13th International Workshop on Slow Positron Beam Techniques and Applications

(SLOPOS13)

Journal of Physics: Conference Series Volume 505

Munich, Germany
15-20 September 2013

Editors:

Christoph Hugenschmidt
Christian Piochacz

ISBN: 978-1-63439-285-3
ISSN: 1742-6588

Printed from e-media with permission by:

Curran Associates, Inc.
57 Morehouse Lane
Red Hook, NY 12571

Some format issues inherent in the e-media version may also appear in this print version.

Printed by Curran Associates, Inc. (2014)

For permission requests, please contact the Institute of Physics
at the address below.

Institute of Physics
Dirac House, Temple Back
Bristol BS1 6BE UK

Phone: 44 1 17 929 7481
Fax: 44 1 17 920 0979

techtracking@iop.org

Additional copies of this publication are available from:

Curran Associates, Inc.
57 Morehouse Lane
Red Hook, NY 12571 USA
Phone: 845-758-0400
Fax: 845-758-2634
Email: curran@proceedings.com
Web: www.proceedings.com

Table Of Contents

Papers

Preface

These proceedings originate from the 13th International Workshop on Slow Positron Beam Techniques and Applications SLOPOS13 which was held at the campus of the Technische Universität München in Garching between 15th–20th September, 2013. This event is part of a series of triennial SLOPOS conferences. In total 123 delegates from 21 countries participated in the SLOPOS13.

The excellent scientific program comprised 50 talks and 58 posters presented during two poster sessions. It was very impressive to learn about novel technical developments on positron beam facilities and the wide range of their applications all over the world. The workshop reflected the large variety of positron beam experiments covering fundamental studies, e.g., for efficient production of anti-hydrogen as well as applied research on defects in bulk materials, thin films, surfaces, and interfaces.

The topics comprised:

- Positron transport and beam technology
- Pulsed beams and positron traps
- Defect profiling in bulk and layered structures
- Nanostructures, porous materials, thin films
- Surfaces and interfaces
- Positronium formation and emission
- Positron interactions with atoms and molecules
- Many positrons and anti-hydrogen
- Novel experimental techniques

The international advisory committee of SLOPOS awarded student prizes for the best presented scientific contributions to a team of students from Finland, France, and the NEPOMUC team at TUM.

The conference was overshadowed by the sudden death of Professor Klaus Schreckenbach immediately before the workshop. In commemoration of him as a spiritus rectus of the neutron induced positron source a minutes' silence was held.

We are most grateful for the hard work of the Local Organising Committee, the help of the International Advisory Committee, and all the students for their friendly and efficient support during the meeting. The workshop could not have occurred without the generous support of the Heinz Maier-Leibnitz Zentrum (MLZ), Deutsche Forschungsgemeinschaft (DFG), and IOP publishing.

Finally we would like to thank all attendees for their outstanding scientific contributions to SLOPOS13, and for the fruitful scientific discussions also in informal atmosphere during the social events.

We are looking forward to SLOPOS14 in Japan in 2016!

Christoph Hugenschmidt and Christian Piochacz

(Guest Editors) Garching, March 2014

Local Organising Committee of SLOPOS13

- Christoph Hugenschmidt (chair)
- Christian Piochacz
- Florian Lippert
- Silvia Valentin-Hantschel
- Hubert Ceeh
- Ina Lommatzsch
- Ramona Bucher
- NEPOMUC team

International Advisory Committee

- M-F Barthe (France)
- C. Corbel (France)
- P. Coleman (UK)
- K. Lynn (USA)
- A. Weiss (USA)
- H. Schut (Netherlands)
- M. Fujinami (Japan)
- R. Suzuki (Japan)
- R. Brusa (Italy)
- R. Krause-Rehberg (Germany)
- G. Amarendra (India)
- F. Tuomisto (Finland)

Journal of Physics: Conference Series **505** (2014) 011001 doi:10.1088/1742-6596/505/1/011001

Delegates of SLOPOS 13, 15—20 September 2014, Garching, Germany
(17/09/2013, T. Gigl)

Peer review statement

All papers published in this volume of *Journal of Physics: Conference Series* have been peer reviewed through processes administered by the proceedings Editors. Reviews were conducted by expert referees to the professional and scientific standards expected of a proceedings journal published by IOP Publishing.

Total reflection high-energy positron diffraction (TRHEPD)

T Hyodo[1], Y Fukaya[2], M Maekawa[2], I Mochizuki[1], K Wada[1], T Shidara[3], A Ichimiya[4] and A Kawasuso[2]

[1]Institute of Materials Structure Science, High Energy Accelerator Research Organization (KEK), 1-1 Oho, Tsukuba, Ibaraki 305-0801, Japan.
[2]Advanced Science Research Center, Japan Atomic Energy Agency, 1233 Watanuki, Takasaki, Gunma 370-1292, Japan.
[3]Accelerator Laboratory, High Energy Accelerator Research Organization (KEK), 1-1 Oho, Tsukuba, Ibaraki 305-0801, Japan.
[4]Nagoya University, Furo-cho, Chikusa-ku, Nagoya 464-8601, Japan.

toshio.hyodo@kek.jp

Abstract. Total reflection high-energy positron diffraction (TRHEPD) promises to be an ideal technique for determining the atomic arrangement of solid surfaces. TRHEPD makes full use of the total reflection of the positron beam from a surface in reflection high-energy positron diffraction (RHEPD), the positron counterpart of reflection high-energy electron diffraction (RHEED). Owing to the development and use of a brightness-enhanced intense positron beam at KEK, it is now possible to obtain clear TRHEPD patterns. It is shown that the TRHEPD pattern from a Si(111)-7×7 reconstructed surface for the total reflection condition does not contain contributions from atoms in the bulk.

1. Introduction

In materials research, X-ray and neutron diffraction has long been the standard technique used to elucidate the bulk structure. Any interpretation of the phenomena found in bulk crystals have to be compatible with the structure determined with these techniques. This is not, however, necessary the case for the surface atomic structures.

Two types of techniques are used to investigate surface structures. One is the imaging technique. Scanning tunneling microscopy (STM) and atomic force microscopy (AFM) are the two most widely used methods. These are sensitive to the topmost surface, and probe the local structure within the area where the probe scanned. It cannot, however, get information on subsurface layers, and in general, it is difficult to distinguish different elements.

The other is the diffraction technique. Low energy electron diffraction (LEED) [1] and reflection high-energy electron diffraction (RHEED) [2] are the most commonly used techniques. As in the 3D diffractions, the symmetry of the patterns gives information on the symmetry of the surface atoms, and the intensity distribution of the spots gives the details of the positions. Unfortunately there is no way for these methods to observe only the topmost surface layers.

As positron counterparts of these electron diffraction techniques, low energy positron diffraction (LEPD) [3] and reflection high-energy positron diffraction (RHEPD)[4,5] have been developed. Here we show that total reflection high-energy positron diffraction (TRHEPD), which is nothing but a refined RHEPD with an intense, brightness-enhanced positron beam, has the promise to be an ideal technique for the surface structure analysis. TRHEPD makes full use of the total reflection of the positron beam from a solid surface.

Though X-ray diffraction in conjunction with its total reflection is also used for the surface structure, the critical angle for X-ray is usually less than 0.2°, not easy to handle. Furthermore, though the penetration depth of X-ray into the sample in the total reflection condition is much smaller than that at a practical angle such as 2°, more than 1000 nm, it is still as large as 3 nm.

2. Total Reflection of a Positron

The positron wave function ψ outside and inside a crystal is described by

$$\left(-\frac{\hbar}{2m}\nabla^2 + qV\right)\psi = E\psi, \qquad (1)$$

where potential V is 0 outside the crystal and V>0 inside in every crystal. The potential energy of a positron, qV, is positive since the charge of the positron $q = +e > 0$, where e is the elemental charge. The abrupt increase in the potential energy at the crystal surface makes the positron reflected at the surface.

The work function of the positron in the same material is a different thing. The work function of the positron is related with the energy eigenvalue E of the ground state of the fully interacting positron (and the effect of the electric dipole layer in the case of a metal).

In a case where a positron beam is directed on a surface with an glancing angle smaller than a certain critical value, the beam is totally reflected [4]. This is not the case for an electron beam. This makes a positron much more sensitive to the topmost atomic layer of the surface than an electron. The wave vector of the incident positron can be decomposed into the component parallel to the surface, $\boldsymbol{k}_{\parallel}$, and that perpendicular, $\boldsymbol{k}_{\perp}$. It follows that its kinetic energy can be expressed with the component parallel to and perpendicular to the surface, $\boldsymbol{E}_{0\parallel}$ and $\boldsymbol{E}_{0\perp}$, respectively:

$$E_0 = \frac{\hbar^2 k_0^2}{2m} = \frac{\hbar^2 k_{0\parallel}^2}{2m} + \frac{\hbar^2 k_{0\perp}^2}{2m} = E_{0\parallel} + E_{0\perp}. \qquad (2)$$

For a beam incident with a glancing angle θ, the perpendicular component is expressed in terms of E_0 and θ as

$$E_{0\perp} = \frac{\hbar^2 k_{0\perp}^2}{2m} = \frac{\hbar^2 k_0^2 \sin^2\theta}{2m} = E_0 \sin^2\theta\ . \qquad (3)$$

If $E_{0\perp}$ is smaller than the average potential energy of the positron inside the crystal, eV, the beam will be totally reflected. The critical angle for the total reflection, θ_c, is expressed as

$$E_0 \sin^2\theta_c = eV \rightarrow \ \theta_c = \sin^{-1}\sqrt{eV/E_0}. \qquad (4)$$

3. Improvements in Intensity and Brightness

RHEPD was proposed by Ichimiya [4] and its usefulness was proved by Kawasuso and Okada [5] using apparatus based on a ^{22}Na positron source. The intensity of the beam was 10^3-10^4 slow- e^+/s. Unfortunately the intensity of the beam was not satisfactory for RHEPD experiments. It was not even possible to orient the sample with the positrons in real time. Yet fruitful determinations of the structures of metal-deposited surfaces of Si and Ge crystals, some of which were impossible or very difficult with other techniques, were performed [6].

Principally used technique in RHEPD is the measurement of a rocking curve of the specular reflection spot in which the intensity of the specular spot is plotted as a function of the incident glancing angle. The experimental pattern is compared with a calculated pattern for an assumed atomic arrangement consistent with the symmetry of the RHEPD pattern. The goodness of fit is expressed by using a reliability factor, or R-factor and the assumed atomic arrangement is modified. The process is repeated until satisfactory fit is obtained.

Now the only one RHEPD station in the world has been moved to the Slow Positron Facility at KEK and connected to the intense slow positron beam there. The facility provides a high-intensity, pulsed slow-positron beam produced by using a dedicated linac (55 MeV, 600 W) [7]. The moderator is kept at high voltage (variable up to 35 kV) that the energy of the positrons is controlled easily. The long (width 1 μs) and short (width 1-10 μs) pulse modes provide 5×10^7 slow-e^+/ s and 5×10^6 slow-e^+/ s, respectively.

The one-dimensional structure of Pt-deposited Ge(001) surface, with which a number of different structure models had been proposed, was determined with the RHEPD rocking curve [8], showing that one of the proposed models was essentially correct.

Then a transmission-type brightness-enhancement unit [9, 10] was installed upstream of the RHEPD chamber. This has resulted in an enhancement of the brightness by ~5000 times over the initial magnetically guided beam in the long pulse mode, with an expense of the beam intensity reduction to 1/100 through the release into the magnetic-field free region and the annihilation in the remoderator foil. This refined beam of energy 10 keV and 5×10^5 slow e^+/s, having a beam diameter ~0.5 mm (FWHM), an energy spread ~40 meV, an angular divergence ~12 mrad, and a normalized emittance ~0.01 cm rad $eV^{1/2}$ [11], yields clear positron diffraction patterns from a crystal surface. The orientation of the sample is now performed in real time observing the phosphor screen behind MCP. A RHEPD pattern can be taken within 1 hour and a full single rocking curve for the specular spot within 3 hours.

The brightness-enhanced beam was recently used to confirmed that silicene constructed on Ag(111) surface has a buckling structure [12] unlike graphene.

4. Form RHEPD to TRHEPD

The pattern in Fig. 1 (1)-(5) show the RHEPD patterns from the Si(111)-7×7 reconstructed surface observed for the indicated glancing angles, θ, observed with the refined beam [13]. The value of the critical glancing angle for the total reflection of 10 keV positrons from this surface is 2.0°. Thus the pattern (1) corresponds to the glancing angle within the total reflection condition, i.e., TRHEPD in a narrow sense. It is to be noted that the integer order spots due to bulk diffraction, commonly seen on

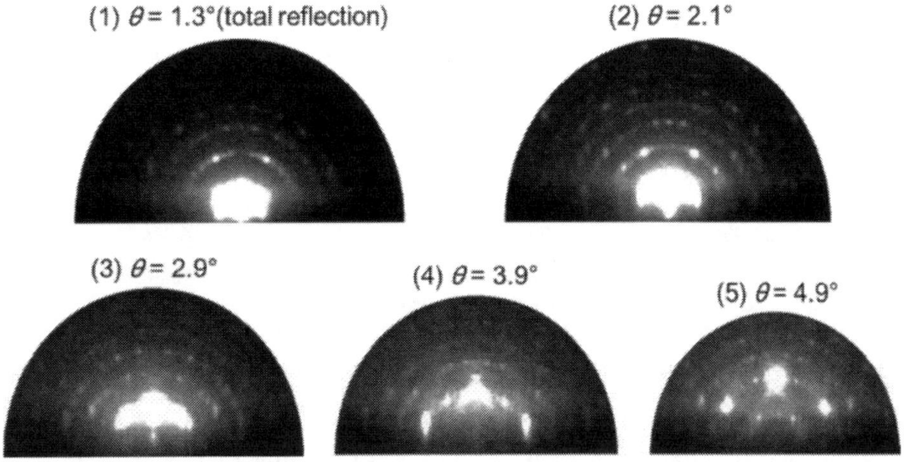

Figure 1. RHEPD patterns observed with a brightness-enhanced high-intensity positron beam.

the RHEED patterns are barely visible on the pattern, indicating that it does not have information on the bulk crystal. Another feature is that Kikuchi lines which are also commonly observed on RHEED patterns are not noticeable.

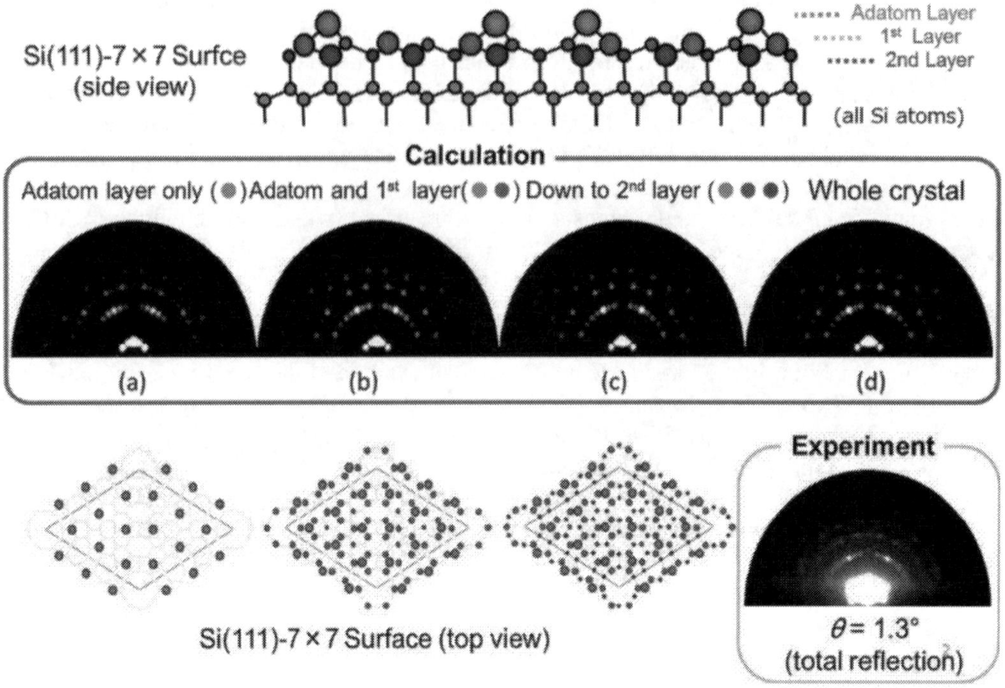

Figure 2. Experimental TRHEPD pattern for the Si(111)-7x7 reconstructed surface structure observed with the glancing angle θ =1.3° (bottom right). The patterns (a)-(c) are those calculated for the adatom sheet, adatom sheet and the first layer, and adatom, first and second layers. The pattern (d) shows that calculated for the bulk sample with the reconstructed surface. The schematic diagram at the top shows the side view of the Si(111)-7x7 structure, while those at the bottom show the top view of the sets of layers corresponding to the calculated patterns. The circles are all silicon atoms.

In order to unambiguously show that the TRHEPD pattern, Fig. 1(1), has only information about the atoms on the surface and not those in the bulk, we calculated contributions of the sheet(s) of the Si(111)-(7×7) surface atoms to the pattern. Figure 2 (a)-(d) show RHEPD patterns for a two dimensional single sheet of the adatom (atoms colored red), the adatoms and first surface layer (atoms colored green) and down to the second surface layer (atoms colored blue) and the whole crystal, respectively. Code developed by Ichimiya for the analysis of RHEED patterns based on dynamical electron diffraction theory [2, 14] was adapted to be applicable to RHEPD by changing the electric charge of the incident particle from negative to positive. Detailed positions of the atoms were taken from the ab initio calculation of this surface, i.e., the rumpling model, by Brommer et al. [15]. It is immediately noticed that the pattern (b) and (d) are in good agreement. This indicates that only the atoms exposed on the surface contribute to the pattern even a bulk sample is observed. Note that the distribution of the Si adatoms in the 7×7 super structure is rather sparse so that most of the atoms in the first surface layer (colored green) are exposed.

Figure 3 shows the experimentally observed RHEED pattern (bottom right) with the same condition as in the TRHEPD pattern (Fig. 2, bottom right) and the calculated patterns for the sheets as for Fig. 2 (a)-(d). The code developed by Inchimiya was used. It is seen that the calculated pattern (c) is appreciably different from the pattern (b) and (d), indicating that (c) contains contributuion from the atom on the layer below the second surface layer.

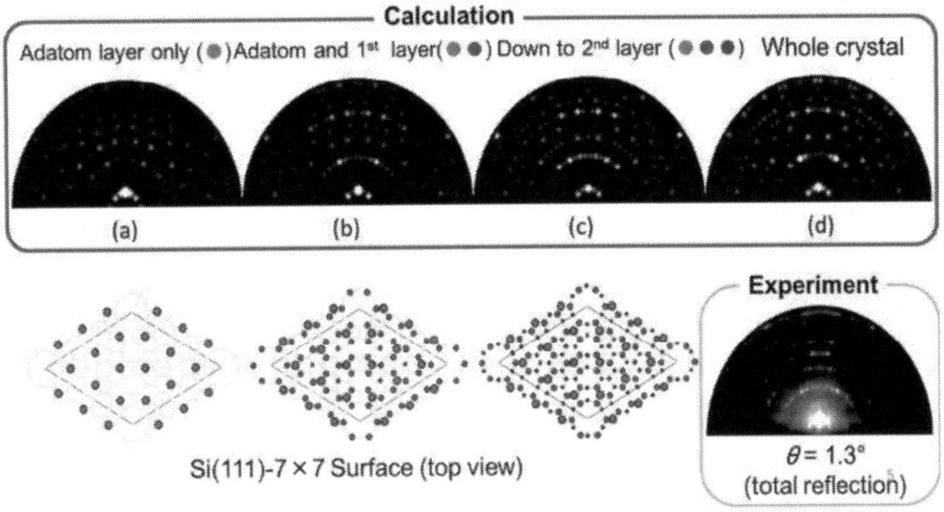

Figure 3. Experimental RHEED pattern for the Si(111)-7×7 reconstructed surface structure observed with the glancing angle θ =1.3° (bottom right). The patterns (a)-(c) are those calculated for the adatom sheet, adatom sheet and the first layer, and adatom, first and second layers. The pattern (d) shows that calculated for the bulk sample with the reconstructed surface. The schematic diagram at the bottom show the top view of the sets of layers corresponding to the calculated patterns.

In summary it has been shown that TRHEPD yields diffraction patterns separating out the contributions from the atoms exposed on the surface alone by simple measurements on a bulk sample. In addition, by analysing the RHEPD patterns across the total reflection range as shown in Fig. 1, it should be possible to determine accurately the crystal structure of the surface from the topmost layer downward. We propose to call hereafter such methods of analyse including the RHEPD data taken in total reflection condition as Total Reflection High-Energy Positron Diffraction (TRHEPD) in a broad sense.

References

[1] for example, van Hove M A and Tong S Y 1979 *Surface Crystallography by LEED, Theory, Computation and Structural Results* (Springer)
[2] Ichimiya A and Cohen P I 2004 *Reflection High-Energy Electron Diffraction* (Oxford Univ. Press)
[3] Coleman P G 2002 *Appl. Surf. Sci.* **194** 264 and references therein
[4] Ichimiya A 1992 *Solid State Phenom.* **28/29** 143
[5] Kawasuso A and Okada S 1998 *Phys. Rev. Lett.* **81** 2695
[6] Fukaya Y et al. 2013 *J. Phys. Conf. Series* **443** 012068 and references therein
[7] Wada K et al. 2012 *Eur. Phys. J. D* **66** 37
[8] Mochizuki I et al. 2012 *Phys. Rev. B* **85** 245438
[9] Chen D, Lynn R, Pareja M K G, Nielsen B 1985 *Phys. Rev. B* **31** 4123
[10] Oshima N et al. 2008 *J. Appl. Phys.* **103** 094916
[11] Maekawa M et al. 2014 to be published in *Eur. Phys. J. D*
[12] Fukaya Y et al. 2013 *Phys. Rev. B* **88** 205413
[13] Fukaya Y et al. 2014 to be published in *Appl. Phys. Express*
[14] Ichimiya A 1983 *Jpn. J. Appl. Phys.* **22** 176
[15] Brommer K D Needels M Larson B E Joannopoulos J D 1992 *Phys. Rev. Lett.* **68** 1355

Accurate description of the van der Waals interaction of an electron-positron pair with the surface of a topological insulator

R Saniz[1], S Vercauteren[1], D Lamoen[1] B Partoens[1] and B Barbiellini[2]
[1] CMT & EMAT, Departement Fysica, Universiteit Antwerpen, Groenenborgerlaan 171, B-2020 Antwerpen, Belgium
[2] Physics Department, Northeastern University, Boston, Massachusetts 02115, USA

Abstract. Positrons can be trapped in localized states at the surface of a material, and thus quite selectively interact with core or valence surface electrons. Hence, advanced surface positron spectroscopy techniques can present the ideal tools to study a topological insulator, where surface states play a fundamental role. We analyze the problem of a positron at a TI surface, assuming that it is a weakly physisorbed positronium (Ps) atom. To determine if the surface of interest in a material can sustain such a physisorption, an accurate description of the underlying van der Waals (vdW) interaction is essential. We have developed a first-principles parameter-free method, based on the density functional theory, to extract key parameters determining the vdW interaction potential between a Ps atom and the surface of a given material. The method has been successfully applied to quartz and preliminary results on Bi_2Te_2Se indicate the existence of a positron surface state. We discuss the robustness of our predictions versus the most relevant approximations involved in our approach.

Topological insulators (TIs) are materials with a bulk electronic structure that is insulating, but which possess at their surface "protected" conducting states. These surface states result, remarkably, from the time-reversal symmetry and topology of the bulk electronic structure, and are protected in the sense that they cannot be altered by impurities or defects if these do not break the time-reversal symmetry. Quite crucially, the surface states of a TI have their spin and momentum coupled [1]. To further our understanding of TIs, there is great interest in probing directly these surface states, and positron spectroscopy offers the ideal tools for this purpose. Indeed, it is known that after irradiation of a material with a positron beam, positrons can diffuse to the surface and may be momentarily trapped, or bound to it, if such a surface state exists [2]. More recently, a method based on an Auger process was devised to directly deposit positrons at a surface bound state using slow positron beams [3]. The signals from the subsequent positron annihilation with core or valence electrons are collected in experiment with various techniques, such as positron annihilation lifetime, Doppler broadening of annihilation radiation, two-dimensional angular correlation of annihilation radiation, or positron annihilation-induced Auger electron spectroscopy [2, 4, 5, 6]. If the annihilating positron is at the surface, the collected signals contain crucial information about the electronic properties, structure, and composition of the surface. Moreover, since spin-polarized positron beams can be used and annihilation occurs only with electrons of opposite spin, the spin of the probed states can be known.

Thus, to study the surface states of a TI with positron spectroscopy, it is of the greatest interest to determine whether positrons can be bound to its surface. In this regard, an important

Journal of Physics: Conference Series **505** (2014) 012002 doi:10.1088/1742-6596/505/1/012002

model for positron-surface interactions was advanced by Platzman and Tzoar [7]. These authors viewed a positron in a low density electron gas as a weakly physisorbed positronium (Ps) atom (i.e, an electron-positron bound state). A few years ago, this idea was applied with success to the Ps-quartz surface interaction. Indeed, using essentially a first-principles approach, Saniz *et al.* [8] found a bound state for Ps on quartz with an energy close to the experimental value[9], solving a long standing disagreement between theory and experiment [10].

Current state-of-the-art *ab initio* techniques, based on density functional theory (DFT) [11], allow an accurate determination of the structural and electronic properties of narrow band gap semiconductors, such as the TIs. This means that one can theoretically determine with reliability the potential seen by an atom at their surface and, thus, whether Ps can be bound to it and what its binding energy would be [12]. In the following we describe how this is achieved. As an example, we consider Bi_2Te_2Se, a well known tetradymite TI [13, 14]. Our preliminary results indicate that Ps should have a bound state on the (111) surface. In this respect, we note that very recently the group of A. Weiss has produced very encouraging evidence for the presence of a positron surface state on the surface of Bi_2Te_2Se [15].

The interaction between a physisorbed atom and a crystal surface is given by an attractive van der Waals potential away from the surface, and a short-ranged repulsive potential close to it. If z denotes the coordinate of the atom with respect to the crystal surface, the interaction is given by [10, 16, 17, 18]

$$\begin{aligned} v(z) &= v_{\rm R}(z) + v_{\rm vdW}(z). \\ &= V_0 e^{-(z-z_0)/\lambda} - \frac{C}{(z-z_0)^3} F_3((z-z_0)/\lambda). \end{aligned} \tag{1}$$

The repulsive part is proportional to the valence electron density near the surface of the crystal. Specifically, define $\langle \rho \rangle_z$ as the electron density at coordinate z, averaged over the section of the unit cell perpendicular to the z axis. $\langle \rho \rangle_z$ will decay exponentially away from the point at which the contribution of the core electrons to the density vanishes. This point, z_0, defines the "background edge". λ is the exponential decay length. The amplitude of V_0 is given by the negative of the Ps work function [10]. The latter is given by $\epsilon_{\rm Ps} = -A^+ - 6.8$ eV, with A^+ the positron affinity given by $A^+ = \mu^- + \mu^+$, where μ^- and μ^+ are the electron and positron chemical potentials, respectively [19].

The van der Waals attraction decreases essentially as the inverse cubic power of the "distance" of the atom to the surface. This distance is measured from the background edge, z_0 [17]. The proportionality constant (van der Waals constant), C, is given by

$$C = \frac{\hbar}{4\pi} \int_0^\infty d\xi\, \alpha(i\xi) \left(\frac{\varepsilon(i\xi) - 1}{\varepsilon(i\xi) + 1} \right), \tag{2}$$

with ε the dielectric function of the material and α the polarizability of the adatom [17]. F_3 is the damping, or regularization function, introduced by Patil and co-workers [18], ensuring that the adatom-surface interaction does not diverge at the reference position, z_0.

Except for the Ps polarizability, all the above quantities can be calculated with first-principles methods, as already indicated. The polarizability of Ps is most easily determined by rescaling the polarizability of the H atom, which is well parametrized [20]. We discuss the calculation of the other quantities here. The most common exchange-correlation functionals used in DFT based calculations, such as the local-density approximation (LDA) or the Perdew-Burke-Ernzerhof approximation (PBE), are typically able to describe with good accuracy the structural properties of surfaces. However, it is important to realize that these functionals suffer from the well-known band gap problem. That is, the band gap and energies of the excited states of semiconductors and insulators are seriously underestimated when using these functionals. In the case of narrow

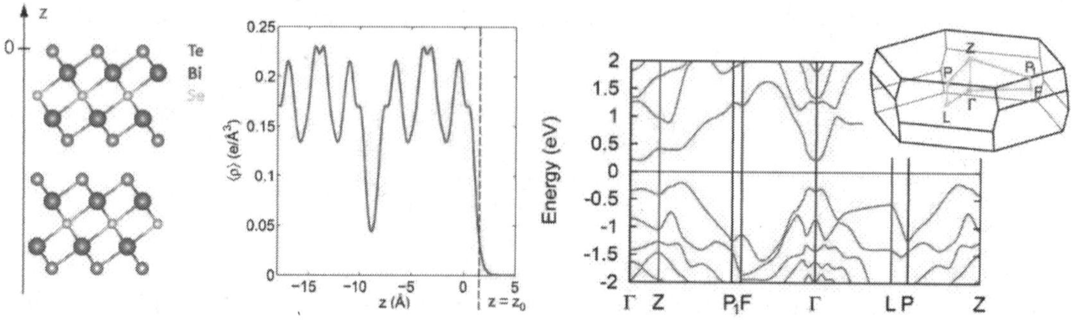

Figure 1. (a) Partial side view of the Bi_2Te_2Se slab and average electron density as a function of z. (b) Bulk band structure of Bi_2Te_2Se (HSE06 hybrid functional), including spin-orbit coupling.

gap semiconductors, such an error can lead to their erroneous description as metals. Given that incorrect gap and excited state values can lead to an incorrect dielectric function, it is important to follow a calculation scheme that does not present this problem. One can obtain a much better description of the electronic structure of semiconductors and insulators within the GW approximation [21, 22]. However, this approximation is computationally very expensive, and an accurate calculation of the dielectric function over sufficiently large frequency range is prohibitive. A less expensive approximation is provided by hybrid functionals, such as the HSE06 [23]. In essence, the latter combines the PBE exchange functional at long range, with Hartree-Fock exchange at short range. This is because the PBE (or the LDA) is known to overscreen the exchange interaction. The HSE06 hybrid functional has proven to provide accurate structural and electronic properties. In most *ab initio* codes, the imaginary part of the dielectric tensor is calculated through [26]

$$\mathrm{Im}\,\varepsilon_{ij}(\omega) = \frac{4\pi^2 e^2}{\Omega} \lim_{q\to 0} \frac{1}{q^2} \sum_{c,v} \sum_{\mathbf{k}} 2w_{\mathbf{k}} \delta(\epsilon_{c\mathbf{k}} - \epsilon_{v\mathbf{k}} - \omega) \langle u_{c\mathbf{k}+\hat{e}_i q} | u_{v\mathbf{k}} \rangle \langle u_{c\mathbf{k}+\hat{e}_j q} | u_{v\mathbf{k}} \rangle^*, \qquad (3)$$

where Ω is the unit cell volume, and $\epsilon_{c\mathbf{k}}$ is the energy of conduction band state c with momentum $\mathbf{k}$, $w_{\mathbf{k}}$ denoting the cell-periodic part of the corresponding wavefunction ($u_{\mathbf{k}}$ is the $\mathbf{k}$-point weight). The real part, $\mathrm{Re}\,\varepsilon_{ij}(\omega)$, is obtained with the Kramers-Kronig relation (see Ref. [26] for details). One final point is that in materials with heavy ions, spin-orbit interactions can contribute importantly to the properties of materials. This is the case of TIs, in which spin-orbit coupling leads to the band inversion that is responsible for the existence of the topologically protected surface states which give the Dirac cone. Thus, it is important to include spin-orbit coupling in *ab initio* studies of topological insulators. Such computations are considerably more expensive, but are still tractable on large parallel computers.

We have applied the above scheme to find out if Ps has a bound state on the surface of Bi_2Te_2Se. For this purpose we used the Vienna ab initio simulation package (VASP) [24]. The calculation of the structure and electron density at the surface of a slab were done using the PBE functional. Figure 1(a) shows a partial side view of the slab used for this calculation (the slab in the calculations is three quintuple layers thick), as well as the average electron density profile in the z-direction. The position of the background edge (which we find to be 1.55 Å) is also indicated. The electron density decay length is 0.35 Å. The calculations of the electronic structure were done with the experimental lattice structure parameters [25], and using the hybrid HSE06 exchange-correlation functional. The band structure along high symmetry lines is shown in Fig.1 (b). The dielectric function was calculated using a $12 \times 12 \times 12$ $\mathbf{k}$-point mesh, and up to 480 bands, resulting in a van der Waals constant $C = 15.95$ eV a_0^3 (a_0 = Bohr radius). Finally, our estimate of the Ps work function in $BiTe_2Se_2$ yields a value close to zero (−0.034 eV). This is based on calculations of the electron and positron chemical potentials with the Abinit code [27], using the functional of Puska *et al.* [28]. With these parameters we find a bound state with energy −0.27 eV. This is a preliminary result because of some of the simplifying assumptions in the calculations. Indeed, in this first study, spin-orbit coupling was not included [except for the band plot in Fig. 1(b)]. This represents heavier calculations, which we are currently performing. Also, we used an approximate value for the Ps work function in $BiTe_2Se_2$, whose small value has yet to be confirmed. It would be of great interest to have this value determined experimentally. Thus, the final theoretical binding energy may deviate somewhat from the value we report here, but our result is robust to small variations of the different parameters determining the physisorption potential, and we do not expect the main conclusion of our work to be different.

Acknowledgments

We thank A. Weiss for very useful conversations. We acknowledge financial support from FWO-Vlaanderen (project G.0150.13). This work was carried out using the HPC infrastructure of the University of Antwerp (CalcUA), a division of the Flemish Supercomputer Center (VSC), funded by the Hercules foundation and the Flemish Government (EWI Department). B. B. is supported by DOE grants Nos. DE-FG02-07ER46352 and DE-AC02-05CH11231 for theory support at ALS, Berkeley, and a NERSC computer time allocation.

References

[1] Ando Y 2013 J. Phys. Soc. Jpn. **82** 102001.
[2] Puska M J and Nieminen R M 1994 Rev. Mod. Phys. **66** 841.
[3] Mukherjee S, Nadesalingam M P, Guagliardo P, Sergeant A D, Barbiellini B, Williams J F, Fazleev N G, and Weiss A H 2010 Phys. Rev. Lett. **104** 247403 (2010).
[4] Weiss A H, Mayer R, Jibaly M, Lei C, Mehl D, and Lynn K G 1988 Phys. Rev. Lett. **61** 2245.
[5] Barbiellini B 2001, in *New Directions in Antimatter Chemistry and Physics*, edited by Surko C M and Gianturco F A (The Nederlands: Kluwer Academic Publishers), p. 127.
[6] Tuomisto F and Makkonen I 2013 Rev. Mod. Phys. (to appear).
[7] Platzman P M and Tzoar N 1986 Phys. Rev. B **33** 5900.
[8] Saniz R, Barbiellini B, Platzman P M and Freeman A J 2007 Phys. Rev. Lett. **99** 096101; *ibid.* 2008 **100** 019902.
[9] Sferlazzo P, Berko S, and Canter K F 1985, Phys. Rev. B **32** 6067.
[10] Ford G W, Sander L M, and Witten T A 1976 Phys. Rev. Lett. **36** 1269.
[11] Martin R M 2008 *Density Functional Theory: Basic Theory and Practical Methods* (Cambridge: Cambridge University Press).
[12] Knowing the Ps wavefunction, furthermore, can allow one to estimate the pick-off annihilation rates with valence and core electrons near the surface.
[13] Ren Z, Taskin A A, Sasaki S, Segawa K, and Ando Y 2010 Phys. Rev. B **82** 241306.
[14] Lin H, Das T, Wray L A, Xu S-Y, Hasan M Z, and Bansil A 2011 New J. Phys. **13** 095005.
[15] A. Weiss *et al.*, to be published.
[16] Nordlander P and Harris J 1984 J. Phys. C **17** 1141.
[17] Zaremba E and Kohn W 1976 Phys. Rev. B **13** 2270.
[18] Patil S H, Tang K T, and Toennies J P 2002 J. Chem. Phys. **116** 8118.
[19] Puska M J, Lanki P, and Nieminen R M, J. Phys.: Condens. Matter **1**, 6081 (1989).
[20] Dalgarno A and Victor G A 1966 Mol. Phys. **10** 333.
[21] Aulbur W G, Jönsson L, and Wilkins J W 1999 Solid State Phys. **54**.
[22] Onida G, Reining L, and Rubio A 2002 Rev. Mod. Phys. **74** 601.
[23] Heyd J, Scuseria G E, Ernzerhof M 2003 J. Chem. Phys. **118** 8207.
[24] See Kresse G and Joubert D 1999 Phys. Rev. B **59** 1758 and references therein.
[25] Nakajima S 1963 J. Phys. Chem. Solids **24** 479.
[26] Gajdoš M, Hummer K, Kresse G, Furthmüller J, and Bechstedt F 2006 Phys. Rev. B **73** 045112.
[27] Gonze X *et al.* 2009 Comput. Phys. Commun. **180** 2582.
[28] Puska M J, Seitsonen A P, and Nieminen R M 1994 Phys. Rev. B **52** 10947.

Journal of Physics: Conference Series **505** (2014) 012003 doi:10.1088/1742-6596/505/1/012003

Upgrade of the Surface Spectrometer at NEPOMUC for PAES, XPS and STM Investigations

S Zimnik[1,2], F Lippert[1] and C Hugenschmidt[1,2]

[1] Technische Universität München, Physik Department, Lehrstuhl E21, James-Franck-Straße, D-85748 Garching, Germany

[2] Technische Universität München, Forschungs-Neutronenquelle Heinz Maier-Leibnitz, Lichtenbergstraße 1, D-85748 Garching, Germany

E-mail: samantha.zimnik@frm2.tum.de

Abstract. The characterization of the elemental composition of surfaces is of great importance for the understanding of many surface processes, such as surface segregation or oxidation. Positron-annihilation-induced Auger Electron Spectroscopy (PAES) is a powerful technique for gathering information about the elemental composition of only the topmost atomic layer of a sample. The upgraded surface spectrometer at NEPOMUC (NEtron induced POsitron source MUniCh) enables a comprehensive surface analysis with the complementary techniques STM, XPS and PAES. A new X-ray source for X-ray induced photoelectron spectroscopy (XPS) was installed to gather additional information on oxidation states. A new scanning tunneling microscope (STM) is used as a complementary method to investigate with atomic resolution the surface electron density. The combination of PAES, XPS and STM allows the characterization of both the elemental composition, and the surface topology.

1. Introduction

The surface of a solid is involved in many of processes of particular interest such as oxidation, surface segregation, or catalysis. These processes significantly affect many macroscopic properties of thin films or nanoparticles. Detailed knowledge of the surface structure and elemental composition is necessary to understand these processes.

PAES is a non-destructive technique to determine the elemental composition of only the topmost atomic layer of a solid. Positrons are implanted into the sample with an energy of only 20 eV. In the solid, the positrons thermalize within a few picoseconds, and the vast majority diffuse back to the surface [1]. There they become trapped in the surface potential, and initiate the Auger process [2], which is a radiationless transition of an excited atom: an electron from a higher level fills the hole at a lower level, and the released energy is transferred to another electron, which leaves the atom with an element-specific energy. In the case of PAES, the hole in the inner shell is created by electron-positron annihilation. Typically, the core annihilation probability for positrons is in the range of 1 to 10% [3]. PEAS exhibits two essential advantages: due to the low energy of the incident positrons, the spectrum is — in contrast to XAES and EAES — free of secondary electron background in the energy range of the Auger transitions, leading to an excellent signal-to-noise ratio. The surface trapping leads to the outstanding surface sensitivity of PAES.

To overcome the main drawback of PAES measurements — the long measurement time when using β^+ lab sources — the surface spectrometer is attached to the positron beam facility NEPOMUC at the research reactor Heinz Maier-Leibnitz. This reactor-based source delivers a 1 keV positron beam with an intensity of 10^9 positrons per second. For brightness enhancement, this primary beam is re-moderated with a tungsten single-crystal with an efficiency of 5% [4]. This high-intensity positron beam yields measurement times of only a few minutes per PAES spectrum, and enables time-dependent observation of e.g. surface segregation processes [5].

A new X-ray source for XPS measurements was recently installed at the spectrometer. XPS, as an established method for elemental characterization with its high chemical specificity, is a particularly suited completion of PAES for e.g. the determination of oxidation states. Due to the high penetration depth of the X-rays, XPS, in contrary to PAES, shows the average of the elemental composition of 3-10 atomic layers, depending on the energy of the photo electrons.

As a complementary method to determine the surface structure, a new scanning tunneling microscope is used. STM images the electron density distribution with atomic resolution. Thus, changes in the crystal structure, such as dislocations or grain boundaries leading to a variation in the surface potential, become visible. These deviations are attractive for positrons, and therefore influence the probability density of positrons. Hence, STM data is of special interest for the interpretation of PAES spectra, in order to determine the annihilation site of the positron. Furthermore, in a first approximation, the mapping of the electron density distribution can be seen as the topology of the surface.

2. Experimental setup of SuSpect

The upgraded Surface Spectrometer (SuSpect) combines sample preparation, a sample storage, and analysis with STM, XPS and PAES under UHV conditions in one setup (see figure 1). For sample cleaning, an Ar-ion sputter gun and electron impact heating can be used. Three electron beam evaporators are installed, to prepare coated samples. The thickness of the coating is controlled by a piezo-oscillator serving as a thickness monitor. Thicknesses from a fraction of an atomic layer (ML) up to several hundred nm are possible.

The samples are transferred by magnetically linked linear feedthroughs to the analysis chambers. In the first analysis chamber the STM Aahrus 150 HT, with a temperature range between 90-1300 K, a sample storage for up to 5 samples, and an electron impact heater for sample preparation are installed.

The next analysis chamber houses the setup for Auger spectroscopy. For conventional Auger spectroscopy a focusable electron gun with a tunable emission energy up to 5 keV and an X-ray source with Mg/Al twin anode are used. For PAES measurements the re-moderated positrons are guided adiabatically to the analysis chamber where they pass a magnetic field termination and are focused electrostatically onto the sample. To avoid the influence of external magnetic fields, the analysis chamber is μ-metal shielded. A NaI(Tl) scintillator is used for detection of γ-radiation from the positron annihilation. For beam adjustment a moveable MCP with a phosphor screen and CCD readout was recently installed. The electron detection is done with a Phoibos 150 hemispherical energy analyzer with a large acceptance angle of up to $\pm$ 13°. The energy analyzer is bipolar, with the result that reemitted or backscattered positrons are also detectable. Readout is done with a cooled CCD camera. In both analysis chambers, an ion getter pump and a titanium sublimation pump are used, in addition to turbo molecular pumps, to obtain a vacuum in the range of 10^{-9} mbar and below. For residual gas analysis a mass spectrometer is used.

3. First measurements

Several measurements were done at the upgraded spectrometer to demonstrate the performance of all components. For the STM a Si(111) wafer was used as reference. Figure 2 shows the

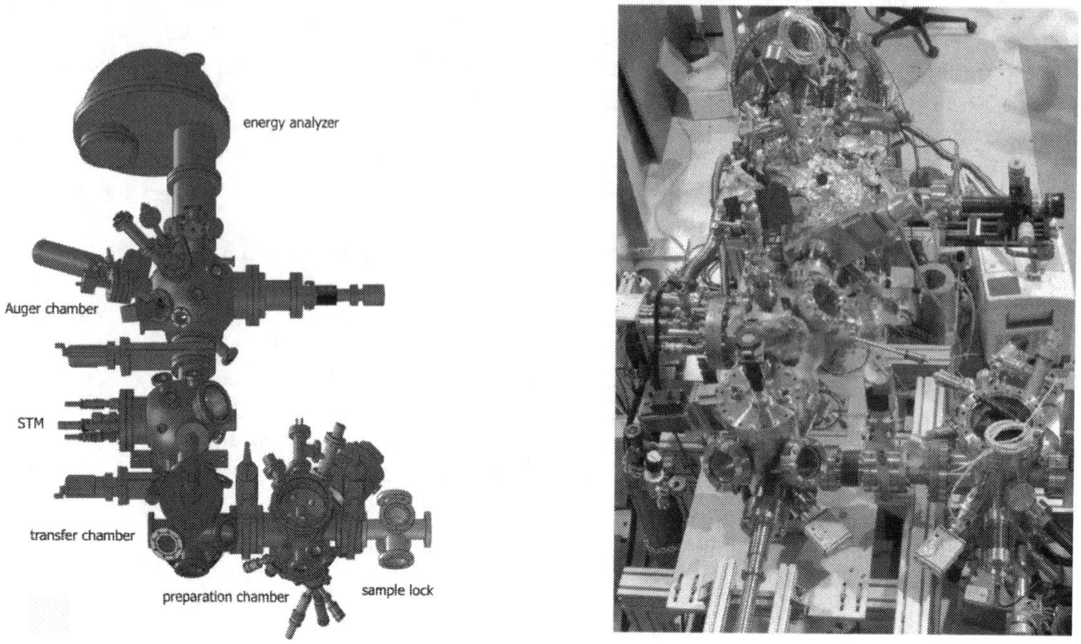

Figure 1: The upgraded surface spectrometer SuSpect at NEPOMUC for PAES, XPS and STM measurements.

occupied and empty states of the Si(111) 7×7 surface reconstruction with atomic resolution, recorded with the new STM. The sample was cleaned with acetone and deionized water, followed by annealing for 30 minutes and flashing using electron impact heating. Afterwards, a Cu(100) crystal was investigated. The STM image in figure 3 shows the $\sqrt{17} \times \sqrt{17}\, R\,14$-reconstruction of S atoms on the Cu(100) surface in atomic resolution.

XPS measurements with the new X-ray source were performed on Cu and Fe-coated Cu foils. For this purpose Cu foils were cleaned with Ar-ion sputtering at 1 keV for 1 hour and annealed with electron impact heating. Subsequently, the foils were coated with 3 ML and 6 ML Fe respectively, evaporated from a Fe rod with a purity of 99.99% using electron beam evaporation. The thickness was controlled with a piezo-oscillator thickness monitor. The XPS spectra of the three samples are shown in figure 4 with different offsets for reasons of clarity.

Recently, a first PAES measurement on Cu was performed, after the modification of the spectrometer and the upgrade of the NEPOMUC source [4]. The re-moderated positron beam with an energy of 20 eV was used. From the count rate of the NaI(Tl) scintillator the positron intensity on the sample was calculated to be $7.5 \cdot 10^6$ e^+/s. Figure 5 shows the PAES spectrum with the Cu MVV Auger transition at 60 eV.

4. Conclusion and outlook

In this work the upgraded surface spectrometer at NEPOMUC was presented. The spectrometer enables a comprehensive surface analysis using PAES as a non destructive, element selective analysis method with outstanding surface sensitivity, XPS with its chemical specificity, and STM for topological imaging of the surface with atomic resolution. In order to reduce the measurement time for PAES spectra and enable time-dependent PAES measurements, a moveable MCP was recently installed and will be used for focusing a precise beam onto the sample. Future studies

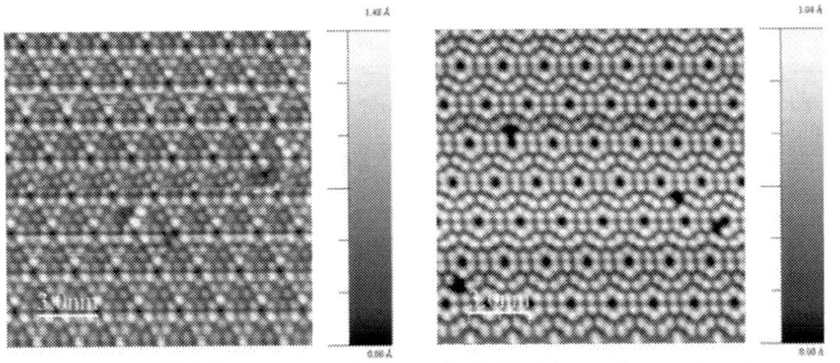

Figure 2: STM image of occupied (left) and empty states (right) of the Si(111) 7 × 7-reconstruction with atomic resolution.

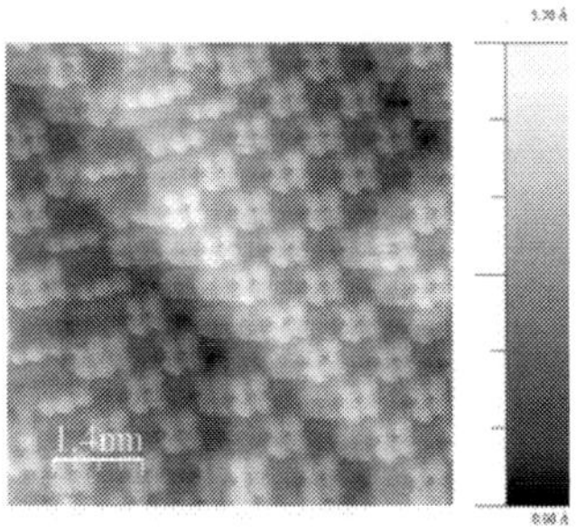

Figure 3: STM image of $\sqrt{17} \times \sqrt{17}$ $R14$-reconstruction of S atoms on the Cu(100) surface with atomic resolution.

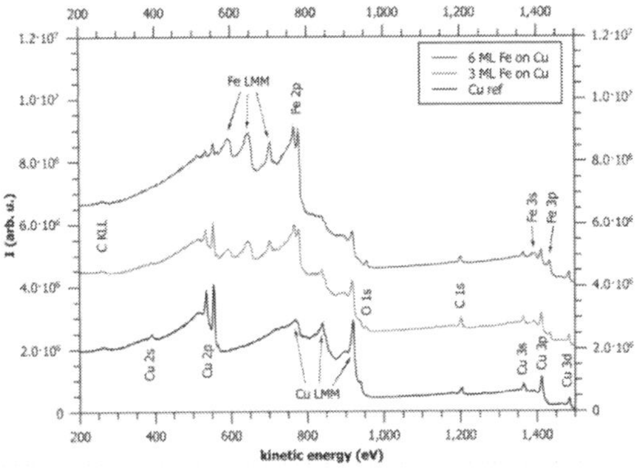

Figure 4: XPS spectra of a Cu reference and Fe coated Cu foils with different offsets.

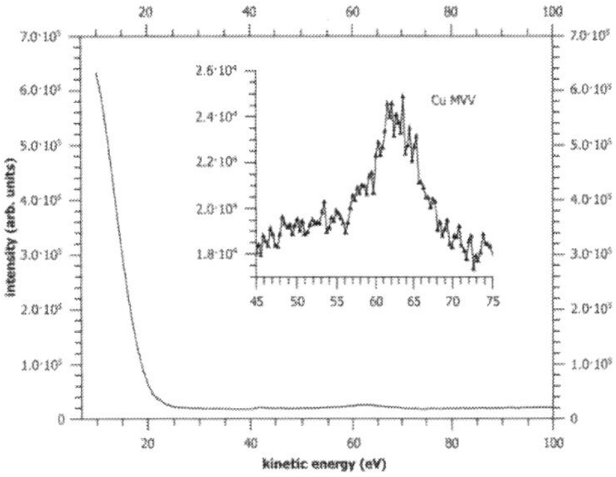

Figure 5: PAES spectrum of Cu with Cu MVV Auger-transition.

will be focused on the detection of surface segregation processes in order to benefit from the surface sensitivity and elemental selectivity of PAES in combination with complementary STM imaging and XPS.

5. Acknowledgments

Financial support provided for the improvement of the instruments PAES and CDBS (project nos. 05KI0WOB and 05K13WO1) by the BMBF is gratefully acknowledged.

References

[1] Schultz P and Lynn K 1988 *Rev. Mod. Phys.* **60** 701–779
[2] Weiss A, Mayer R, Jibaly M, Lei C, Mehl D and Lynn K G 1988 *Phys. Rev. Lett.* **61** 2245–2248
[3] Jensen K O and Weiss A 1990 *Phys. Rev. B* **41** 3928–3936
[4] Hugenschmidt C, Piochacz C, Reiner M and Schreckenbach K 2012 *New Journal of Physics* **14** 055027
[5] Mayer J, Hugenschmidt C and Schreckenbach K 2010 *Phys. Rev. Lett.* **105** 207401

Positron reemission from clean and LiF coated W(100): Effect of oxygen exposure

K Sudarshan[1,3,§], SN Samarin[1], PJ Wilkie[1], P Guagliardo[1], VN Petrov[1,2] and JFWilliams[1]

[1]ARC Centre of Excellence for Antimatter and Matter Studies, School of Physics, The University of Western Australia, Crawley, Perth, Western Australia 6009. Australia.
[2]St. Petersburg State Polytechnical University, Polytechnicheskaya 29, St. Petersburg 195251, Russia
[3]Radiochemistry Division, Bhabha Atomic Research Centre, Mumbai, India 400085.

[§]E-mail: kathis@barc.gov.in

Abstract. Positron reemission has been measured from clean and oxygen exposed W(100), as well as W(100) with LiF deposited on the surface at an incident positron energy of 600eV. The reemitted positron yield was found to be much lower from LiF deposited W(100) than from clean W(100). Oxygen exposure caused a drastic reduction in the reemission yield from W(100) but did not cause any major changes in the LiF deposited sample.

1. Introduction

Tungsten is one of a number of materials with a negative work function for positrons, emitting thermalised positrons with a kinetic energy equivalent to its work function. Its simplicity of use has led frequently to its use as a moderator in slow positron beams for many years in many laboratories [1]. The experimental results for various parameters that influence the positron reemission yield and other associated parameters of reemission vary considerably among different samples. For example, oxygen adsorption on GaAs(100) was shown to enhance the reemitted positron yield because of reduction in the Ps yield by the conversion of Ps atoms to positrons at the surface [2]. Lynn and Lutz, in measurements on Al(111) and Al(100), showed that oxygen exposure increases positronium formation and reduces the positron reemission yield [3].Suzuki *et al.* reported a significantly higher positron reemission yield from W(110) and polycrystalline tungsten foil when the samples were exposed to 10^{-7} mbar of O_2 and held at a temperature of 900-1000 °C [4], and in contrast at room temperature, oxygen exposure caused a reduction in the positron reemission yield in W(100) [5]. Recipes for conditioning of moderators such as tungsten have been optimized over the years by trial and error, however, the physics governing the conditioning procedures is not fully understood and calls for further investigation. In the present experiments, the influence of oxygen exposure on the reemitted positron yield and energy spectrum have been measured for clean W(100) and W(100) with over layers of LiF.

2. Experiment

Reemission measurements were performed using the electrostatic slow-positron beam at The University of Western Australia. Details of the positron beam and associated experimental facilities are described in Ref. 6. The W(100) single crystal was cleaned by heating the crystal to about ~1200 ^{0}C in the presence of oxygen (10^{-8} mbar) for 5 min and followed by a high-temperature flash at ~2400 ^{0}C using electron bombardment. The procedure was repeated until no further improvements in the thermal-positron yield could be observed. For oxygen exposure studies, the sample was allowed to cool for 1.5 hours before admitting oxygen into the chamber. LiF was evaporated onto the W(100) surface and the amount of LiF was measured using a microbalance under identical conditions. The reemitted positron yield was measured as a function of time using a large acceptance angle retarding field analyser in order to ensure good collection efficiency. Since our detection system is at a fixed angle of 45° with respect to the beam direction, the angle of the sample was changed so that positrons emitted in the direction normal to crystal surface reached the detector. In that way the positrons from the beam were incident on the crystal surface at an angle of 45°.

3. Results and discussion

The retarding curves of reemitted positrons in the direction normal to the crystal surface were recorded for clean W(100) and W(100) with over layers of LiF; these are shown in Fig.1. The negative of the derivative of this curve gives the energy spectrum of the emitted positrons (also given in Fig.1). As seen from the figure, the reemitted positron intensity in the direction normal to the surface reduces with increasing thickness of LiF deposited on the surface. There were no new peaks in the LiF reemitted positron energy spectra showing that LiF has a positive or near zero work function for positrons. The reemitted positron energy spectrum from LiF coated W(100) is broader, has a marginally higher contribution in the low energy range and a lower mean energy than clean W(100). These features might be due to either greater contribution from inelastic scattering of positrons in the LiF over layers or modified work function of the material due to the change in the surface dipole.

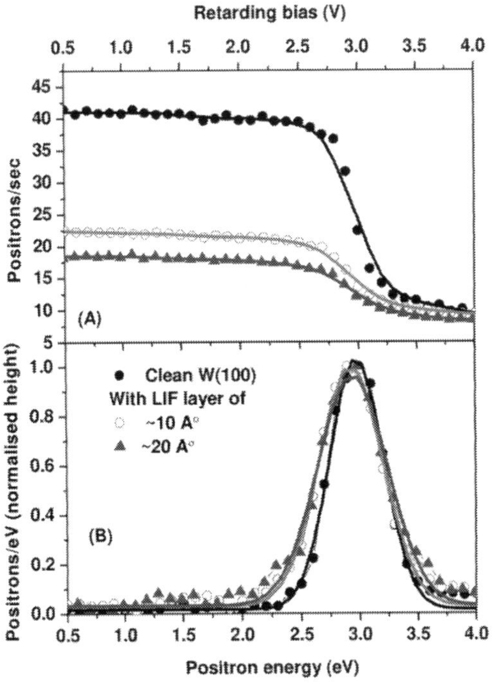

Fig.1.(A) Retarding curves of positrons and(B) reemitted positron energy spectra in the direction normal to the surface from clean W(100) and from W(100) with over layers of LiF. The incident positron energy is 600 eV.

Clean W(100) and about 20 Å of LiF coated W(100) were exposed to O_2 at a pressure of 1×10^{-8} mbar and retarding curves of the reemitted positrons were recorded at voltage steps of 0.5 eV. Positrons in the energy range of 2 to 4 eV were taken as reemitted positrons. The yield as a function of oxygen exposure time from clean and from LiF coated W(100) are shown in Fig.2. The yield at zero exposure time has been normalized to unity. It can be seen that the reemitted positron yield decreased exponentially with oxygen exposure time in case of clean W(100) while there was no significant change for the LiF coated W(100).

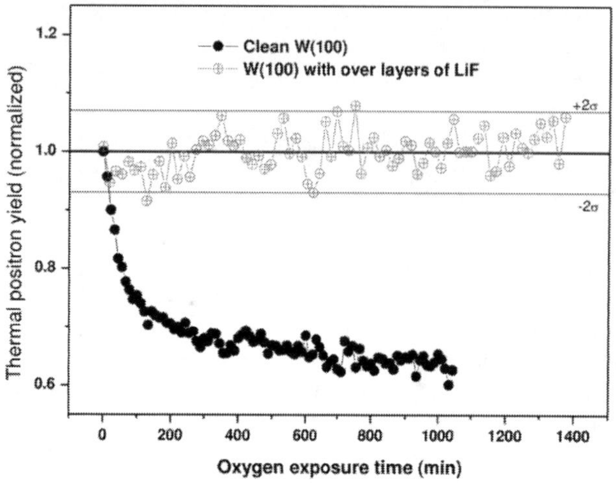

Fig.2. Change in positron reemission yield with oxygen exposure time. The yield is normalized to the yield without oxygen exposure. Oxygen pressure is 1×10^{-8} mbar.

To monitor changes in the reemitted positron energy spectra due to oxygen exposure, retarding curves of remitted positrons in the direction normal to the surface were recorded on samples exposed to oxygen for about 15 hours.

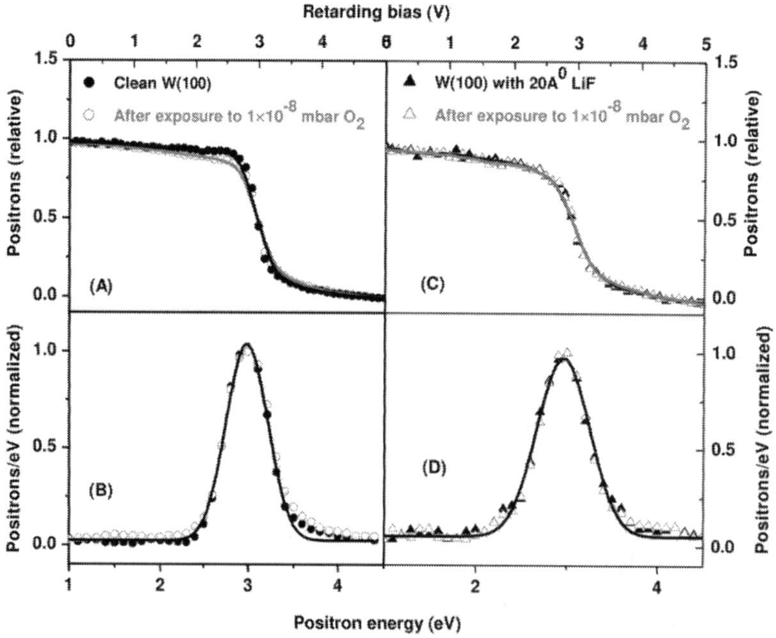

Fig. 3 (A) Retarding curves of reemitted positrons from clean and oxygen exposed W(100). (B) Energy spectra of reemitted positrons from clean and oxygen exposed W(100). (C) Retarding curves of reemitted positrons from LiF coated W(100) with and without oxygen exposure. (D) Energy spectra of reemitted positrons from LiF coated W(100) with and without oxygen exposure.

Retarding curves of reemitted positrons from oxygen exposed samples of W(100) and LiF coated W(100) are shown in Figs 3A and 3C, along with those taken before oxygen exposure. The corresponding reemitted positron energy spectra are also shown in Figs 3B and 3D. For W(100), oxygen exposure caused a reduction in the yield of reemitted positrons and increased the contribution from inelastically scattered positrons. However, for the LiF covered sample, oxygen exposure did not change either the positron reemission yield or any other parameters associated with the energy spectra. This might be due to a low sticking coefficient for oxygen on LiF compared to W(100).

4. Conclusion

Positron reemission spectra were measured from W(100) and LiF coated W(100), before and after exposure to oxygen at a pressure of 1×10^{-8} mbar. The positron yield was also determined as a function of oxygen exposure time. In the case of clean W(100), oxygen exposure caused an exponential decrease in the reemission positron yield. Oxygen exposure also caused an increase in the fraction of inelastically scattered positrons. Although the LiF coated sample showed a lower reemission yield than clean W(100), the yield was insensitive to oxygen exposure. Also the clean W(100) surface gave higher reemission yields, the yield was very sensitive to surface cleanliness. Although the presence of oxygen or LiF on clean W(100) decreased the reemission yield, they appeared to be helpful in maintaining constant reemission yield.

5. References

[1] Schultz P J, Lynn K G 1988 *Rev. Mod. Phy.* **60** 701.
[2] Kim JH, Nangia A, Weiss AH 2000 *Rad. Phys. Chem.* **58** 655.
[3] Lynn KG, Lutz H 1980 *Phys. Rev.* **B. 22** 4143.
[4] Suzuki R et al. 1998 *Jpn. J. Apl. Phys.* **37** 4636.
[5] Sudarshan K et al. 2013 *J. Appl. Phys.* **A. 113** 633.
[6] Sudarshan K et al. 2013 *Phys. Rev.* **B. 87** 085418.

Reflection high-energy positron diffraction study on the first surface layer

Y Fukaya[1], M Maekawa[1], I Mochizuki[2], K Wada[2], T Hyodo[2] and A Kawasuso[1]

[1]Advanced Science Research Center, Japan Atomic Energy Agency, 1233 Watanuki, Takasaki, Gunma 370-1292, Japan
[2]Institute of Materials Structure Science, High Energy Accelerator Research Organization (KEK), 1-1 Oho, Tsukuba, Ibaraki 305-0801, Japan

E-mail: fukaya.yuki99@jaea.go.jp

Abstract. Reflection high-energy positron diffraction (RHEPD) is a powerful tool for studying surface structure. In particular, the topmost surface layer can be observed, facilitated by the characteristic of total reflection for positrons. A previous RHEPD study on a Si(111)-√21×√21-Ag surface, using a ^{22}Na-based beam, is revisited and the analysis detailed.

1. Introduction

The structure of the surface of a crystal is, in general, different from that of the bulk. A typical example is a 7×7 reconstructed structure on a clean Si(111) surface [1]. Crystal surfaces can exhibit electronic and magnetic properties which may significantly change with slight displacements of atomic position in the first surface layer. Therefore, a precise knowledge of the atomic arrangement of this layer is needed to fully understand surface properties.

Reflection high-energy positron diffraction (RHEPD) is a powerful method used to investigate the structure of the first surface layer [2,3]. When the glancing angle of a positron beam directed onto a surface is below a particular value, positrons are totally reflected from the material, as depicted in Fig. 1, due to the positive crystal potential. This critical angle is given by $\theta_c = \arcsin(eV/E)^{1/2}$, where eV and E denote the mean potential energy of the positron in the crystal and the energy of the incident positron beam [2]. For example, for a beam of energy E = 10 keV incident on Si for which eV = 12 eV, θ_c is 2.0°. Under the total reflection condition, the intensities of the diffraction spots are very sensitive to the structures and thermal vibration of the first surface layers. A RHEPD apparatus using a ^{22}Na positron source was developed in 1998 [3,4] and used in surface structure investigations [5] until recently. In this paper, we describe details of the study of a √21×√21 superstructure made with this apparatus as an example of surface structure analysis using RHEPD.

2. Structure analysis by RHEPD

2.1. Si(111)-√21×√21 surface

Adsorption of one monolayer (ML) of Ag atoms on a clean Si(111)-7×7 surface leads to the formation of a √3×√3-Ag structure. The basic framework of the Si(111)-√3×√3-Ag structure was deduced by pioneering work using surface x-ray diffraction [6]. Subsequently, using low-temperature scanning tunneling microscopy and first-principles calculations, the ground-state structure was confirmed to be an inequivalent triangle (IET) structure [7]. In 1994, it was discovered that a √21×√21 superstructure having a large unit cell appeared when a small amount of Au atoms were deposited onto the √3×√3-Ag surface [8, 9]. Moreover, it was found that adsorption of other noble and alkali metal atoms also led to the formation of a √21×√21 superstructure, accompanied by a significant increase of surface electrical conductivity [10].

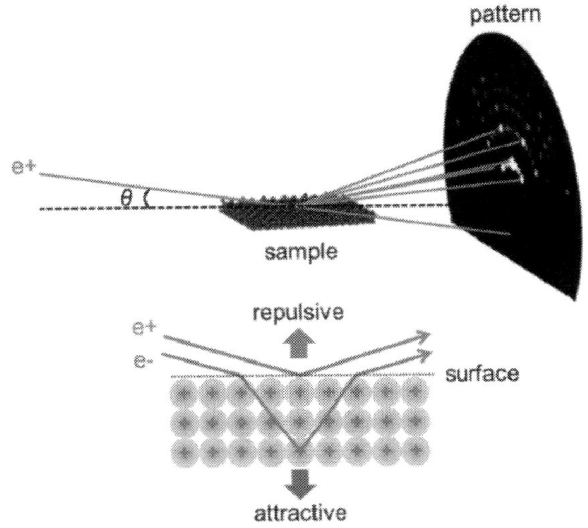

Figure 1. Experimental setup of RHEPD (top) and schematic drawings of positron and electron incidences on a crystal surface (bottom). When the positron beam is incident on the surface, the positron is repelled by the positive crystal potential and is totally reflected when the glancing angle is below a certain critical angle.

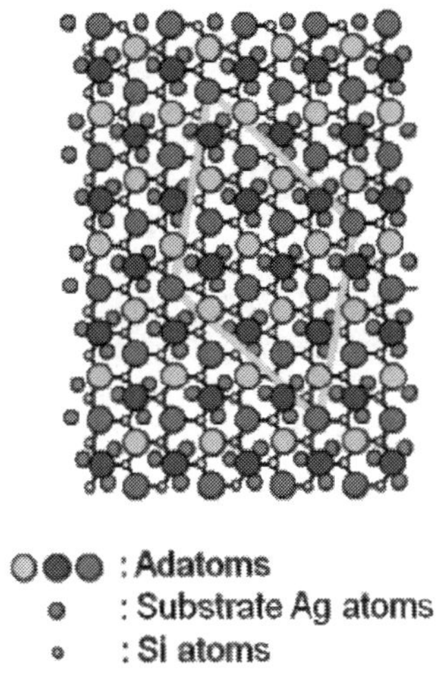

Figure 2. The twenty-one possible adatom adsorption sites of the √21×√21 superstructure: the centers of seven large Ag triangles, the centers of seven small Ag triangles, and the centers of seven Si trimers are denoted by orange, blue, and green circles, respectively. Red and open circles denote the substrate Ag and Si atoms. The rhombus indicates a √21×√21 unit cell.

Various models for the √21×√21 superstructure were proposed by a number of researchers [8,9,11,12], however, it could not be ascertained which one was correct. As depicted in Fig. 2, there are 21 possible adsorption sites on the Si(111)-√3×√3-Ag surface [13]: (i) the centers of seven large Ag triangles, (ii) the centers of seven small Ag triangles, and (iii) the centers of seven Si trimers. Since seven equivalent sites exist for each of these candidates in the √21×√21 unit cell, the number of possible structure models is ${}_{21}C_3$ = 1330 when the number of adatoms is three (~0.14 ML coverage). To determine the surface structure, we performed a screening analysis of this structure using RHEPD rocking curves and intensity profiles of the diffraction spots [14].

2.2. *Determination of the vertical components of atomic positions*

A RHEPD rocking curve is a plot of the intensity of a certain diffraction spot versus glancing angle. Figure 3 illustrates two different incident directions used in the analysis. Figure 3(a) shows a view along the $[11\bar{2}]$ direction, neglecting the small glancing angle, where periodicity of the atomic distribution along the perpendicular directions of [111] and $[1\bar{1}0]$ is seen. This incidence condition is called the many-beam condition [15]. Figure 3(b) shows a view from 7.5° off the $[11\bar{2}]$ direction, where the atoms appear to be randomly distributed in plane and only the periodicity along [111], i.e., the direction perpendicular to the surface, is seen. This is called the one-beam condition [16]. The

Journal of Physics: Conference Series **505** (2014) 012005 doi:10.1088/1742-6596/505/1/012005

rocking curve analysis is conducted in two steps. First, by analyzing the rocking curve in the one-beam condition, we determine the vertical components of the atomic positions ignoring the accurate in-plane configurations but taking only the atomic density in the plane into account. Second, we determine the in-plane components by analyzing the rocking curves in the many-beam condition with fixed vertical components determined by the analysis in the one-beam condition.

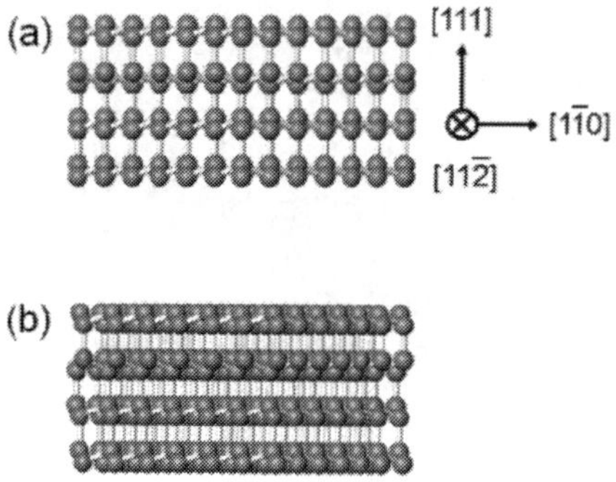

Figure 3. Diamond structures when viewed (a) along the [11$\bar{2}$] direction (many-beam condition) and (b) along 7.5° off the [11$\bar{2}$] direction (one-beam condition).

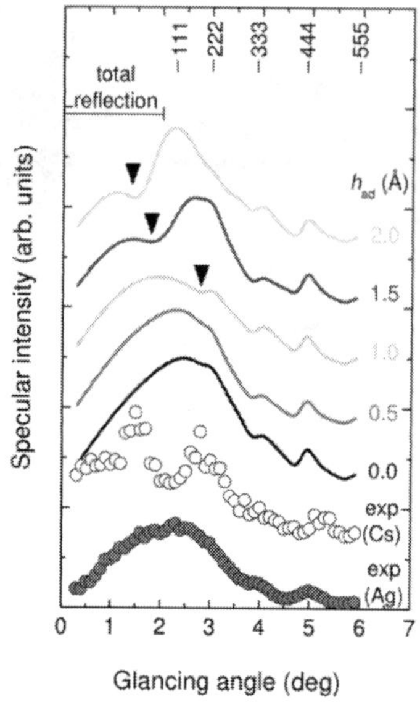

Figure 4. RHEPD rocking curves from the Si(111)-√21×√21 super-structures under the one-beam condition. Red closed circles show the experimental curves from Si(111)-√21×√21-Ag. For compareson, the open circles show those from a Si(111)-√21×√21-(Ag,Cs) surface. Solid lines indicate the rocking curves calculated with various adatom heights (h_{ad}) up to 2.0 Å.

Red filled circles in Fig. 4 show a measured RHEPD rocking curve of the specular reflection spot for a Si(111)-√21×√21-Ag surface at 7.5° off the [11$\bar{2}$] direction (one-beam condition [16]). The solid lines show RHEPD rocking curves at various Ag adatom heights (h_{ad}) from the substrate Ag layer calculated using dynamical diffraction theory [15]. The parameters used in the calculations are described in reference [14]. From the curve fitting procedure using the reliability factor (*R*-factor) defined in reference [17], the adatom height for the Si(111)-√21×√21-Ag surface was determined to be 0.53 Å [14].

In order to illustrate the sensitivity of the rocking curve in the one-beam condition, Fig. 4 includes the calculations for a range of values of h_{ad} which illustrate that the shape of the rocking curve varies with h_{ad}. As h_{ad} increases from zero, the broad peak rising in the total-reflection region shifts towards lower glancing angles. This is due to the decrease in the critical angle of total reflection caused by a reduction in the crystal potential at the surface because of less overlap between the adatoms and the substrate Ag atoms. At higher h_{ad}, a dip structure appears in the broad peak, as shown by the arrowheads. The position of the dip gradually shifts towards lower glancing angles with increasing h_{ad}. The tendency of the shift can be explained by the interference of the positron waves reflected from the adatom and the substrate Ag layers [18]. The value of h_{ad} for the Si(111)-√21×√21-Ag surface is small so there is no significant dip structure observed. For comparison, the rocking curve measured from the Cs atoms adsorbed √21×√21 superstructure (Si(111)-√21×√21-(Ag,Cs)) [19] is also plotted in Fig. 4. In this case, a distinct dip structure is seen because of the large atomic radius of the Cs atoms.

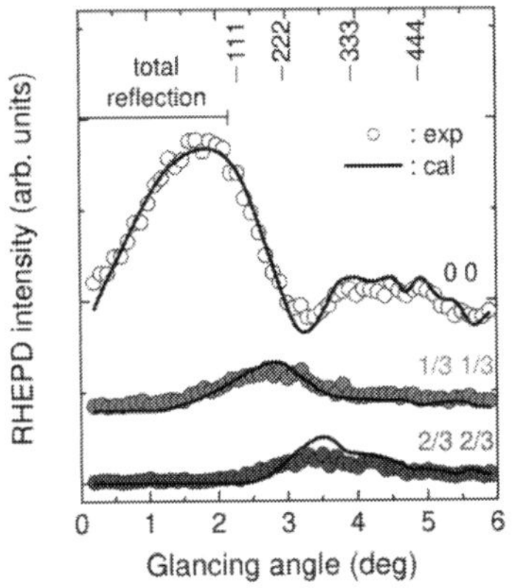

Figure 5. RHEPD rocking curves along the [$11\bar{2}$] direction (many-beam condition) for the indicated spots from the Si(111)-√21×√21-Ag surface. Circles show the experimental curves. Solid lines indicate the rocking curves calculated with optimized structure parameters.

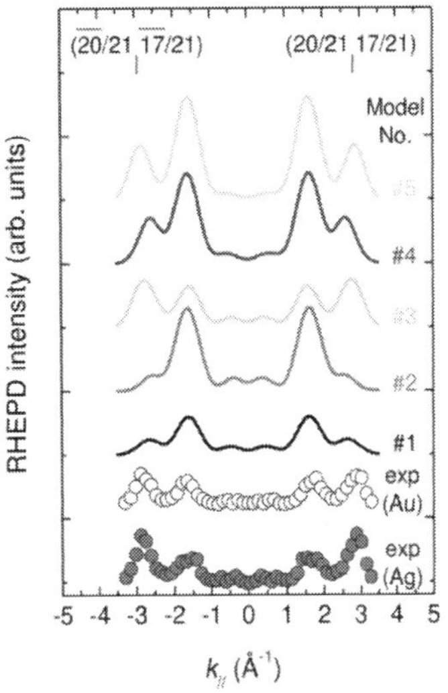

Figure 6. Diffraction intensity profiles along the 1/7th Laue zone for the Si(111)-√21×√21 superstructure. $k_{//}$ is the position in the 1/7 Laue zone. Red and open circles indicate the intensity profiles measured from Si(111)-√21×√21-Ag and -(Ag,Au) surfaces, respectively. Solid lines indicate the profiles calculated for the remaining five possible structures.

2.3. Determination of in-plane components by rocking curve and intensity profiles

To determine the in-plane components of the atomic positions of the Si(111)-√21×√21-Ag surface, the rocking curves along the [$11\bar{2}$] direction (many-beam condition) were measured. Results obtained are shown by the circles in Fig. 5. The rocking curves for the 00, 1/3 1/3, and 2/3 2/3 spots (spots on the 0th Laue zone) were used in the analysis because the intensities of these spots were greater than those of the higher Laue zones. The solid lines show the best fitted rocking curves. As a result, it was found that all of the three adatoms were located on sites at the center of a large Ag triangle. However, the rocking curves for these spots could not discern which three sites out of the seven were occupied. Nevertheless, the possible structure of Si(111)-√21×√21-Ag was narrowed down by this process to five (See Fig. 6 of reference [14]), i.e., the possible symmetrically independent distribution of three atoms among the seven possible sites.

The fractional-order spots resulting from the formation of the √21×√21 superstructure carries information about the in-plane distribution of adatoms. However, the intensities of these spots measured with the ^{22}Na-based slow-positron beam in this study were too weak to give reliable rocking curves. Therefore, the intensity profiles of the spots in the 1/7th Laue zone were used to deduce the correct arrangement out of the five possibilities. The red filled circles in Fig. 6 show the intensity profile extracted from the observed pattern. The solid lines are the profiles calculated for the five possible structures. It is clear that only one graph, #3, matches the intensity profile observed. This is a structure where all the three adatoms are sitting on sites at the center of a large triangle of Ag atoms (orange sites in Fig. 2). More precisely, the occupied sites form a triangle surrounding the Si trimer in the √21×√21 unit cell (#3 in Fig. 6 of reference [14]). It was later shown that the intensity profile from the Au-adsorbed Si(111)-√21×√21 structure (open circles in Fig. 6) resembles that of the Si(111)-√21×√21-Ag surface [20]. This demonstrates that the Si(111)-√21×√21-(Ag,Au) surface has the same atomic consignations as the Si(111)-√21×√21-Ag surface.

3. RHEPD with an intense beam and future prospects

Although it was possible to determine the surface structure using the weaker ^{22}Na-based positron beam at JAEA, Takasaki, owing to the high reflectivity of positrons at surfaces, the RHEPD apparatus was recently moved to the Slow Positron Facility at the Institute of Materials Structure Science, KEK, and connected to a linac-produced, brightness-enhanced intense positron beam [21]. Weak fractional-order spots in the RHEPD pattern from a Si(111)-7×7 surface were successfully observed with this improved apparatus. The intense positron beam makes the structural analysis much more efficient and accurate. A study of a more complex surface structure will be made and even a Pattern function analysis without assuming a possible atomic arrangement will be attempted.

Acknowledgments

This work was conducted under PF Proposal No. 2012G653 and the auspices of Joint Development Research at the High Energy Accelerator Research Organization (KEK). The present work was partly supported by a Grant-in-Aid for Scientific Research (S) 24221007 and for Young Scientists (B) 25800182 from JSPS.

References

[1] Takayanagi K, Tanishiro Y, Takahashi S and Takahashi M 1985 *Surf. Sci.* **164** 367
[2] Ichimiya A 1992 *Solid State Phenom.* **28&29** 143
[3] Kawasuso A and Okada S 1998 *Phys. Rev. Lett.* **81** 2695
[4] Kawasuso A, Ishimoto T, Maekawa M, Fukaya Y, Hayashi K and Ichimiya A 2004 *Rev. Sci. Instrum.* **75** 4585
[5] Fukaya Y, Maekawa M, Mochizuki I, Wada K, Hyodo T and Kawasuso A 2013 *J. Phys.: Conf. Series* **443** 012068
[6] Takahashi T, Nakatani S, Okamoto N, Ishikawa T and Kikuta S 1988 *Jpn. J. Appl. Phys., Part 2* **27** L753
[7] Aizawa H, Tsukada M, Sato N and Hasegawa S 1999 *Surf. Sci.* **429** L509
[8] Ichimiya A, Nomura H, Horio Y, Sato T, Sueyoshi T and Iwatsuki M 1994 *Surf. Rev. Lett.* **1** 1
[9] Nogami J, Wan K J and Lin X F 1994 *Surf. Sci.* **306** 81
[10] For review, Hasegawa S, Tong X, Takeda S, Sato N and Nagao T 1999 *Prog. Surf. Sci.* **60** 89
[11] Tong X, Sugiura Y, Nagao T, Takami T, Takeda S, Ino S and Hasegawa S 1998 *Surf. Sci.* **408** 146
[12] Tajiri H, Sumitani K, Yashiro W, Nakatani S, Takahashi T, Akimoto K, Sugiyama H, Zhang X and Kawata H 2001 *Surf. Sci.* **493** 214
[13] Aizawa H and Tsukada M 1999 *Phys. Rev.* B **59** 10923
[14] Fukaya Y, Kawasuso A and Ichimiya A 2006 *Surf. Sci.* **600** 3141

[15] Ichimiya A 1983 *Jpn. J. Appl. Phys.* **22** 176
[16] Ichimiya A 1987 *Surf. Sci.* **192** L893
[17] Fukaya Y, Kawasuso A, Hayashi K and Ichimiya A 2004 *Phys. Rev.* B **70** 245422
[18] Fukaya Y, Kawasuso A and Ichimiya A 2006 *Surf. Sci.* **600** 4086
[19] Fukaya Y, Matsuda I, Yukawa R and Kawasuso A 2012 *Surf. Sci.* **606** 1918
[20] Fukaya Y, Kawasuso A and Ichimiya A 2007 *Surf. Sci.* **601** 5187
[21] Wada K, Hyodo T, Yagishita A, Ikeda M, Ohsawa S, Shidara T, Michishio K, Tachibana T, Nagashima Y, Fukaya Y, Maekawa M and Kawasuso A 2012 *Eur. Phys. J.* D **66** 37

Vacancies in NiTi shape memory alloys

MH Weber[1], **T Ablekim, KG Lynn**
Center for Materials Research
Washington State University, Pullman, WA 99164-2711, USA

E-mail: m_weber@wsu.edu

Abstract. NiTi shape memory alloys were examined with a mono-energetic positron beam. The thermal vacancy formation enthalpy was found to be 0.77±0.08 eV. Upon heating NiTi the martensitic S value is retained to temperatures significantly above the phase transition temperature and then decreases towards the austenitic value only gradually. In contrast upon cooling, a narrow transition region at 81.5±1.5 °C is observed. The magnitude of the transition strongly depends on the annealing temperature.

1. Introduction

The implementation of smart materials called shape memory alloys offer the promise of extensive savings in weight and fuel consumption as well as lower failure rates in commercial and military aircraft. Shape Memory Alloys (SMA) can oscillate between pre-trained shapes in response to temperature variations. SMA actuators operate without power lines, mechanical linkages or hydraulic systems. They can be implemented into relatively small devices such as rotor blades of vertical take-off and landing craft or into jet engine air intakes as well as exhausts. Tremendous flexibility in shapes can be achieved to reduce environmental noise or improve aerodynamic performance. Nickel titanium (NiTi)-based alloys commonly known as Nitinol™ offer great promise in these areas. The transition temperature for the shape change can be adjusted in a wide range from 0 °C to about 100 °C by minute changes of about 1.3 wt% in the ratio of nickel to titanium. At low temperatures NiTi maintains a martensitic structure, which is soft and easy to deform. Upon heating beyond the transition temperature it transforms into the harder austenitic structure. A sizable temperature hysteresis between the heating and cooling cycles occurs. Upon transition into the austenitic phase a specimen will return to its trained and "memorized" shape. Shape training can be accomplished by clamping the specimen into the desired shape during a brief anneal at about 550 °C.[1,2] More elaborate procedures lead to much better memory retention – the details of which are confidential vendor property.

Commercial implementation of NiTi SMAs is currently not quite possible because of a lack of understanding of the detailed and microscopic materials processes that underlie the shape memory effect. When SMA raw materials are processed and trained to perform as actuators, not every specimen performs within the desired specifications. Often training has to be repeated many times at great cost in time and manpower. A broad range of standard diagnostic tests failed to predict what treatments produce the desired results. The root cause for these variations may stem from atomic sized defects involving vacancies. Positron annihilation spectroscopies are applied to explore these

[1] Correspondence should be addressed to MH Weber at m_weber@wsu.edu.

properties. Only a small number of lifetime measurements have been published since early work by Würschum et al. and Araki et al.[3,4]

2. Samples and techniques

A number of sample materials in sheet form have been acquired from several vendors. The work discussed here focuses on sheets purchased from Johnson Matthey with a high transition temperature A_f of 110 °C where A_f is the temperature where the transition to the austenitic phase is complete. The sheets are received in a cold rolled and "flat annealed" state. All were coated with a layer of native black TiO_x oxide. The samples were etched in a mixture of HF-nitric and H_2O in the approximate ratio of 1:1:2 respectively to remove the oxide. Samples of 2 by 2 cm^2 or 1 by 3 cm^2 are cut from 0.1 mm thick sheet. As a test a sample was "trained" into a simple U-shape strip. Following the manufacturer's instructions the sample was clamped into the desired shape and annealed at 500 °C for about 10 minutes. More reliable shape memory behaviour requires much more elaborate procedures which are proprietary information of commercial vendors. After random deformation at room temperature it returns to the "trained" U-shape at ~100 °C as expected. Repeating the deformation and heating cycle in air reveals a rather rapid "memory loss" after a few cycles. Clearly, the more elaborate training regimen performed by commercial users are required. Positrons from the beam at Washington State University were implanted about 5 um into the sample to obtain bulk properties. Annealing was carried out in-situ in the positron beam.

3. Measurements and discussion

To investigate the formation rate and accumulation of vacancy like defects a sample was bent multiple times 180° and again 180° in the opposite direction with a radius of about 5 mm. Figure 1 shows the change in S parameter as a function of the number of bending pairs. With increasing number of bending pairs the S parameter rapidly increases and saturates after about 100 pairs with a value of 3% above the as received "flat annealed" starting state. This is consistent with the formation of vacancies.

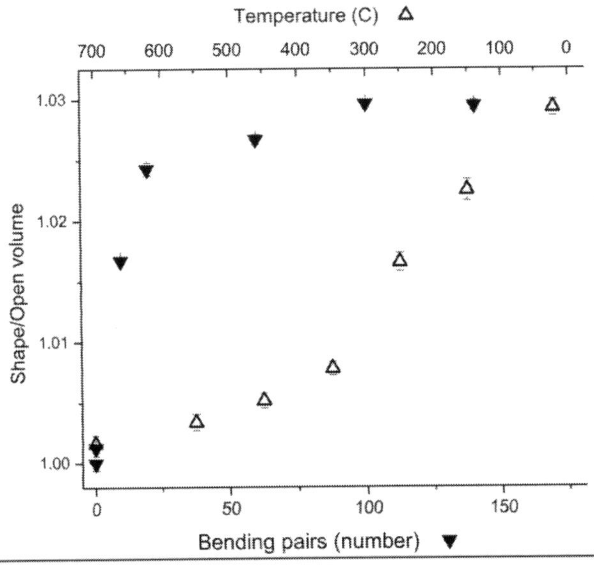

Figure 1: Change in S parameter as a function of bending pairs (solid symbols, bottom scale) and subsequent annealing to a temperature listed on the top scale (open symbols). Data are normalized to the as-received flat annealed state.

Subsequently the sample was annealed to increasing temperatures followed by a measurement upon cooling to room temperature. These data are also shown in figure 1. The sample returned to approximately the as received state only after annealing as high as 700 °C. It should be noted that every annealing step occurred above the A_f transition temperature. In fact, typical training conditions do not involve temperatures above 150 °C. Clearly the actuators accumulate a large concentration of vacancies over the course of the training period as well as during regular operation.

To evaluate the enthalpy to form thermal vacancies H_v, an as received (and etched) sample was heated in situ to 700 °C and cooled slowly at ~2 °C/min while measuring the S parameter rapidly with relatively poor statistical uncertainty. They are shown in figure 2 while the sample was in the austenitic phase only. The insert shows a frequency count plot for data between 85 and 150 °C. The Gaussian distribution is consistent with an error bar of about 0.005 in S. The enthalpy for thermal vacancy formation was obtained by fitting the data with the function.[5]

$$S(T) = \frac{S_f(T) + S_v \xi \exp(\frac{-H_v^f}{k_b T})}{1 + \xi \exp(\frac{-H_v^f}{k_b T})}, \qquad S_f(T) = S_b(300K) + \Delta S(T - 300K)$$

Here $S_b(300K)$ and ΔS are the S values for bulk austenitic NiTi and the change in S due to thermal expansion respectively. The equation $\xi = C_v^0 \mu \lambda_b^{-1}$ contains the frequency factor and the vacancy concentration and trapping rate while k_b is the Boltzmann constant. A vacancy formation enthalpy of H_v = 0.77±0.08 eV results in the best fit. The defect type (Ni or Ti vacancies) was not identified and remains subject of future studies.

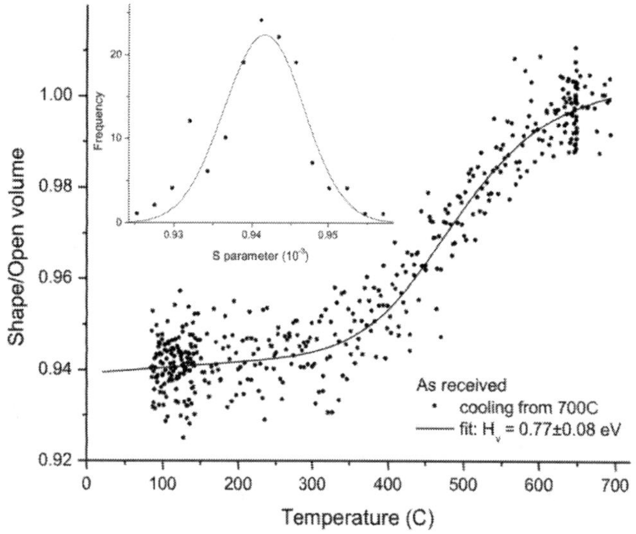

Figure 2: Change in S parameter while the sample is cooling from 700 °C. Each data point was accumulated for 1 min. The spread of the S values for T<150 °C is shown in the insert with a Gaussian fit. The statistical error is about 0.005 in S. The solid line is a fit of the thermal vacancy formation enthalpy H_v.

Figure 3 shows the data collected during temperature increases and decreases as well as across the phase transition from martensitic to austenitic and back. A "flat annealed" etched sample was used. While the sample is heated above the transition temperature and beyond, S drops gradually up to a temperature around 350 °C and then follows the cooling trend as shown in figure 2. While cooling past the phase transition a rapid change in S back the martensitic value at room temperature is observed. The transition occurs within less than 2 °C at 81.5±1.5 °C. During increasing temperatures the initial S values is retained to temperatures significantly above the A_f phase transition temperature. If the temperature rise is halted at an intermediate temperature below 300 °C (not shown here) the S value of that temperature is observed until the sample is cooled below ~81 °C. At that point S rises back to the starting value at room temperature. A detailed understanding of the data awaits further work. The relation between S and W follows a single straight line for the full temperature range. This indicates the presence of only a single type of defect.

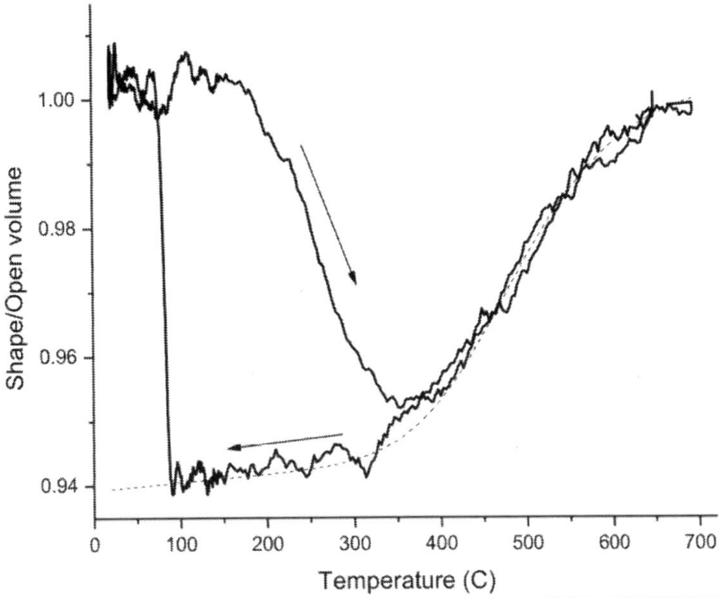

Figure 3: Change in S parameter during increasing and decreasing temperature from room temperature to 700 °C. The solid line represents a running average of 10 short measurements as in fig. 2. The arrows indicate the sequence. The dashed line is the fit of the thermal vacancy formation enthalpy.

4. Conclusions

NiTi shape memory alloy with a high $A_f \approx 110$ °C was examined by with positron Doppler Broadening studies. A vacancy formation enthalpy of 0.77±0.08 eV was measured. S and W parameters show hysteresis with a much wider temperature spread than expected from martensitic to austenitic structure transitions. SW-data follow a single line, indicating a single defect type.

Acknowledgements

The authors acknowledge valuable discussions with Dan Clingman. This work was supported by the Joint Center for Aerospace Technology and Innovation of Washington State, USA.

[1] Robertson SW, Pelton AR and Ritchie RO (2012) *Int. Mater. Reviews* **57** 1
[2] for example the online presence of Johnson Matthey.
[3] Würschum R, Donner P, Hornbogen E and Schaefer H-E 1992 *Mater. Sci. Forum* **105-110** 1333
[4] Araki H, Matake N, Chalermkarnnon P and Shirai Y 2000 *Mater. Sci. Forum* **327-3** 437 2000
[5] Matter H, Winter J and Triftshäuser W 1980 *J. Nuclear Materials* **88** 273-278

Journal of Physics: Conference Series **505** (2014) 012007 doi:10.1088/1742-6596/505/1/012007

Vacancy defects study in Fe based alloys induced by irradiation under various conditions

K Dawi[1], C-W He[1], M-F Barthe[1], P Desgardin[1] and A Volgin[1,2,3,4]

[1] CNRS, CEMHTI UPR3079, Univ. Orléans, F-45071 Orléans, France
[2] GPM UMR-CNRS 6634, Saint-Etienne du Rouvray, France
[3] CSNSM UMR-CNRS 8609, Orsay, France
[4] EDF, R&D, France

E-mail: kamel.dawi@cnrs-orleans.fr

Abstract. In this study, a pure FeNiCr (D302) model alloy has been investigated in parallel with an industrial 316L alloy. Both alloys have been irradiated in different conditions with 5 MeV Ni^{2+} at high temperature (HT) 450°C, and to damage doses: 0.5 or 1 dpa (displacement per atom), or with 1.5 MeV $^4He^{2+}$ at room temperature (RT) and to damage dose between $1x10^{-4}$ and $1x10^{-2}$ dpa. The positron annihilation spectroscopy (PAS) implemented on a slow positron beam has been used to characterize vacancy defects. The results are compared to other analysis methods such as: TEM (Transmission Electron Microscopy) and APT (Atomic Probe Tomography) reported in [1]. Main results show that only single vacancies have been observed in the case of 316L alloy regardless the irradiation conditions, while vacancy clusters seem to be formed in the case of D302 alloy under HT irradiation conditions and from $1x10^{-3}$ dpa at RT. The difference observed between 316L and D302 alloys could be linked to a role played by the alloying elements such as silicon.

1. Introduction

Austenitic stainless steels are widely used in nuclear industry as internal structures. The exposure of these elements to high irradiation doses (the accumulated dose after 40 years of operation can reach 80 dpa), at temperature close to 350°C, modifies the macroscopic behavior of the steel and thus limits the operational lifetime of the reactor. In the framework of PERFORM60[1] project, twenty European organizations and universities involved in the nuclear field are engaged in the development of experimentally validated multi-scale modeling tools with the main concern to predict the operational lifetime of the reactor pressure vessel (RPV). CEMHTI laboratory is involved in this program to provide experimental data on the evolution of the microstructure of studied materials as a function of irradiation conditions (temperature, fluence, etc.). Then, results obtained will be used by the other program teams to validate their simulation models.

In this work, the Slow positron beam Doppler broadening spectroscopy (SPBDBS) has been used to study the vacancy defects induced by irradiation of a pure FeNiCr (D302) model alloy and an industrial 316L alloy. Results obtained will be interpreted with the help of those obtained on a pure Fe by He and *al.* [2]. Materials were irradiated in different conditions: either with heavy Ni^{2+} ions to

[1]PERFORM 60: Prediction of the Effects of Radiation For RPV and in-vessel Materials using MSM-60 years foreseen plant life time.

simulate reactor neutron irradiations, or with light $^4He^{2+}$ ions to obtain based data complementary to those with Ni^{2+}. The results obtained by PAS will be compared to APT and TEM observations performed in [1].

2. Experimental

The metal alloys (fcc) were provided by OCAS company. Their composition is given in Table 1. Before irradiation, samples (5x5 mm^2) have undergone several stages of preparation (mechanical polishing, electro-polishing, and annealing) in the aim to get "virgin" samples with fewer defects as possible checked by SPBDBS measurements. Samples have been irradiated in different conditions: either with 5 MeV Ni^{2+} at high temperatures (HT) 450°C and to 0.5 or 1 dpa, or with 1.5 MeV $^4He^{2+}$ at RT and with damage dose between $1x10^{-4}$ and $1x10^{-2}$ dpa (fluence between $1x10^{14}$ and $1x10^{16}$ $ions.cm^{-2}$). The pure Fe has been also irradiated with 1.5 MeV $^4He^{2+}$ at RT and to damage dose between $1x10^{-5}$ and 0.1 dpa (fluence between $1x10^{13}$ and $1x10^{17}$ $ions.cm^{-2}$) and studied. Irradiations at HT have been performed on the JANNUS platform using ARAMIS Tandem (Orsay France), and the ones at RT using the Van de Graaff accelerator (CEMHTI Orléans, France).

Table 1. Chemical composition of 316L stainless steel and D302 model alloy (wt.%).

Composition	Cr	Ni	Mo	Mn	Si	C	S	P	Cu	Fe
316L	17.6	12	2.39	1.78	0.59	0.028	0.01	0.01	0.066	bal.
D302	18.27	12.4	-	-	-	0.008	0.003	-	-	bal.

Figure 1 (a) and (b) shows the calculated depth profiles of implanted ions concentration and the corresponding damage calculated for 316L alloy implanted with 1.5 MeV He ions and 5 MeV Ni ions, respectively. These calculations are performed by using SRIM software [3] with a displacement threshold energy of 40 eV (Fe), 44 eV (Cr), and 28 eV (Ni), a density = 7.99 $g.cm^{-3}$ (316L), and for the fluence = $1x10^{14}$ $ions.cm^{-2}$. In the same figure, the slow positron implantation profile is also illustrated for different energies [4]. We point out that the implantation with heavy Ni ions generates dpa greater than that observed with light He ions. The theoretical projected range (Rp) for $^4He^{2+}$ implanted ions is 2310 nm, while Rp for Ni^{2+} implanted ions is lower and equal to 1660 nm. Positrons probe mainly the low damage ($1x10^{-4}$ dpa) and low He concentration (< $4x10^{-6}$ at.%) region for the He implantations, whereas the damage dose becomes strongly higher (7.8 $x10^{-2}$ dpa) for Ni implantations (as calculated at depth of 810 nm). The calculations of the number of displaced Ni atoms by the pka (primary knock-on atoms) for 316L by using SRIM software (ignoring the other atoms of the lattice, Fe and Cr) showed that 56% of pka created by 1.5 MeV He ions is only be able to create Frenkel-pairs (number of displaced atoms ≤ 1), against 39% of pka in the case of 5 MeV Ni implantation, and 99% of pka created by 1.5 MeV He ions participates in the creating of displacement cascades containing less than 30 Ni atoms compared to displacement cascades of 518 Ni atoms when the target is irradiated with 5 MeV Ni ions. These calculations suggest that, positrons will probe a greater concentration of complex defects when irradiation is performed with Ni ions. Knowing that, the area of nuclear cascades may not be detected by SPBDBS technique, since, even the most energetic positrons rarely reach this zone especially when irradiation is done with He ions (Figure 1).

The Slow positron beam Doppler broadening spectroscopy (SPBDBS) [5] at the CEMHTI laboratory has been used to probe vacancy defects in the first μm of metals. This technique is based on the measurement of the low (S) and the high (W) momentum fractions of the 511 keV positron–electron annihilation line. S corresponds to annihilations in the momentum range (0 - $|2.49|$)$x10^{-3}m_0c$, and W in the windows ($|9.64|$ - $|24.88|$)$x10^{-3}m_0c$, respectively. The measurements were performed at RT. The S and W parameters were recorded as a function of the positron energy E which changes by 0.5 steps in the range 0.5 to 25 keV. The SPBDBS uses a high purity germanium (Ge) detector with a good resolution (< 1.3 keV at 511 keV) and a high efficiency (> 25% at 1.33 MeV).

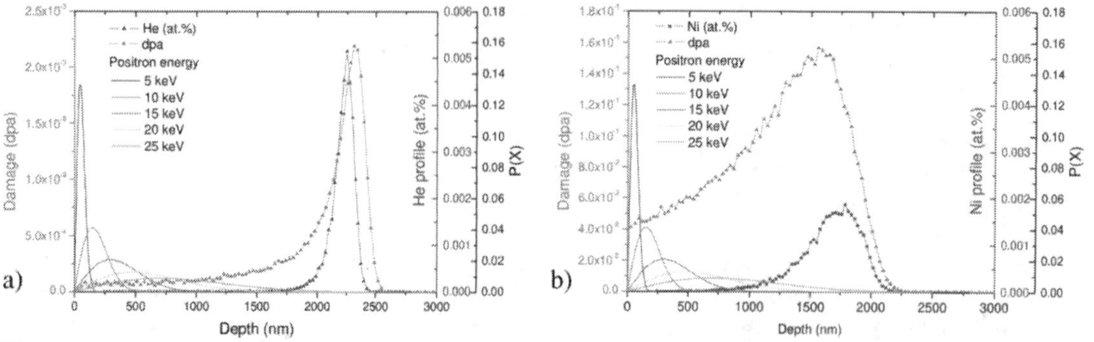

Figure 1. SRIM damage profile of 316L implanted with 1.5 MeV He ions (a) and 5 MeV Ni ions (b) at fluence of $1x10^{14}$ ions.cm^{-2} compared to positron implantation profile.

3. Results and discussion

3.1. Irradiation with light He ions at RT

In this case, metal alloys (316L and D302) were irradiated with 1.5 MeV $^4He^{2+}$ at RT up to damage dose between $1x10^{-4}$ and $1x10^{-2}$ dpa. Figure 2 (a) shows the average values of S and W parameters of 316L and D302 obtained for positron energy between 20 and 22 keV compared to results obtained on pure Fe irradiated with 1.5 MeV $^4He^{2+}$ ($1x10^{-5}$ to 0.1 dpa) at RT [2]. This energy range is considered to be representative of the annihilation characteristics in the bulk. The error bars are limited in the points and thus are not plotted in this figure ($\Delta W = \pm 7.5x10^{-4}$; $\Delta S = \pm 13.6x10^{-4}$).

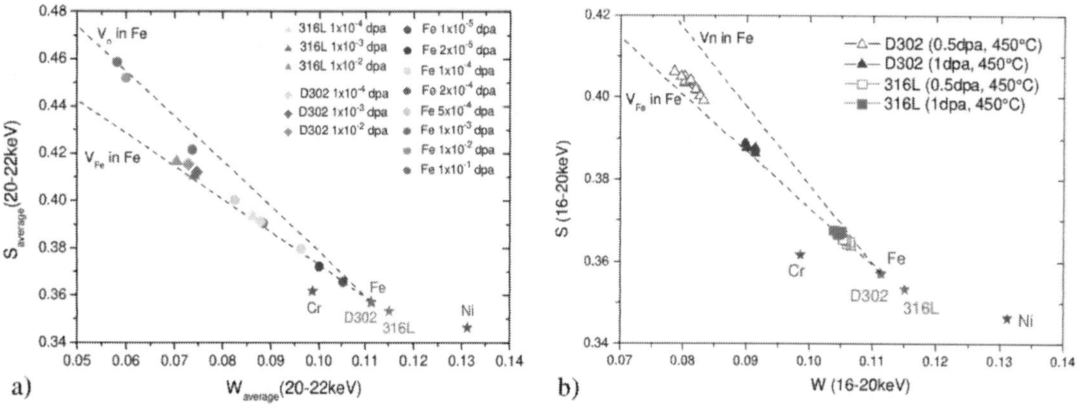

Figure 2. Variation of low momentum annihilation fraction S as a function of high momentum annihilation fraction W: (a) for irradiated Fe, 316L, and D302 with 1.5 MeV He ions to different damage doses and (b) for 316L and D302 alloys irradiated with 5 MeV Ni ions at 450°C and to two damage doses: 0.5 and 1 dpa. Stars points are the lattice points obtained on virgin samples.

For the metal alloys, we notice different behaviour depending on its composition. For 316L, the S,W points lie at the same "V_{Fe} in Fe" line, the single vacancy characteristic line V_{Fe} as determined in [1] whatever the irradiation conditions are (low and high dpa, at RT). In the case of D302 alloy, we notice that from $1x10^{-3}$ dpa, the S,W points are slightly above the "V_{Fe} in Fe" line. It could be due to the detection either of small vacancy clusters, or of Stacking Faults Tetrahedral (SFTs) which are typical irradiation induced defects in fcc metals. These types of defects are special arrangements of vacancy defects essentially in 2D morphology leading to annihilation characteristics very different of the ones for 3D arrangement of the same number of vacancies. In fact, as shown in Ni by Kuramoto [6], calculated positron lifetimes for different numbers of vacancy are found strongly lower values for

large number. The size of these clusters observed in D302 alloy is smaller than that observed in pure Fe from the same damage dose ($1x10^{-3}$). It could be related to the different migration properties of vacancy in both fcc and bcc lattices limiting the clustering. In the case of 316L alloy, the migration temperature has been estimated at about 230 to 270°C, compared to ~ -53 to 5°C in the case of pure iron [7]. Moreover in D302 or other fcc materials such as Ni, SFTs are the predominant form of vacancy clusters.
For both alloys and at the highest damage dose, the S and W remain close for an increase of a factor of 10 in damage dose. It suggests that a saturation of vacancy detection is reached or close to be reached. However, for pure Fe such saturation is not reached yet because of 3D-V_n formation and the increase of the size of vacancy clusters with the fluence [2].

3.2. Irradiation with heavy Ni ions at HT

Under high temperature conditions, metal alloys have been irradiated with 5 MeV Ni^{2+} at 450°C, and to 0.5 or 1 dpa damage doses. Figure 2 (b) shows the variation of S parameter versus W parameter, obtained for positron energy between 16 and 20 keV which correspond to a depth zone also observed by APT and TEM. First of all, we notice that the defect concentration in 316L is drastically lower than in D302. Moreover for 316L, only one type of defect is detected: single vacancy (V), independently of damage dose (0.5 or 1 dpa) like it was observed under RT irradiation conditions, while in the case of D302, vacancy clusters seem to be the major defect detected (bulk points are lying above the "V_{Fe} in Fe" line). The size of clusters is smaller than that observed in the case of pure iron. In addition, for D302 we notice a lower defect concentration at 1 dpa dose compared to 0.5 dpa dose: decreasing of S parameter and increasing of W parameter. Such difference in defect concentration can be explained by an annealing effect of the induced defects during the irradiation which is two times longer for 1 dpa compared to 0.5 dpa (same flux, but fluence (irradiation time) is two times higher).

A. Volgin has performed APT and TEM analyses for 316L and D302 irradiated at 450°C and to 1 dpa damage dose [1]. APT shows, in all bulk samples irradiated at high temperature, the formation of Cottrell clouds. These clouds were enriched in Ni and depleted in Cr (Size: 5 nm; Nd = $(6.8\pm1.3)x10^{22}$ m^{-3}) in the case of D302 alloy, and in Ni, Si, and P in 316L. In the latter case, the Cottrell clouds are along linear dislocation lines (disl. density: $(2.7 \pm 0.2)x10^{14}$ m^{-2}), and the presence of some small objects (clusters and clouds) NiSiP-rich also is noticed (Figure 3). TEM observations show the formation of Frank loops for both alloys D302 and 316L. The mean size of loops is however different, the diameter of loops is bigger in the D302 alloy, 15.3 nm compared to 6 nm in the case of 316L. Frank loops were distributed homogeneously in D302 alloy while they were preferentially located in the vicinity of the dislocation network in the 316L. SFTs are also detected in D302 but not in 316L. Their number density (Nd) is $(2.8\pm0.6)x10^{22}$ m^{-3} with mean size of 4 nm (450°C, 1 dpa). These SFTs could correspond to the clusters detected by SPBDBS from 0.5 dpa dose. Neither cavities, nor precipitates are detected in both alloys [1].

To explain this difference observed between 316L and D302, especially as regarding the absence of vacancy clusters in the case of 316L, it was suggested that the presence of Si in 316L could play a role. In fact, it is known that the addition of silicon in pure nickel, Ni-Cr alloys, and Fe-Ni-Cr alloys, raises the diffusivity of each of the alloy components [8]. The resultant increase in the effective vacancy diffusion coefficient causes large reductions in the nucleation rate of voids during irradiation. The concept of existing a correlation between the suppression of void formation and the presence of fast-diffusing solute or solvent elements initially proposed by Venker, Ehrlich, and Giesecke [9, 10] appears to be an alternative to the concept that defect trapping at solute elements reduces void formation. The latter possibility has been proposed and analyzed by several authors [11]. Indeed, the trapping of interstitials and/or vacancies at solute atoms or impurities enhances the probability of recombination, and thereby lowers the supersaturation of vacancies. Silicon in 316L was observed by APT after irradiation to be especially segregated along linear dislocation lines. The dislocations as known are a major sink of defects; this can be added to the idea of existing Si-interstitial complexes and can contribute in the recovering of the alloy.

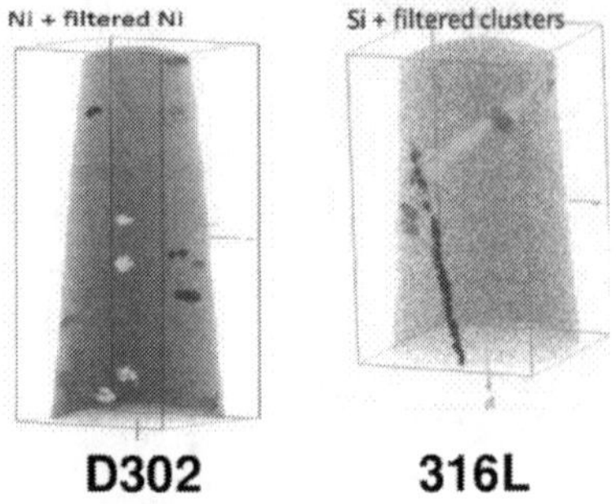

Figure 3. APT 3D reconstruction of typical volumes of D302 and 316L implanted with Ni ions: 1 dpa, 450°C. Detected elements are filtered and imaged in colour; in this paper we only show the Ni and Si enrichment in D302 and 316L, respectively [1].

4. Conclusions

The principal results obtained show that only single vacancies have been observed in the 316L alloy independent of the irradiation conditions studied in this work (low and high dpa, at RT or HT), while vacancy clusters seem to be formed in the D302 alloy under HT irradiation conditions and from $1x10^{-3}$ dpa at RT. The size of clusters is smaller than that observed in the case of pure iron. The difference observed between 316L and D302 alloys could be linked to a role played by the alloying elements such as silicon on the nature of created defects.

References

[1] Volgin A 2012 Characterization and understanding of ion irradiation effect on the microstructure of austenitic stainless steels. University of Rouen)

[2] He C-W, Dawi K, Platteau C, Barthe M-F, Desgardin P and Akhmadaliev S 2013 Vacancy type defect formation in irradiated α-iron investigated by positron beam Doppler broadening technique *Journal of Physics: Conference Series (JPCS), IOP Publishing* in press

[3] Ziegler J F, Ziegler M D and Biersack J P 2010 SRIM - The stopping and range of ions in matter *Nuclear Instruments and Methods in Physics Research, Section B: Beam Interactions with Materials and Atoms* **268** 1818-23

[4] Krause-Rehberg R and Leipner H S 1999 *Positron Annihilation in Semiconductors* vol 127 (Berlin: Springer-Verlag)

[5] Desgardin P, Liszkay L, Barthe M-F, Henry L, Briaud J, Saillard M, Lepolotec L, Corbel C, Blondiaux G, Colder A, Marie P and Levalois M 2001 Slow Positron Beam Facility in Orléans *Materials Science Forum* **363-365** 523-5

[6] Kuramoto E, Tsutsumi T, Ueno K, Ohmura M and Kamimura Y 1999 Positron lifetime calculations on vacancy clusters and dislocation in Ni and Fe *Computational Materials Science* **14** 28-35

[7] Takaki S, Fuss J, Kugler H, Dedek U and Schultz H 1983 The resistivity recovery of high purity and carbon doped iron following low temperature electron irradiation *Radiation effects* **79** 87-122

[8] Esmailzadeh B, Kumar A and Garner F A 1985 The influence of silicon on void nucleation in irradiated alloys *Journal of Nuclear Materials* **133-134** 590-3

[9] Venker H and Ehrlich K 1976 Relation Between Partial Diffusion Coefficients in Alloys and their Swelling Behavior under Fast Neutron Irradiation *Journal of Nuclear Materials* **60** 347-9

[10] Venker H, Giesecke P and Ehrlich K 1977 The influence of fast diffusing substitutional elements on the swelling behavior of Ni- and Cu- alloys. In: *International conference on radiation effects in Breeder reactor structural materials,* ed M L Bleiberg and J W Bennett (Scottsdale, Arizona pp 415-20

[11] Mansur L K Effects of point defect trapping and solute segregation on irradiation-induced swelling and creep *Journal of Nuclear Materials* **83** 109-27

Journal of Physics: Conference Series **505** (2014) 012008 doi:10.1088/1742-6596/505/1/012008

Corrosion-related defects in Zircaloys: a preliminary study with slow positron beam

Z J Zhu[1], M Y Yao[2], X D Xue[1], Y C Wu[1] and B X Zhou[2]

[1] School of Physics and Technology, Hubei Nuclear Solid Physics Key Laboratory, Wuhan University, Wuhan 430072, P.R. China

[2] Institute of Materials, Shanghai University, Shanghai 200072, P.R. China

E-mail address: ycwu@whu.edu.cn

Abstract. Corrosion-related microstructure and defects in Zircaloy-4 and N18 alloys were investigated by variable energy positron annihilation spectroscopy. The specimens were corroded in 0.01mol/L LiOH aqueous solution at 360 ℃/18.6 MPa and in super heated steam at 400 ℃/10.3 MPa, respectively. Defect profiles were analyzed by measuring the *S* parameter as a function of incident positron energy from 0.25 to 27 keV. Results indicated that Zircaloys corroded in LiOH aqueous solution contained more defects in the oxide layer than that in superheated steam, which implies that formation of defects in oxide layer may relate to the effects of Li^+ ions in corrosion solution.

Introduction

Zirconium based alloy (Zircaloy) has been widely used as cladding materials in nuclear reactors due to its low thermal neutron capture cross section, good corrosion resistance and adequate mechanical properties [1]. During its service in reactors, Zircaloy was corroded in high temperature high pressure condition. Corrosion-induced defects of Zircaloy are one of the important degradation phenomena in long term operation in nuclear reactors. Research indicated that the corrosion process led to a volumetric expansion of the oxide layer as zirconium (Zr) near the oxide-metal interface region were oxidized and transformed to zirconia (ZrO_2). It is widely known that high compressive stresses were built up during oxidation due to a high Pilling-Bedworth ratio (1.56 for Zr). Under these high compressive stresses, the dense oxide was formed near the interface which controlled the corrosion behavior of Zircaloys. However, the relaxation of the stress in companion with the destabilization might happen while the oxides were formed, which led to defects in the oxide layer [2, 3, 4]. Analyzing the details of these defects will contribute to the understanding of the corrosion mechanism of Zircaloys. Positron annihilation spectroscopy (PAS) has been widely used as a probe for defect studies at the atomic dimension in solids, such as vacancies, vacancy clusters, dislocations, and nanometer-scale voids. The use of variable energy positron beams allows one to probe the depth profile of solids, and is particularly suitable for analysis of surface and near-surface defects [5]. In this work, corrosion-related microstructure and defects in Zircaloy-4 and N18 alloys in 0.01mol/L LiOH aqueous solution at 360 ℃/18.6 MPa and in super heated steam at 400 ℃/10.3 MPa were investigated by slow positron beam-based Doppler broadening spectroscopy.

Experiment

Specimens with 20mm × 10mm ×1.0 mm in size were cut from Zircaloy-4 (denoted as Zr-4) and N18 alloy plates for corrosion tests and Doppler broadening measurements. Chemical composition of both alloys is shown in table 1. One group of specimens was corroded in an autoclave with 0.01 mol/L LiOH aqueous solution at 360 ℃/18.6 MPa for 30 days (denoted as Zr-4-360 and N18-360, respectively), while another group of specimens was corroded in super heated steam at 400 ℃/10.3 MPa for 30 days (denoted as Zr-4-400-s and N18-400-s). The thicknesses of the oxide layer induced by corrosion were about 1~2 μm, which were estimated by measuring the weight gain of these specimens before and after the corrosion tests. Prior to the corrosion test, the specimens were annealed at 580 ℃ for 4h in a vacuum furnace at about 5×10^{-3} Pa, and cleaned using a pickling solution (a mixture of 10 vol.% HF, 30 vol.% HNO_3, 30 vol.% H_2SO_4 and 30 vol.% H_2O), sequentially rinsed in cold tap water, boiling deionized water and then blow-dried with warm air.

Table 1: Chemical composition of Zr-4 and N18 alloys (wt.%).

	Sn	Fe	Cr	Nb	Zr
Zr-4	1.5	0.2	0.1	—	Balanced
N18	1.0	0.3	0.1	0.35	Balanced

Doppler broadening measurements of uncorroded and corroded specimens were carried out by the slow positron beam with incident positron energies E from 0.25 to 27 keV, and a high purity-Ge detector with an energy resolution of 1.64 keV at 511keV peak was used to collect the annihilation γ energy spectra. 10^6 annihilation events with a count rate of 1000 cps near the 511 peak were collected in each Doppler broadening spectrum.

Results and Discussion

Figure 1 shows *S* parameter as a function of incident positron energy in uncorroded and corroded specimens. The *S-E* curves show that there exists a peak value of *S* parameters in uncorroded Zr-4 and N18 specimens when positron incident energy is from 0.25 to around 8 keV, which indicates an interface defect layer near the oxide-metal interface region. Then with further increasing positron incident energy, the *S* parameters tend to a plateau where most positrons are implanted in the Zr substrate and annihilate therein.

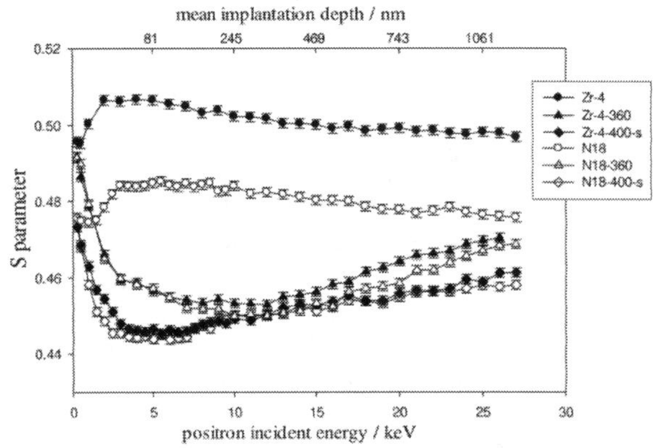

Figure 1. S parameter as a function of incident positron energy in uncorroded and corroded specimens.

The *S-E* curves of uncorroded Zr-4 and N18 specimens from figure 1 were fitted by VEPFIT using a two-state model (interfacial defect layer and substrate or bulk). Table 2 shows fitted values of *S* parameter, positron diffusion length in the interface layer and in the Zr substrate, and the thickness of interface layer. In uncorroded Zircaloys, defects were mainly located in the oxide-metal interface layer with thickness of 100~150 nm. In addition, the *S* parameter in Zr-4 substrate is larger than that in N18 substrate. This is due to Nb element in N18 alloy whose outer electronic configurations are $4d^45s^1$, which gives positron more chances to annihilate with high-momentum electrons, leading to a smaller *S* value.

Table 2: Fitted S values of the diffusion length in interface layer (L_d) and in the Zr substrate (L_B), *S* parameter in interface layer (S_d) and in the Zr substrate (S_B), and thickness of interface layer (B_d).

	S_d	L_d (nm)	B_d (nm)	S_B	L_B (nm)	Chi-square
Zr-4	0.5078±0.023	10±2	112±8	0.4996±0.012	103±6	32
N18	0.4842±0.034	11±2	148±9	0.4798±0.017	105±9	42
Zr-4-360	0.4552±0.025	115±8	—	—	—	45
N18-360	0.4523±0.032	118±8	—	—	—	48
Zr-4-400-s	0.4463±0.026	76±5	—	—	—	34
N18-400-s	0.4435±0.037	73±6	—	—	—	36

The *S-E* profiles of corroded specimens shown in figure 1 are evidently different from those of the uncorroded ones. The *S* parameters of Zr-4-360, Zr-4-400-s, N18-360, N18-400-s specimens decrease rapidly as the positron incident energies increasing from 0.25 to 3 keV, then keep unchanged as the positron incident energies increasing from 3 to 15 keV and finally increase slowly with positron incident energy. For the corroded specimens, *S* parameters and diffusion lengths of the oxide layer were fitted by VEPFIT using positron incident energy from 0.25 to 15 keV, and the fitted *S* values and diffusion lengths are also shown in table 2. As most incident positrons were implanted and annihilated in the oxide layer. The smaller *S* values of corroded specimens are due to the broadened momentum distribution of electrons from the oxygen *p* state, which can attributed to an origin of the small value of *S* for ZrO_2 [6].

It is interesting to observe in figure 1 that *S* parameters of Zr-4-360 and N18-360 are larger than Zr-4-400-s and N18-400-s, which indicates that more defects were induced in the oxide layer when Zircaloys corroded in 0.01 mol/L LiOH aqueous solution at 360 ℃/18.6 MPa. This difference is due to the action of Li^+ ions in corrosion aqueous solution. Previous studies showed that when Zircaloys corroded in LiOH aqueous solutions, the Li^+ ions can easily enter the oxide layer. The Li^+ ions could be absorbed by vacancies and holes in the oxide layer and led to a decrease of the surface free energy of vacancies, which would accelerate the procedure of the diffusion of defects and the formation of pores and micro-cracks [7]. Meanwhile, as the radius of Li^+ ions and Zr^{4+} ions are approximately equal (the radius of Li^+ ions are 0.076 nm, the radius of Zr^{4+} ions are 0.072 nm), Zr^{4+} ions can be easily displaced by Li^+ ions and produce an amount of anion vacancies. It is noted here that these vacancies can make O^{2-} enter the oxide layer easier, and accelerate the corrosion rate of Zircaloys [8].

Furthermore, the *S* parameter of Zr-4-360 is larger than N18-360, which indicates that fewer defects were induced in N18-360 specimen, in agreement with the better corrosion resistance of N18 alloys in LiOH aqueous solution [9]. Besides, there does not exist an evident difference of *S-E* curves in Zr-4-400-s and N18-400-s specimens, indicating that total defect concentrations in both Zircaloys were similar. However, for specimens corroded in super heated steam for long-term corrosion (475 days, about 20 μm thickness of oxide layer), positron annihilation lifetime measurements found that defect size of oxide layer in N18 is smaller, but higher concentration of larger-scale defects than that of Zr-4 specimens. Further investigations are in progress and will be reported in the future.

Information about the number of defect types could be revealed by plotting the *S* parameter as a function of the *W* parameter (*S-W* plots). All the data for uncorroded and corroded specimens are shown in figure 2. For uncorroded specimens, each *S-W* plot shows two positron annihilation states, which are in metal-oxide interface and in Zr substrate. The inflection points represent annihilation of positrons trapped in open-volume defects near the surface oxide-metal interface. However, as shown in figure 2, the *S-W* plots of Zr-4 and N18 specimens are not in the same straight line from interface to substrate, which is due to the different chemical composition between two kinds of specimens. For corroded specimens, the *S* and *W* values all fall on a common locus, which consists of two straight-line segments connecting three verticals. As *E* is increased the points move along the bottom segment from right to left and then along the top from left to right. Positrons annihilate from surface to oxide layer as *E* is increased, and then from oxide layer to interface layer and substrate at higher energy. Positron annihilation states in oxide layer and interface layer are very complicated, further investigations have to be done.

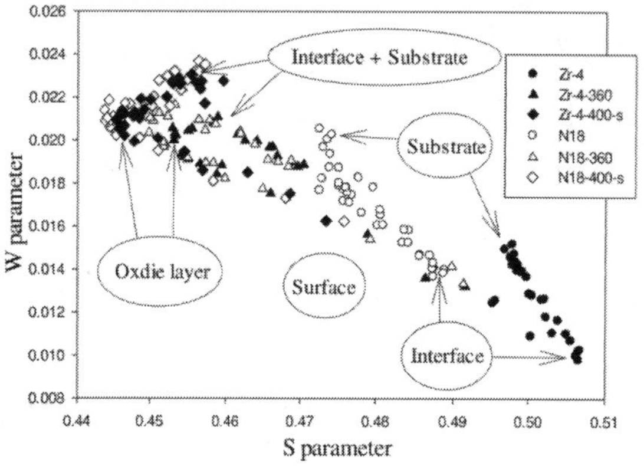

Figure 2. *W* parameter as a function of the *S* parameter for uncorroded and corroded Zircaloys.

Conclusions

Slow positron beam Doppler broadening measurements of corroded Zircaloys showed that open-volume defects were induced during the corrosion. When Zircaloys corroded in 0.01 mol/L LiOH aqueous solution at 360 ℃/18.6 MPa, more defects were introduced in the oxide layer compared to the corrosion in superheated steam 400 ℃/10.3 MPa. This should be the effect of Li^+ ions in the corrosion solution. Furthermore, the *S* parameter of Zr-4-360 is larger than N18-360 which indicates that fewer defects were induced in N18-360 specimens during the corrosion, in agreement with better corrosion resistance of N18 alloys in LiOH aqueous solution.

Acknowledgments

This research is supported by the National Natural Science Foundation of China under grants 11175136, 51071111 and J1210061, and is also performed under the contract between Shanghai University and Wuhan University.

References

[1] Wilson P D 1996 *The Nuclear Fuel Cycle from Ore to Waste* (Oxford: Oxford Science Publication) p 89
[2] Yao M Y, Shen Y F, Li Q, Peng J C, Zhou B X and Zhang J L 2013 *J. Nucl. Mater.* **435** 63
[3] Benali B, Herbst and G M, Gallet I 2006 *Appl. Surf. Sci.* **253** 1222
[4] Park J Y, Choi B K, Jeong Y H and Jung Y H 2005 *J. Nucl. Mater.* **340** 237
[5] Schultz P J and Lynn K G 1988 *Rev. Mod. Phys.* 60 701.
[6] Uedono A, Ikeuchi K, Yamabe K, Ohdaira T and Muramatsu M 2005 *J. Appl. Phys.* **98** 203506
[7] Liu W Q, Li Q and Zhou B X 2004 *Rare Metal Materials and Engineering* **33** 728
[8] Jeong Y H, Baek J H and Kim S J 1999 *J. Nucl. Mater.* **270** 322
[9] Zhou B X, Yao M Y, Li Z K, Wang X M, Zhou J, Long C S, Liu Q and Luan B F 2012 *J. Mater. Sci. Technol.* **28** 606

Defects in nitride-based semiconductors probed by positron annihilation

A Uedono[1], M Sumiya[2], S Ishibashi[3], N Oshima[4] and R Suzuki[4]

[1] Division of Applied Physics, Faculty of Pure and Applied Science, University of Tsukuba, Tsukuba, Ibaraki 305-8573, Japan
[2] Wide Bandgap Material Group, National Institute for Materials Science, Tsukuba, Ibaraki 305-0044, Japan
[3] Nanosystem Research Institute (NRI) "RICS", National Institute of Advanced Industrial Science and Technology (AIST), Tsukuba, Ibaraki 305-8568, Japan
[4] Research Institute of Instrumentation Frontier, National Institute of Advanced Industrial Science and Technology, Tsukuba, Ibaraki, 305-8568, Japan

E-mail: uedono.akira.gb@u.tsukuba.ac.jp

Abstract. Point defects in $In_xGa_{1-x}N$ grown by metal organic chemical vapor deposition were studied by a monoenergetic positron beam. Measurements of Doppler broadening spectra of the annihilation radiation as a function of incident positron energy for $In_xGa_{1-x}N$ ($x = 0.08$ and 0.14) showed that vacancy-type defects were introduced with increasing InN composition. From comparisons between coincidence Doppler broadening spectra and the results calculated using the projector augmented-wave method, the major defect species was identified as the complexes between a cation vacancy and nitride vacancies. The concentration of the defects was found to be suppressed by Mg doping. An effect of Mg-doping on the positron diffusion properties in GaN and InN was also discussed. The momentum distribution of electrons at the $In_xGa_{1-x}N$/GaN interface was close to that in defect-free GaN or $In_xGa_{1-x}N$, which was attributed to the localization of positrons at the interface due to the electric field caused by polarizations.

1. Introduction

Growth techniques of GaN and Ga-rich $In_xGa_{1-x}N$ have been extensively studied for the past two decades, and this has led to the realization of present GaN-based solid state lighting technologies [1,2]. The $In_xGa_{1-x}N$ material system has also attracted great research interest for use as high-efficiency thin-film photovoltaic device materials [3] because of its potential in controlling the band gaps spanning from 0.65 eV to 3.43 eV, which can cover the wavelength from infrared to ultraviolet regions. The theoretical conversion efficiency of the solar cell using $In_xGa_{1-x}N$ has been reported to be greater than 60% [4]. The efficiency of solar cells using $In_xGa_{1-x}N$/GaN multilayers, however, has been reported to be lower than the theoretical value. The origin of the degradation has been explained by low $In_xGa_{1-x}N$ crystal quality which mainly due to the difficulty of its epitaxial growth. This is mainly associated with large differences in the growth temperatures, equilibrium vapor pressures, and lattice constants of InN and GaN. As a result, the film growth process gives rise to unusually high strain and high defect densities. Dislocations and stacking faults, misorientations, residual strains, and phase separations of $In_xGa_{1-x}N$ have been studied [5], but knowledge regarding point-defects, such as monovacancies and vacancy composites, is also crucial to improve the conversion efficiency. Positron annihilation is a powerful technique for evaluating vacancy-type defects in semiconductors [6], and the defects in group-III nitrides have been investigated using this method [7-9]. In the present study, we have used a monoenergetic positron beam to probe native vacancies in undoped and Mg-doped $In_xGa_{1-x}N$ grown by metal organic chemical vapor deposition (MOCVD).

2. Experiment

The samples investigated were $In_xGa_{1-x}N$ ($x = 0.08$ and 0.14) grown using MOCVD. The growth method and the optical properties of the samples are described elsewhere [10,11]. 1.4-μm-thick GaN films (+*c*-face) were grown on sapphire substrates at 1000°C with buffer layers, and then 300-nm-thick $In_xGa_{1-x}N$ films were grown. For $In_{0.14}Ga_{0.86}N$, the film was deposited on Mg-doped GaN(120 nm)/GaN/sapphire substrate. The growth temperatures of the $In_xGa_{1-x}N$ layers were 790-800°C, respectively. The electron and hole concentrations of undoped and Mg-doped $In_{0.08}Ga_{0.92}N$ were $n_e = 5\times10^{17}$ cm^{-3} and $n_p = 5\times10^{18}$ cm^{-3}, respectively. InN and GaN grown by MOCVD and hydride vapor phase epitaxy (HVPE) were also studied [12,13].

With a monoenergetic positron beam, the Doppler broadening spectra of the annihilation radiation were measured as a function of the incident positron energy *E*. The low-momentum part of the spectra was characterized by the *S* parameter, defined as the number of annihilation events over the energy range of 511 keV $\pm \Delta E_\gamma$ (where $\Delta E_\gamma = 0.76$ keV) around the centre of the peak, and the high-momentum part was characterized using the *W* parameters, defined as the annihilation events in the range of 3.4 keV $\leq |\Delta E_\gamma| \leq 6.8$ keV. The relationship between *S* and *E* was analysed by VEPFIT [14].

Doppler broadening spectra corresponding to the annihilation of positrons were theoretically calculated using QMAS (Quantum MAterials Simulator) code, which uses valence-electron wavefunctions determined by the projector augmented-wave (PAW) method [15]. Details of the calculation method and its application to group-III nitrides are described elsewhere [11-13]. The structural optimization was done for a cell containing about 128 atoms by means of *ab initio* quenched molecular dynamics. For $In_{0.5}Ga_{0.5}N$, the initial bulk structure containing 128 atoms has been generated by means of the special-quasirandom-structure (SQS) approach [16]. The simulation was also perfumed for ordered $In_{0.5}Ga_{0.5}N$, where In- and Ga-faces were assumed to locate in *c*-faces alternately.

3. Results and Discussion

The *S* values of HVPE-GaN and Mg-doped GaN before and after annealing as a function of *E* are shown in Fig. 1. The mean implantation depth of positrons is shown on the upper horizontal axis. For HVPE-GaN, the *S* value saturated at $E > 20$ keV, suggesting that almost all positrons annihilate in the bulk in this energy range. The solid curve is a fit to the experimental data, and the diffusion length of positrons L_d was obtained as 72±1 nm. The derived value is the typical one for defect-free GaN [11-13]. For Mg-doped GaN after annealing, the *S* value near the surface (< 1 keV) decreased rapidly with increasing *E*. For this sample, the L_d value was derived as 2.3±0.1 nm. This very short diffusion length indicates the suppression of thermalized positrons diffusing toward the surface. A schematic of the band structure for p-type GaN is shown as an inset of Fig. 1. Because of the electric field caused by the band bending, the positrons implanted into the depletion region are dragged toward the inside of the sample, and this decreases the L_d value for p-GaN after annealing.

For p-GaN before annealing, the *S* values at $E \approx 2$ keV were almost the same as those after annealing; above this energy, however, they increased with increasing *E* (<12 keV), suggesting the trapping of positrons by vacancy-type defects. When the Fermi level position inside the sample was higher than that near the surface, the diffusion of positrons toward the surface would be enhanced. According to this model, the region sampled by the positrons was divided into three blocks; the first and second blocks respectively correspond to the positron annihilation in the Mg-doped layer, and the third block corresponds to that in the Si-doped layer. The derived *S* values for the first and second layers were close to the *S* values at $E = 2$ keV and 12 keV, respectively. The interface between the first and second layers was obtained as 6.4±0.4 nm. The L_d value in the first layer was fixed to the value for the sample after annealing, and that in the second layer was derived as 127±5 nm. The long diffusion length of positrons in the second layer can be attributed to the enhanced diffusion toward the surface due to the band structure. The *S* value for the first layer was close to that for the sample after annealing, suggesting that the vacancy concentration in the subsurface region (<6 nm) were close to those in the film after annealing. The phenomena observed for the sample before annealing can be attributed to the partial activation of Mg in the subsurface region. Mg is known to segregate on the growth surface of GaN. Thus, because of a combination between the high Mg concentration and an enhanced reaction between Mg and vacancies at the surface, the Fermi level position in the subsurface region could decrease without annealing.

Figure 2 shows the *S–E* curves for undoped InN and Mg-doped InN (p-InN, [Mg] = 3.1×10^{19} cm^{-3}) after annealing, where the activation of Mg in p-InN was confirmed using electrolyte-based capacitance-voltage analysis[12]. For the undoped sample, the L_d value was derived as 34±4nm, and this short diffusion length is considered to be related to the surface band bending for InN [17]. The *S–E* curve for p-InN can be fitted using the model applied to p-GaN before annealing. The width of the first region and the L_d value in the second block were obtained as 70±7 nm and 127±8 nm, respectively. Thus, a similar band structure discussed for p-type GaN before annealing is expected for p-InN. Because of the electron accumulation at the surface of InN, the confirmation of carrier species in InN is difficult [17]. The present results suggests that the analysis of the *S–E* curve can provide useful information about the band structure near the surface.

Figure 3 shows undoped and Mg-doped $In_xGa_{1-x}N$ as a function of incident positron energy *E*. For Mg:$In_{0.08}Ga_{0.92}N$, the *S* value near the surface decreased rapidly with increasing *E*. This is the typical behavior of *S* for *p*-type GaN and InN, and was attributed to the suppression of the positron diffusion toward the surface due to the band bending near the surface. For $In_{0.14}Ga_{0.86}N$, the *S* value was almost flat at *E* = 1-3 keV, which suggests that almost all positrons annihilated in the $In_{0.14}Ga_{0.86}N$ layer without diffusing back to the surface. For all samples, the dips in the *S–E* curves were observed at *E* = 10-15 keV, which can be attributed to the annihilation of positrons near the $In_xGa_{1-x}N$/GaN interface.

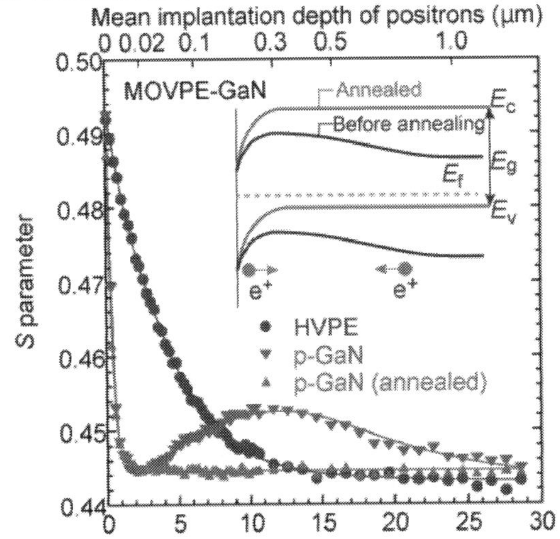

Figure 1. *S* as a function of incident positron energy *E* for undoped HVPE-GaN, Mg-doped GaN (p-GaN) before and after annealing. The solid curves are fits to the experimental data. An inset shows a schematic viewgraph for the band structure for *p*-type GaN.

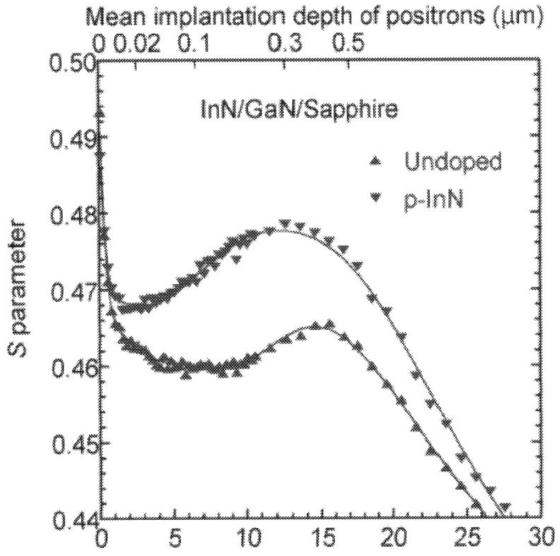

Figure 2. *S* parameters as a function of *E* for undoped and Mg-doped InN (p-InN) after activation of Mg.

Journal of Physics: Conference Series **505** (2014) 012009 doi:10.1088/1742-6596/505/1/012009

Annihilation characteristics of positrons in layered structures can be discussed by graphical analysis of the relationship between the *S* and *W* parameters. Figure 4 shows the *S*–*W* relationship for $In_{0.14}Ga_{0.86}N$, $In_{0.08}Ga_{0.92}N$, and Mg:$In_{0.08}Ga_{0.92}N$. The (*S*,*W*) value for HVPE-GaN (defect-free GaN) is also shown. Arrows show directions of the increase of *E*. For Mg:$In_{0.08}Ga_{0.92}N$, the (*S*,*W*) trajectory can be represented by a straight line [an identical line was also shown in Fig. 4(a) and (b)]. The value approached the defect-free (*S*,*W*) for GaN with increasing *E* ($E \leq 10$ keV). For $In_{0.14}Ga_{0.86}N$, although the (*S*,*W*) value saturated at $E = 2.9$-5.4 keV (corresponding (*S*,*W*) value was shown in dotted lines), above $E = 5.9$ keV, it started to approach the defect-free (*S*,*W*) value. For undoped $In_{0.08}Ga_{0.92}N$, although no clear indication of the annihilation of positrons in the $In_{0.08}Ga_{0.92}N$ layer observed in the *S*–*E* curve, the deviation from the line was observed in the *S*–*W* plot. This suggests that positrons implanted into the $In_xGa_{1-x}N$ layer were under strong influence of the $In_xGa_{1-x}N$/GaN interface. Because the (*S*,*W*) value tend to approach the value for HVPE-GaN, the momentum distribution of electrons near the $In_xGa_{1-x}N$/GaN interface is considered to be close to that for defect free $In_xGa_{1-x}N$ or GaN.

The *S*–*E* curves for $In_xGa_{1-x}N$/GaN were fitted, and the region sampled by the positrons was divided into four blocks; they respectively correspond to the positron annihilation in the $In_xGa_{1-x}N$ layer, the $In_xGa_{1-x}N$/GaN interface, the GaN layer, and the sapphire substrate. The widths of the blocks and the corresponding *S* values were determined by the fitting. The derived depth distributions of *S* are shown

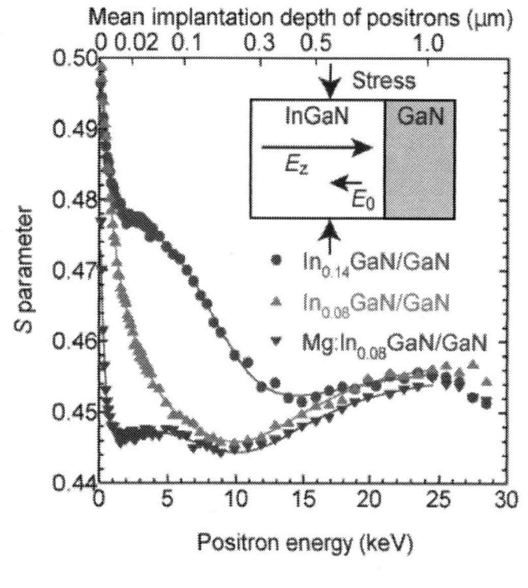

Figure 3. *S* parameters as a function of *E* for undoped In_xGaN_{1-x} (x=0.08 and 0.14) and Mg-doped $In_{0.08}Ga_{0.92}N$. Dips at $E = 9$-15 keV in the curves were attributed to the annihilation of positrons at the $In_xGaN_{1-x}N$/GaN interface. An inset shows the piezoelectric field E_z and the electric field generated from the spontaneous polarization E_0.

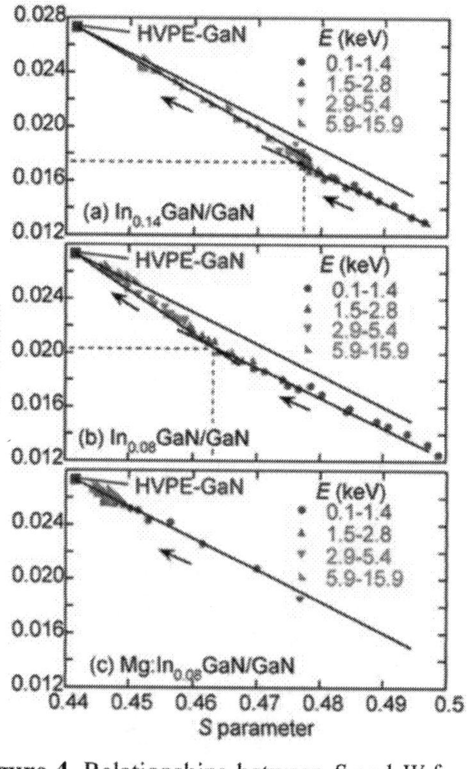

Figure 4. Relationships between *S* and *W* for (a) $In_{0.14}Ga_{0.86}N$, (b) $In_{0.08}Ga_{0.92}N$, and (b) Mg-doped $In_{0.08}Ga_{0.92}N$. The (*S*,*W*) values for HVPE-GaN (defect-free) is also shown. Arrows show the increase in the incident energy of positrons.

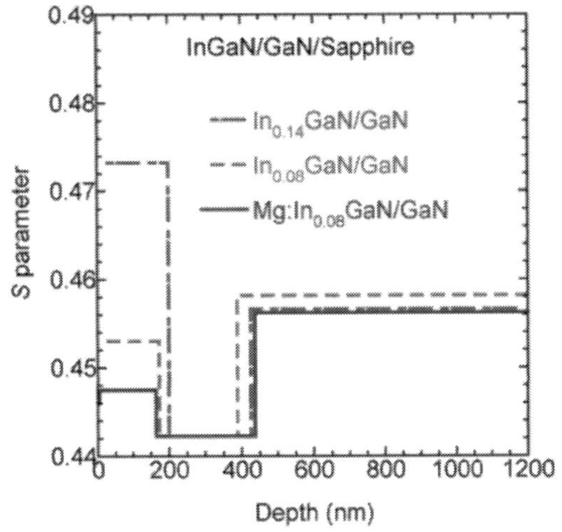

Figure 5. Depth distributions of S for undoped $In_xGa_{1-x}N$ (x = 0.08 and 0.14) and Mg-doped $In_{0.08}Ga_{0.92}N$ derived from the fitting (Fig. 3).

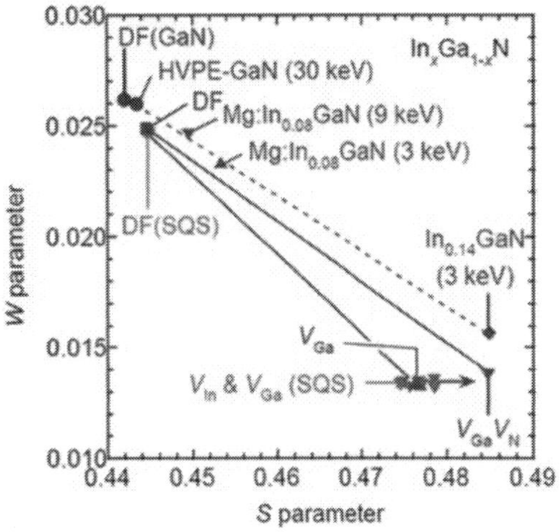

Figure 6. (S,W) values (blue) corresponding to the annihilation of positrons in the delocalized state (defect-free: DF) and that of positrons trapped by V_{Ga} as calculated for $In_{0.125}Ga_{0.875}N$. The result for DF-GaN is also shown. Arrows show the effect of V_N coupled with V_{Ga}. The (S,W) values for $In_{0.14}Ga_{0.86}N$, Mg-doped $In_{0.08}Ga_{0.92}N$, and HVPE-GaN are also shown.

in Fig. 5. For all samples, although the second block represents the $In_xGa_{1-x}N$/GaN interface, their widths were obtained as about 200 nm. Makkonen *et al.* [9] reported the localization of positrons at the InN/GaN interface due to the built-in electric field caused by the piezoelectric effect. Because this effect was not included in the fitting model, the accumulation of positrons near the interface appeared as the wide width of the second block. Because a compressive stress is introduced in the $In_xGa_{1-x}N$ layer grown on the +c-plane GaN template, the positrons in the layer are dragged toward the interface by the piezoelectric field E_z, where the spontaneous electric field E_0 is far smaller than E_z (inset of Fig. 3). This is the major cause of the lack of clear signals for the positron annihilation in the $In_{0.08}Ga_{0.92}N$ layer.

Figure 6 shows the (S,W) values calculated from the coincidence Doppler broadening spectra using the PAW method. The values correspond to the annihilation of positrons in the delocalized state (defect-free: DF) in ordered $In_{0.125}Ga_{0.875}N$ and those of positrons trapped by Ga-vacancy V_{Ga} and divacancy ($V_{Ga}V_N$) (blue symbol) are shown. The result for DF-GaN is also shown. Arrows show the effect of N-vacancy V_N coupled with V_{Ga}. The (S,W) values calculated using SQS for DF-$In_{0.125}Ga_{0.875}N$ and cation vacancies are also shown. The (S,W) values calculated using the ordered structure and SQS are close to each other, suggesting that the (S,W) values calculated using ordered-$In_{0.125}Ga_{0.875}N$ can be used to compare with experimentally obtained values.

Using the coincidence system, the (S,W) values were measured at E = 3 keV for $In_{0.14}Ga_{0.86}N$, at E = 3 and 9 keV for Mg:$In_{0.08}Ga_{0.92}N$, and at E = 30 keV for HVPE-GaN. For $In_{0.14}Ga_{0.86}N$, the obtained value is located on the right side of the line connecting the values of the defect-free and $V_{Ga}V_N$. Because the formation of vacancy complexes with V_N causes a rightward shift in the S–W plot, the major trapping center of positrons can be determined as complexes between a cation vacancy and V_Ns. For Mg:$In_{0.08}Ga_{0.92}N$, the measured (S,W) values are located on the line connecting the values of $In_{0.14}Ga_{0.86}N$ and HVPE-GaN. Thus, the major defect species is expected to be the same as that for $In_{0.14}Ga_{0.86}N$, but their concentration was suppressed by Mg-doping.

Using first-principles calculation [18], it was reported that the formation energy of V_N decreased as the number of In adjacent to V_N increased. Thus, both the bond-length/angle distortions and the decrease in the formation energy of V_N are expected to promote the introduction of V_N with increasing InN content. Although positrons are not trapped by isolated V_N, some of them preferentially couple with cation vacancies and form stable complexes. Thus, we can conclude that the observed behaviour of defects in in $In_xGa_{1-x}N$ is considered to be the introduction of V_N and related formation of cation-vacancy-V_N complexes.

4. Summary

We studied vacancy-type defects in $In_xGa_{1-x}N$ grown by MOCVD using a monoenergetic positron beam. The major defect species in the $In_xGa_{1-x}N$ film was identified as cation vacancies coupled with V_Ns, and their concentration increased with increasing InN content. The defect concentration, however, was decreased by Mg-doping. The intrinsic (S,W) value corresponding to the annihilation of positrons at the $In_xGa_{1-x}N$/GaN interface was close to that for defect-free GaN or $In_xGa_{1-x}N$, suggesting the localization of positrons at the interface due to the piezoelectric field. From the measurements of Mg-doped InN, $In_xGa_{1-x}N$, and GaN, it was found that the diffusion property of positrons is sensitive to variation in the band structures in the subsurface region. We have shown that the positron annihilation parameters are sensitive to vacancy-type defects in $In_xGa_{1-x}N$ films, meaning that this technique can be a useful tool for evaluating $In_xGa_{1-x}N$-based devices.

References

[1] Nakamura S and Chichibu S F 2000 *Introduction to Nitride Semiconductor Blue Lasers and Light-Emitting Diodes* (London: Taylor and Francis)
[2] Akasaki I and Amano H 2006 *Jpn. J. Appl. Phys.* **45** 9001
[3] Wu J, Walukiewicz W, Yu K M, Shan W, Ager III J W, Haller E E, Lu H, Schaff W J, Metzger W K and Kurtz S 2003 *J. Appl. Phys.* **94**, 3967
[4] Devos A 1992 *Endoreversible Thermodynamics of Solar Energy Conversion* (New York: Oxford University Press).
[5] Wu J 2009 *J. Appl. Phys.* **106** 011101
[6] Krause-Rehberg R and Leipner H S 1999 *Positron Annihilation in Semiconductors, Solid-State Sciences* (Berlin: Springer-Verlag)
[7] Saarinen K, Laine T, Kuisma S, Nissilä J, Hautojärvi P, Dobrzynski L, Baranowski J M, Pakula K, Stepniewski R, Wojdak M, Wysmolek A, Suski T, Leszczynski M, Grzegory I and Porowski S, 1997 *Phys. Rev. Lett.* **79** 3030
[8] Tuomisto F, Pelli A, Yu K M, Walukiewicz W, Schaff W J 2007 *Phys. Rev. B* **75** 193201
[9] Makkonen I, Snicker A, Puska M J, Mäki J-M and Tuomisto F 2010 *Phys. Rev. B* **82** 041307(R)
[10] Sang L, Liao M, Ikeda N, Koide Y and Sumiya M 2011 *Appl. Phys. Lett.* **99** 161109
[11] Uedono A, Tsutsui T, Watanabe T, Kimura S, Zhang Y, Lozac'h M, Sang L W, Ishibashi S and Sumiya M 2013 *J. Appl. Phys.* **113**, 123502
[12] Uedono A, Nakamori H, Narita K, Suzuki J, Wang X, Che S-B, Ishitani Y, Yoshikawa A and Ishibashi S 2009 *J. Appl. Phys.* **105** 054507
[13] Uedono A, Ishibashi S, Tenjinbayashi K, Tsutsui T, Nakahara K, Takamizu D and Chichibu S F 2012 *J. Appl. Phys.* **111** 014508
[14] Van Veen A, Schut H, De Vries J, Hakvoort R A and Ijpma M R 1990 *AIP Conf. Proc.* **218** 171
[15] Ishibashi S 2004 Materials Science Forum **445-446** 401. Up-to-date information of QMAS is reported in http://qmas.jp/.
[16] Zunger A, Wei S-H, Ferreira L G and Bernard J E 1990 *Phys. Rev. Lett.* **65** 353
[17] Yoshikawa A, Wang X Q, Ishitani Y and Uedono A 2010 *Physica Status Solidi A* **207** 1011
[18] Obata T, Iwata J, Shiraishi K and Oshiyama A, 2009 *J. Crystal Growth* **311** 2772

First-principles calculation of positron states and annihilation parameters for group-III nitrides

S Ishibashi[1] **and A Uedono**[2]

[1]Nanosystem Research Institute (NRI) "RICS", National Institute of Advanced Industrial Science and Technology (AIST), Tsukuba, Ibaraki 305-8568, Japan
[2]Division of Applied Physics, Faculty of Pure and Applied Science, University of Tsukuba, Tsukuba, Ibaraki 305-8573, Japan

E-mail: shoji.ishibashi@aist.go.jp

Abstract. We calculated theoretically positron states and annihilation parameters for V_{Ga}-C complexes in GaN. Positron lifetimes and Doppler-broadening S–W parameters were evaluated. The obtained results are compared other various types of defects in group-III nitrides. A random alloy $In_{0.5}Ga_{0.5}N$ and its multilayer system together with GaN are investigated also. The results of the present study are expected to be useful for interpretation of experimental results on practical materials.

1. Introduction

Group-III nitrides are thought to be important and promising materials for various applications such as optoelectronics etc. Their properties and functions can be controlled by forming alloys and/or superlattices. For practical materials and devices, often, defects such as vacancies and dislocations play a critical role in their performances. We have been successfully applying the positron annihilation technique in characterizing defects for group-III nitrides [1]-[10].

In the present study, we calculated positron states and annihilation parameters for V_{Ga}-C complexes, which are thought to be closely related to photoluminescence, in GaN and compared with the previous results for various kinds of defects in AlN, GaN and InN. We also investigated a random alloy $In_{0.5}Ga_{0.5}N$. For the purpose of more realistic modeling of alloys, we have used the special-quasirandom-structure (SQS) approach [12, 13]. Distributions of Doppler S and W parameters as well as positron lifetimes for cation vacancies are evaluated. Positron states and annihilation parameters for a superlattice system $In_{0.5}Ga_{0.5}N$/GaN are also investigated.

2. Computational details

To calculate positron states and annihilation parameters in systems containing defects or in alloys, we used orthorhombic supercells which are equivalent to $4 \times 4 \times 2$ hexagonal unit cells of the wurzite structure. As for the alloys, we utilized the special quasirandom structure (SQS) approach [12, 13], which provides realistic structural modeling for random alloys using periodic supercells. With this approach, two kinds of atoms are arranged so that pair-correlation functions become as close to those for an infinite random alloy as possible. Practically, an atomic arrangement, which gives the lowest Warren-Cowley parameter for several neighboring shells $\alpha_i = 1 - P_B(i)/x_B$, is chosen, where $P_B(i)$ is the probability of finding a B atom on the

i-th nearest neighbor shell about an A tom as the origin. In the present study, shells up to the 7-th are taken into account and the quantity $\sum_{i=1}^{7} \alpha_i^2$ should be minimized.

Electronic structures have been calculated with our computational code QMAS [14], which adopts the projector augmented-wave (PAW) method [15] and the planewave basis. As for the electronic exchange-correlation energy, we used the generalized gradient approximation [16]. Although lattice parameters can be determined with first-principles calculations, we used the experimental ones in the present study for convenience. Internal atomic positions were optimized with evaluating the Hellmann-Feynman force. Effects of the positron localization to atomic configurations were not considered. For each system, we assumed a neutral charge state. The planewave energy cutoff and the force convergence criterion were set to be 20 Ha and 5×10^{-5} Ha/bohr, respectively. As for $\boldsymbol{k}$-point sampling, we used the Γ point only. Positron states were calculated with constructing the positron effective potential using core electron wavefunctions from isolated atom calculations and valence electron wavefunctions described in the PAW formalism [17]. With the resultant positron and electron wavefunctions, positron annihilation Doppler broadening spectra $\rho_{1D}(p_z)$ and positron lifetimes τ were calculated with the following equations:

$$\rho_{1D}(p_z) = \int\int \rho(\boldsymbol{p})\mathrm{d}p_x\mathrm{d}p_y, \quad \rho(\boldsymbol{p}) \propto \sum_j \left| \int \psi_+(\boldsymbol{r})\psi_{-j}(\boldsymbol{r})\sqrt{\gamma\left(n_-(\boldsymbol{r})\right)}\exp(-i\boldsymbol{p}\cdot\boldsymbol{r})\mathrm{d}\boldsymbol{r} \right|^2, \quad (1)$$

$$1/\tau = \pi r_0^2 c \int n_+(\boldsymbol{r})n_-(\boldsymbol{r})\gamma\left(n_-(\boldsymbol{r})\right)\mathrm{d}\boldsymbol{r}, \quad (2)$$

where ψ_+, ψ_{-j}, n_+ and n_- are the positron wavefunction, the j-th electron wavefunction, the positron density and the electron density, respectively. r_0 represents the classical electron radius and c is the speed of light. γ is the enhancement factor. Here, we used the Boroński-Nieminen enhancement factor $\gamma(r_s) = 1 + 1.23r_s + 0.8295r_s\sqrt{r_s} - 1.26r_s^2 + 0.3286r_s^2\sqrt{r_s} + r_s^3/6$ [18], where $r_s = [3/(4\pi n_-)]^{1/3}$ is the density parameter, with a small modification to deal with semiconductors [19]. In analogy with the GGA enhancement factor [20], and to be consistent with the positron-electron correlation energy, we used the expression $\gamma_{\mathrm{semi}} = 1+(\gamma-1)(1-1/\varepsilon_\infty)$. ε_∞ is the dielectric constant of the target system. Although it can be determined by first-principles calculations, we used the experimental values for convenience. For alloys, we used linearly-interpolated values. As for the positron-electron correlation potential, we used that proposed by Boroński and Nieminen [18]. It is also modified for semiconductors by scaling [19].

3. Results and discussion

In Table 1, calculated positron lifetimes for V_{Ga}-C complexes in GaN together with the previous results for various kinds of defects in AlN, GaN and InN are listed. For example, the calculated lifetimes for the bulk and the Ga vacancy in GaN are 159.5 ps and 237.9 ps. They are in good agreement with the experimental values of 166 ps and 235 ps [21]. Positrons are not completely trapped at the N vacancy. Although these predicted lifetime values are already useful to distinguish defect species, so-called S–W plots are more powerful for this purpose. In Fig. 1, an S–W plot for bulk GaN and defects in GaN is shown. Systematic variations are observed in the two-dimensional space while the lifetime varies in the one-dimensional space. As for V_{Ga}-impurity complexes, the direction of variation differs depending on the impurity species. For several representative ones, ratio curves against the bulk are plotted in Fig. 2.

As for the $\mathrm{In_{0.5}Ga_{0.5}N}$ alloy, the supercell used in the present study contains 128 atoms. Its lattice parameters and dielectric constant are given as average values of those for InN and GaN. There are 64 different cation sites. We made calculations for single vacancies on these sites. In Fig. 3, obtained S and W parameters are plotted. It is clearly shown that the S parameter shows a significant variation while the W parameter shows a smaller variation. We have found that the larger Ga concentration on the nearest 12 cation sites tends to bring the larger W value.

Journal of Physics: Conference Series **505** (2014) 012010 doi:10.1088/1742-6596/505/1/012010

Table 1. Calculated positron lifetimes (ps)

AlN $(a,c) = (3.11, 4.98)$ Å, $\varepsilon_\infty = 4.7$		GaN $(a,c) = (3.189, 5.18625)$ Å, $\varepsilon_\infty = 5.5$		InN $(a,c) = (3.538, 5.7029)$ Å, $\varepsilon_\infty = 8.4$	
bulk	165.3	bulk	159.5	bulk	175.8
V_{Al}	244.6	V_{Ga}	237.9	V_{In}	256.8
V_N	166.9	V_N	160.2	V_N	176.6
$V_{Al}V_N$	253.0, 267.0	$V_{Ga}V_N$	244.9, 245.7	$V_{In}V_N$	258.4, 260.9
		$V_{Ga}(V_N)_2$	252.6		
$V_{Al}(V_N)_3$	281.5	$V_{Ga}(V_N)_3$	259.6	$V_{In}(V_N)_3$	265.6
		$(V_{Ga})_2V_N$	274.2		
		$(V_{Ga})_2(V_N)_2$	295.4, 259.9(line)	$(V_{In})_2(V_N)_2$	294.6, 258.9(line)
		$(V_{Ga})_3(V_N)_3$	329.0		
		$(V_{Ga})_6(V_N)_6$	414.2		
		$V_{Ga}O_N$	238.5, 239.0		
$V_{Al}(O_N)_3$	248.2	$V_{Ga}(O_N)_3$	241.2	$V_{In}(O_N)_3$	263.4
$V_{Al}(O_N)_4$	245.5	$V_{Ga}(O_N)_4$	239.8	$V_{In}(O_N)_4$	264.1
		$V_{Ga}C_N$	230.7, 231.7		
		$V_{Ga}(C_N)_3$	211.5		
		$V_{Ga}(C_N)_4$	203.9		

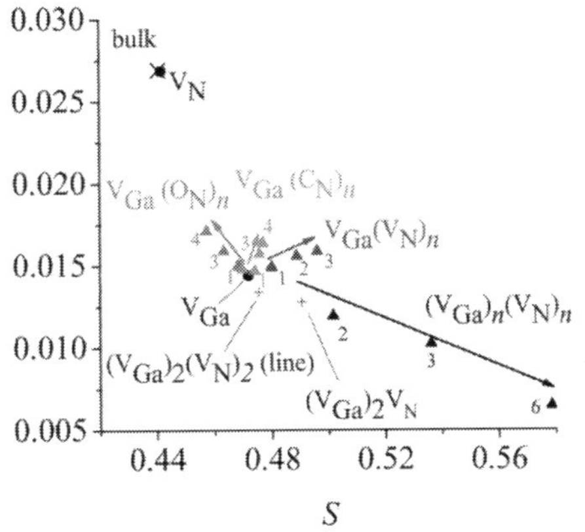

Figure 1. S–W plot for GaN bulk and defects therein (S: $-0.419 \sim +0.419$ a. u., W: $\pm(1.828 \sim 3.688)$ a.u.).

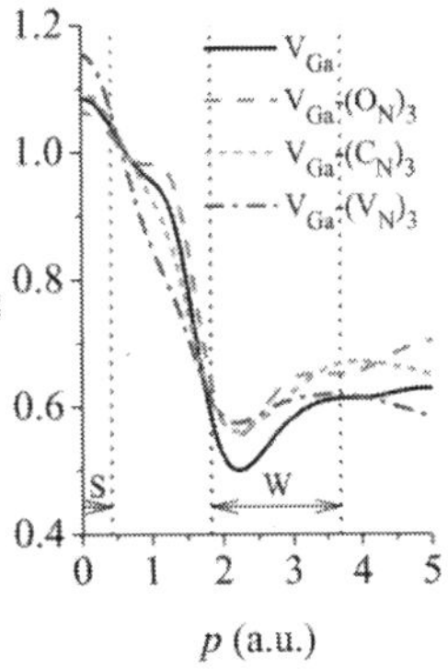

Figure 2. Ratio curve for representative defects in GaN against its bulk.

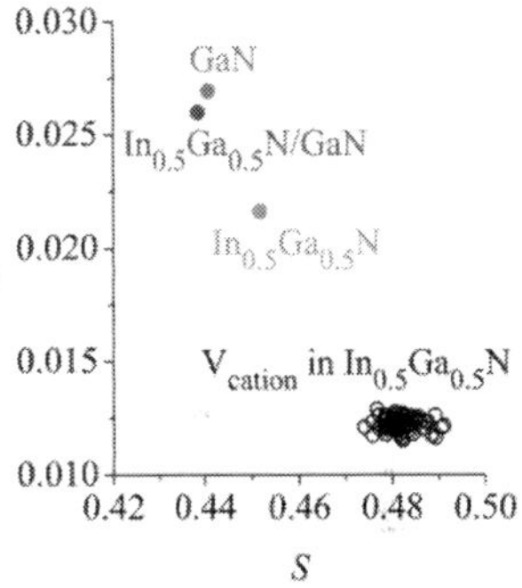

Figure 3. S–W plot for a random alloy $In_{0.5}Ga_{0.5}N$ bulk and defects therein.

We also investigated an $In_{0.5}Ga_{0.5}N/GaN$ superlattice system. The $4 \times 4 \times 2$ $In_{0.5}Ga_{0.5}N$ unit and the $4 \times 4 \times 2$ GaN unit are stacked along the wurtzite c axis as show in Fig. 4 (a). The unit cell of the superlattice system contains 256 atoms. The lattice parameters of the supercell perpendicular to c are fixed to those of the GaN unit. As for the lattice parameter along c, it is set to $c(In_{0.5}Ga_{0.5}N) + c(GaN)$, where $c(In_{0.5}Ga_{0.5}N)$ and $c(GaN)$ are the lattice parameters along c for the units mentioned above. Thus, the Poisson effect in the $In_{0.5}Ga_{0.5}N$ region was not taken into account. It has been revealed that positrons are predominantly distributed in a part of the GaN region, which is close to one type of the interfaces, as show in Fig. 4 (a). This is due to the variation of the positron effective potential shown in Fig. 4 (b) and (c) as the planar average $\tilde{V}$ and the macroscopic average $\tilde{\tilde{V}}$. This is similar to the situation for an InN/GaN superlattice studied by Makkonen *et al.* [11]. The obtained S and W parameters are also plotted in Fig. 2. Naturally, it is close to the location of the plot for the GaN bulk.

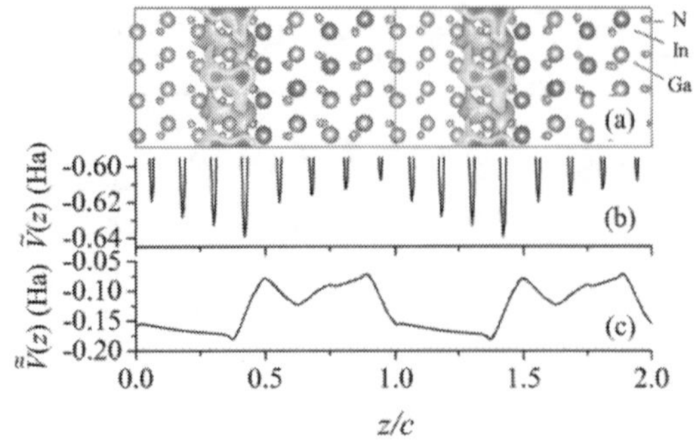

Figure 4. Positron density distribution in $In_{0.5}Ga_{0.5}$/GaN superlattice system. (a) structural model and positron density distribution, (b) planar-average effective potential for positrons and (c) macroscopic-average effective potential for positrons. Here, c represents one of the lattice vectors of the supercell unit cell.

4. Summary

The positron annihilation S and W parameters as well as the positron lifetime have been calculated for V_{Ga}-C complexes in GaN and been compared with other various types of defects in AlN, GaN and InN. They are expected to be useful in interpreting experimental results. A random alloy $In_{0.5}Ga_{0.5}N$ was modeled with the SQS approach and the S, W and lifetime for cation vacancies were calculated and analyzed. An $In_{0.5}Ga_{0.5}N$/GaN superlattice system is also investigated. Modeling for practical materials is now possible. Calculations for $In_{0.125}Ga_{0.875}N$/GaN, which is closer to the situation of existing devices, are now in progress.

References

[1] Uedono A, Ito K, Nakamori H. Mori K, Nakano Y, Kachi T, Ishibashi S, Ohdaira T and Suzuki R 2007 *J. Appl. Phys.* **102** 084505
[2] Uedono A, Shaogiang, Jongwon C S, Ito K, Nakamori H, Honda N, Tomita S, Akimoto K, Kubo H and Ishibashi S 2008 *J. Appl. Phys.* **103** 104505
[3] Uedono A, Ishibashi S, Keller S, Moe C, Cantu P, Katona T M, Kamber D S, Wu Y, Letts E, Newman S A, Nakamura S, Speck J S, Mishra U K, DenBaars S P, Onuma T and Chichibu S F 2009 *J. Appl. Phys.* **105** 054501
[4] Uedono A, Nakamori H, Narita K, Suzuki J, Wang X, Che S-B, Ishitani Y, Yoshikawa A and Ishibashi S 2009 *J. Appl. Phys.* **105** 054507
[5] Chen S, Uedono A, Ishibashi S, Tomita S, Kudo H and Akimoto K 2010 *Appl. Phys. Lett.* **96** 05190
[6] Uedono A, Ishibashi S, Tenjinbayashi K, Tsutsui T, Nakahara K, Takamizu D and Chichibu S F 2012 *J. Appl. Phys.* **111** 014508
[7] Uedono A, Ishibashi S, Watanabe T, Wang X Q, Liu S T, Chen G, Sang L W, Sumiya M and Shen B 2012 *J. Appl. Phys.* **112** 014507
[8] Uedono A, Tsutsui T, Watanabe T, Kimura S, Zhang Y, Lozac'h M, Sang L W, Ishibashi S and Sumiya M 2013 *J. Appl. Phys.* **113** 123502
[9] Uedono A, Yonenaga I, Watanabe T, Kimura S, Oshima N, Suzuki R, Ishibashi S and Ohno Y 2013 *J. Appl. Phys.* **114** 084506
[10] Uedono A, Watanabe T, Kimura S, Zhang Y, Lozac'h M, Sang L, Ishibashi S, Oshima N, Suzuki R and Sumiya M 2013 *J. Appl. Phys.* **114** 184504
[11] Makkonen I, Snicker A, Puska M J, Mäki J-M and Tuomisto F 2010 *Phys. Rev.* B **82** 041307(R)
[12] Zunger A, Wei S-H, Ferreira L G and Bernard J E 1990 *Phys. Rev. Lett.* **65** 353
[13] Mäder K A and Zunger A 1995 *Phy. Rev.* B **51** 10462
[14] http://qmas.jp
[15] Blöchl P E 1994 *Phys. Rev.* B **50** 17953
[16] Perdew J P, Burke K and Ernzerhof M 1996 *Phys. Rev. Lett.* **77** 3865
[17] Ishibashi S 2004 *Mater. Sci. Forum* **445-446** 401
[18] Boroński E and Nieminen R M 1986 *Phys. Rev.* B **34** 3820
[19] Puska M J, Mäkinen S, Manninen M and Nieminen R M 1989 *Phys. Rev.* B **39** 7666
[20] Barbiellini B, Puska M J, Torsti T and Nieminen R M 1995 *Phys. Rev.* B **51** 7341
[21] Saarinen K, Laine T, Kuisma S, Nissila J, Hautojärvi P, Dobrzynski L, Baranowski J M, Pakula K, Stepniewski R, Wojdak M, Wysmolek A, Suski T, Leszczynski M, Grzegory I and Porowski S 1997 *Phys. Rev. Lett.* **79** 3030

Formation of Cu precipitates in a high-energy-particle-irradiated and thermally aged Fe-0.6%Cu alloy

K Sato[1], Q Xu[1], X Z Cao[2], P Zhang[2], B Y Wang[2] and H Tsuchida[3]
[1] Research Reactor Institute, Kyoto University, Kumatori-cho, Osaka 590-0494, Japan
[2] Key Laboratory of Nuclear Analysis Techniques, Institute of High Energy Physics, Chinese Academy of Sciences, Beijing 100049, China
[3] Quantum Science and Engineering Center, Kyoto University, Uji, Kyoto 611-0011, Japan

E-mail: ksato@rri.kyoto-u.ac.jp

Abstract. The stability of Cu precipitates in thermally aged Fe-Cu alloys after high-energy particle irradiation was investigated using a mono-energetic positron beam apparatus. The S-parameters did not change with increasing incident positron energy above 15 keV, which means positron annihilation at the surface does not influence the Doppler broadening (DB) spectra. In a non-aged sample, the S-parameter increased, while the W-parameter did not change upon ion irradiation at above 10 keV. By contrast, in the thermally aged sample, the S-parameter increased, while the W-parameter decreased upon ion irradiation. The DB spectra of thermally aged and non-aged samples have nealy the same shape after high-energy particle irradiation. In both thermally aged and non-aged samples, almost all positrons annihilate with electrons in Cu, and vacancy clusters covered with Cu atoms are formed upon neutron irradiation. This agrees with the results of previous studies. However, the defect growth process may be different in aged and non-aged samples.

1. Introduction

Cu precipitates form in Fe-Cu alloys during thermal aging at high temperatures, as well as upon irradiation with high-energy particles [1,2], since Cu atoms are almost insoluble in α-Fe. Cu precipitates obstruct dislocation motion during deformation, which increases the hardness and decreases the ductility of the alloy, thus inducing embrittlement in Fe-based alloys containing Cu impurities, such as old commercial reactor pressure vessel (RPV) steels. Besides Cu precipitates, defect clusters, such as interstitial clusters and vacancy clusters, are also formed during irradiation, which further increase the hardness and decrease the ductility of RPV steels.

Positron annihilation spectroscopy can detect vacancy clusters and even single vacancies. Doppler broadening (DB) of positron annihilation radiation is a nondestructive technique for testing defect clusters. In the dominant decay mode of a thermal positron and electron, two gamma rays are emitted. In the laboratory frame, the energies of the two photons emitted by the annihilation of a positron and electron are different. This difference in photon energy is proportional to the longitudinal component of the electron-positron momentum in the direction of gamma emission. Through measurements of the photon energies, information about the momentum distribution of core electrons can be obtained. Thus, DB measurements can provide useful information about the distribution of elements around the annihilation site [3–6]. Positron annihilation spectroscopy using slow-positron beam techniques is very effective for investigating the near-surface microstructure of ion-irradiated materials [7].

It is well known that Cu atoms aggregate by a vacancy mechanism even in thermally aged Fe-Cu alloys. However, details of the process of precipitate formation in thermally aged Fe-Cu alloys and irradiated Fe-Cu alloys remain unclear. In the present study, samples were irradiated with ions and neutrons, and the stability of Cu precipitates in thermally aged Fe-Cu alloys was investigated.

2. Experimental

Fe-0.6 wt% Cu was prepared from Fe (99.99%, purity) and Cu (99.9%, purity) in vacuum using a high-frequency induction furnace. Two types of heat treatments were performed before irradiation. In one case, the samples were aged at 1223K for 0.5 h in a vacuum, and quenched in water. In the other case, specimens were aged at 773 K for 100 h in vacuum after being quenched in water, to obtain small Cu precipitates. Samples consisted of 10×10×0.1 mm sheets. Fe ion irradiation was performed at the Quantum Science and Engineering Center, Kyoto University, using a tandem Pelletron accelerator (model: 6SDH-2). The ion energy was 2.5 MeV, and the irradiation temperature was 573 K. The irradiation dose at damage peak area was 1 dpa. Figure 1 shows simulated depth distributions of damage in Fe irradiated by 2.5 MeV Fe ions. In order to investigate the irradiation depth dependence of the microstructural evolution of ion-irradiated samples, the DB measurements were performed using a mono-energetic positron beam apparatus at the Institute of High Energy Physics, Chinese Academy of Sciences. The incident positron energy was 0.18-20 keV. Figure 2 plots the depth profile of the positron stopping probability for energies of 5, 10, 15, and 20 keV. Both the irradiated and opposite side were measured to investigate the effect of irradiation. The parameters S and W are defined as the ratio of the low-momentum ($|P_L| < 3\times10^{-3}mc$) and high-momentum ($10\times10^{-3}\ mc < |P_L| < 27\times10^{-3}\ mc$) areas in the DB spectrum to the total area, respectively. Fission neutron irradiation was performed at Research Reactor Institute, Kyoto University. The irradiation dose was 4.3×10^{-3} dpa (2.0×10^{18} n/cm^2), and the irradiation temperature was 573 K. Coincidence Doppler broadening (CDB) measurements and positron annihilation lifetime measurements were carried out at room temperature using a Na-22 source for neutron-irradiated samples.

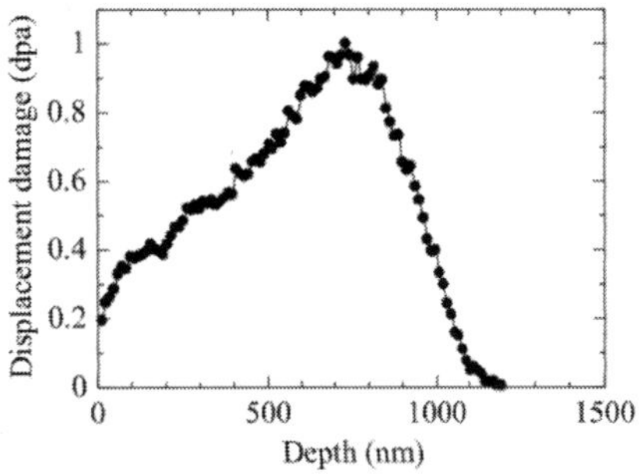

Figure 1. Depth distribution of damage in Fe irradiated by 2.5 MeV Fe ions.

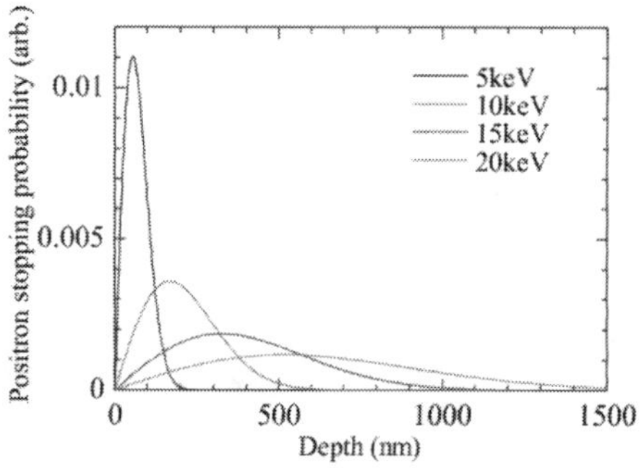

Figure 2. Depth profile of positron stopping probability for energies of 5, 10, 15, and 20 keV.

3. Results and discussion

Figures 3 and 4 show the dependence of the S- and W-parameter on incident positron energy, respectively. Figure 5 shows S–W plots at all incident positron energies. 1#, 2# and 3# denote the quenched sample irradiated with Fe ions, the sample irradiated with Fe ions after aging at 773 K and the sample aged at 773 K (non-irradiated), respectively. "B" denotes measurements made on the non-irradiated side. The S-parameter decreased and the W-parameter increased with increasing incident positron energy. S-parameters did not change with increasing incident positron energy above 15 keV. Positron annihilation at the surface did not affect the DB spectra above 15 keV. Comparison of 1# and 1#B reveals the saturation value of the S-parameter increased upon Fe ion irradiation while that of the

W-parameter did not change. This means that irradiation defects and Cu precipitates were formed during ion irradiation. If Cu precipitates had not formed during irradiation, the increase in the S-parameter due to the formation of irradiation defects would have led to a decrease in the W-parameter. Comparing 2# with 2#B shows that for the aged samples, the saturation value of the S-parameter increased while that of the W-parameter decreased upon Fe ion irradiation. The increase in the S-parameter indicates the formation of irradiation defects. As mentioned in the next paragraph, Cu precipitates can grow by irradiation even in an aged sample. Some of the irradiation defects promote the growth of aging-induced Cu precipitates, while the rest contribute to the formation of new Cu precipitates. The fact that the W-parameter is the same for both irradiated-quenched and aged samples (1# and 2#) indicates that positrons favour annihilation in irradiation-induced newly formed Cu precipitates over those formed by aging. Thus, if only vacancy clusters are formed and Cu precipitates do not grow during irradiation, the W-parameter would decrease more. In S-W plots, we can see the same trend in 2#B and 3#, and 1# and 2#. On the non-irradiated side of ion-irradiated samples without aging (1#B), Cu precipitates are not formed. Thus, these samples exhibit the lowest W-parameter. The same heat treatment is applied to 2#B and 3#.

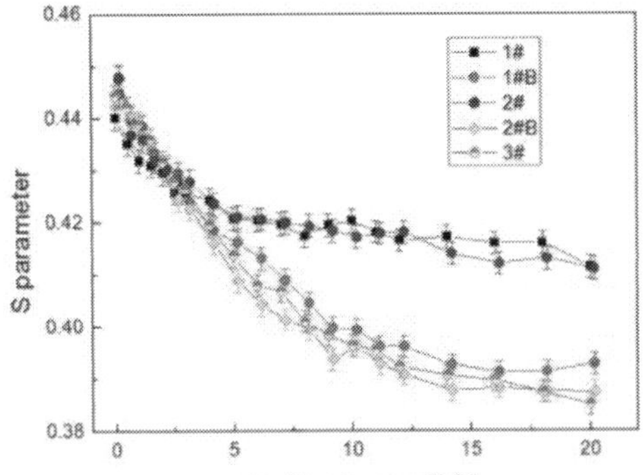

Figure 3. Dependence of S parameter on incident positron energy.

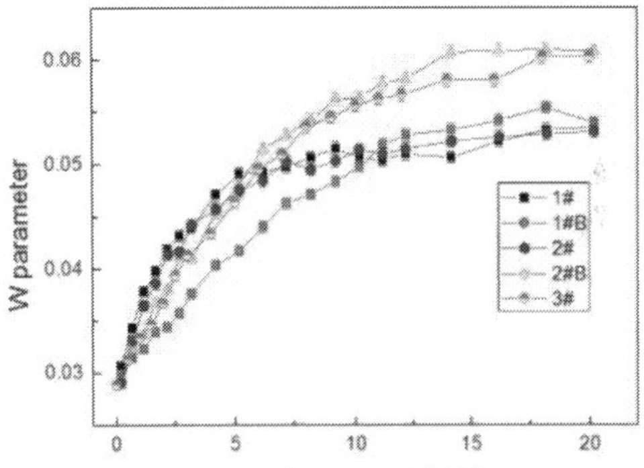

Figure 4. Dependence of W parameter on incident positron energy.

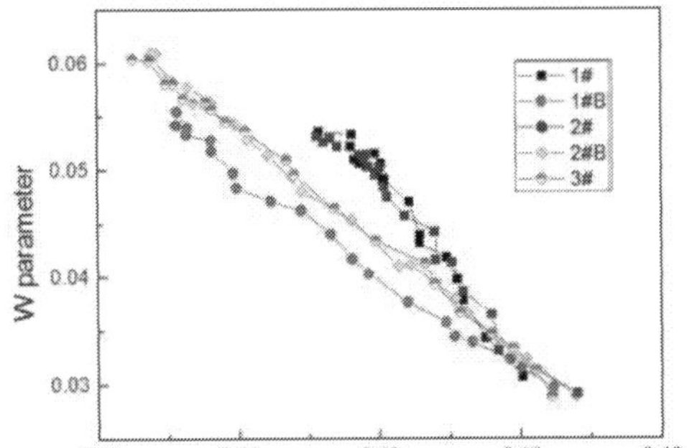

Figure 5. . S–W plots in all incident positron energies.

Figures 6 and 7 plot the CDB ratio curve of neutron-irradiated Fe-Cu alloys with respect to pure Fe and Cu, respectively. The CDB ratio curve of the non-aged Fe-Cu alloy peaks in the high-momentum region after neutron irradiation. As seen in figure 7, the CDB ratio curve of the Fe-Cu alloy aged at 773 K for 100 h has a small valley in the high-momentum region (15×10^{-3} $mc < |P_L| < 30\times10^{-3}$ mc). Thus, in the aged sample, some of the positrons annihilate with electrons in Fe [3]. But, after neutron irradiation, the CDB ratio curves of the irradiated-quenched and aged Fe-Cu alloys are almost flat in the high-momentum region. Cu precipitates grow as a result of neutron irradiation and almost all positrons annihilate with electrons in Cu. Vacancy clusters covered with Cu atoms are formed by neutron irradiation not only in quenched sample, but also the aged sample. This result agrees with references 4 and 6. It is expected that they are also formed under Fe ion irradiation, and more positrons annihilate in them than in Cu precipitates formed by aging, because the latter do not contain vacancy-type defects. The size and density of Cu precipitates cannot be determined by DB measurements alone; transmission electron microscopy and other characterization techniques are necessary. Table 1 shows positron annihilation lifetimes of neutron-irradiated quenched and aged Fe-Cu alloys. The vacancy cluster size and density are slightly larger in the aged sample than in the quenched sample. In the aged sample, the number of Cu atoms that do not form Cu precipitates is smaller than in the quenched sample before irradiation. This may influence the growth of Cu precipitates. The defect growth process may be different for the aged and non-aged samples. Different irradiation doses are required to investigate these possibilities.

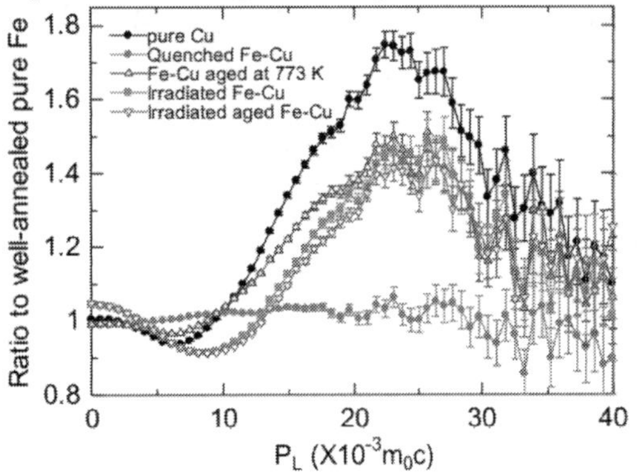

Figure 6. Ratio curves of CDB spectra of pure Cu, quenched Fe-Cu alloy, Fe-Cu alloy aged at 773 K, irradiated quenched Fe-Cu alloy and irradiated aged Fe-Cu alloy with respect to that of pure Fe.

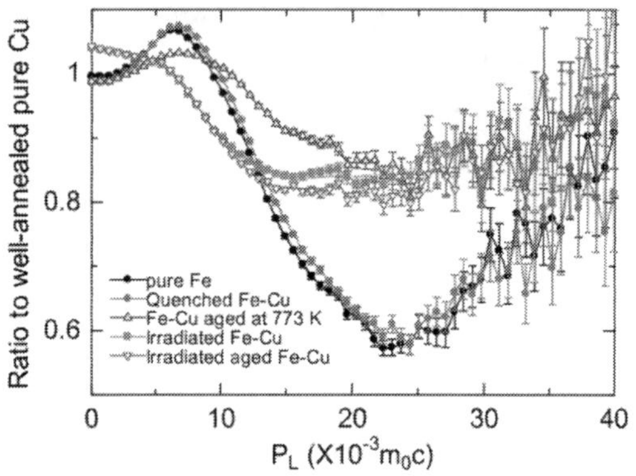

Figure 7. Ratio curves of CDB spectra of pure Fe, quenched Fe-Cu alloy, Fe-Cu alloy aged at 773 K, irradiated quenched Fe-Cu alloy and irradiated aged Fe-Cu alloy with respect to that of pure Cu.

Table 1. Positron annihilation lifetimes of neutron-irradiated Fe-Cu alloys.

	Mean lifetime (ps)	Short lifetime (ps)	Long lifetime (ps)	Long lifetime intensity (%)
Quenched	172 ± 1	129 ± 1	463 ± 5	13 ± 1
Thermally aged	186 ± 1	126 ± 1	479 ± 4	17 ± 1

4. Summary and future plan

The stability of Cu precipitates in thermally aged Fe-Cu alloys was investigated. DB spectra of thermally aged and non-aged samples are almost the same after both ion and neutron irradiation. Almost all positrons annihilate with electrons in Cu, and vacancy clusters covered with Cu atoms are formed upon neutron irradiation in both aged and non-aged samples. Determining the exact size and density of Cu precipitates requires detailed observation of the defect structure by transmission electron microscopy, three-dimensional atom probes, etc. Further experiments using lower irradiation doses and shorter aging times will be conducted in the near future.

References

[1] Buswell J T, English C A, Hetherington M G, Phythian W J, Smith G D W and Worrall G M 1990 Effects of radiation on materials *in 14th International Symposium, ASTM STP, Vol. II*

Journal of Physics: Conference Series **505** (2014) 012011 doi:10.1088/1742-6596/505/1/012011

(American Society for Testing and Materials, Philadelphia, PA) ed Packan N H, Stoller R E and Kumar A S p127
[2] Othen P J, Jenkins M L, Smith G D W and Phythian W J 1991 *Phil. Mag. Lett.* **64** 383
[3] Nagai Y, Hasegawa M, Tang Z, Hempel A, Yubuta K, Shimamura T, Kawazoe Y, Kawai A and Kano F 2000 *Phys. Rev. B* **61** 6574
[4] Nagai Y, Tang Z, Hasegawa M, Kanai T and Saneyasu M 2001 *Phys. Rev. B* **63** 134110
[5] Xu Q, Yoshiie T and Sato K 2006 *Phys. Rev. B* **73** 134115
[6] Xu Q, Yoshiie T and Sato K 2007 *Philos. Mag. Lett.* **87** 65
[7] Fujinami M, Miyagoe T, Sawada T, Suzuki R, Ohdaira T and Akahane T 2003 *Physica* B **340–342** 724 Glowinski L, Fiche C and Lott M 1973 *J. Nucl. Mater.* **47** 232

He implantation induced defects in InN.

F Linez[1], M Ritt[1], C Rauch[1], L Kilanski[2], S Choi[4], J Räisänen[3], J S Speck[4] and F Tuomisto[1]

[1] Department of Applied Physics, Aalto University, Otakaari 1M, 02150 Espoo, Finland
[2] Institute of Physics, Polish Academy of Sciences al. Lotnikow 32/46, 02-668 Warsaw, Poland
[3] Department of Physics, University of Helsinki, P.O. Box 43, FI-00014, University of Helsinki, Finland
[4] Materials Department, University of California, Santa Barbara, California 93106 USA

E-mail: florence.linez@aalto.fi

Abstract. InN layers have been implanted with helium at 158 keV at various fluences to study the nature of the generated defects. The defects have been probed using positron annihilation spectroscopy. The first measurements showed that at least two different kinds of defects are created depending of the implantation fluence. The second measurements performed two years later gave different results suggesting that at least one of these defects is not stable at room temperature.

1. Introduction

InN is a significant semiconductor with many interesting properties such as low direct band gap (0.7 eV) [1–5] and high electron mobility promising applications in optoelectronics and high-speed electronics [6–8]. The presence of intrinsic point defects can change the electrical properties of the material in an advantageous way or not: In vacancies (V_{In}) and N vacancies (V_N) are expected to be the dominant intrinsic acceptors and donors, respectively, according to latest density functional theory calculations. The understanding and control of these defects is consequently an important element in the development of the industrial applications. Despite major research efforts in that area, relatively little is known about the properties of defects in InN especially their annealing temperature.

High-energy particle irradiation has been demonstrated as an effective way to control the conductivity of undoped InN by creating defects. The implantation of InN with helium could be an effective tool for precise control of In vacancy concentration without the risk of unintentional doping. In the frame of a more global work on the effects of the different defects on the InNs electrical properties, we have studied the defects generated by He-ions implantation in this material. The damaged zones of the implanted samples have been probed using the positron annihilation spectroscopy. The samples have been probed twice with a delay of 2 years. The measurements show an evolution of the defects detected by positrons. These never before observed changes will be discussed.

2. Experimental techniques

The studied samples are undoped 2 μm layer of InN grown on GaN wafer by plasma-assisted molecular beam epitaxy (n=5×10^{17} cm^{-3}, μ=1500 cm^{2}V^{-1}s^{-1}), implanted with 158 keV-He ions at three fluences 8×10^{12}, 8×10^{14} and 8×10^{15} cm^{-2} at room temperature in the University of Helsinki. Both damage and He-implantation depth profiles calculated using SRIM 2013 in full damage cascade calculation mode are plotted figure 1. The projected range is at about 580 nm. The Doppler broadening measurements of the positron-electron annihilation radiation were performed with a variable-energy positron beam on an energy range from 0.5 to 35 keV with high-purity Ge detectors. The positron implantation depth profiles drawn figure 1-a show that up to 15 keV positrons probed only damaged area then they start probing the undamaged one behind the He peak. From 25 keV, positrons can reach the GaN substrate. The conventional low-momentum (S) and high-momentum (W) annihilation parameters have been used to analyse the data [9].

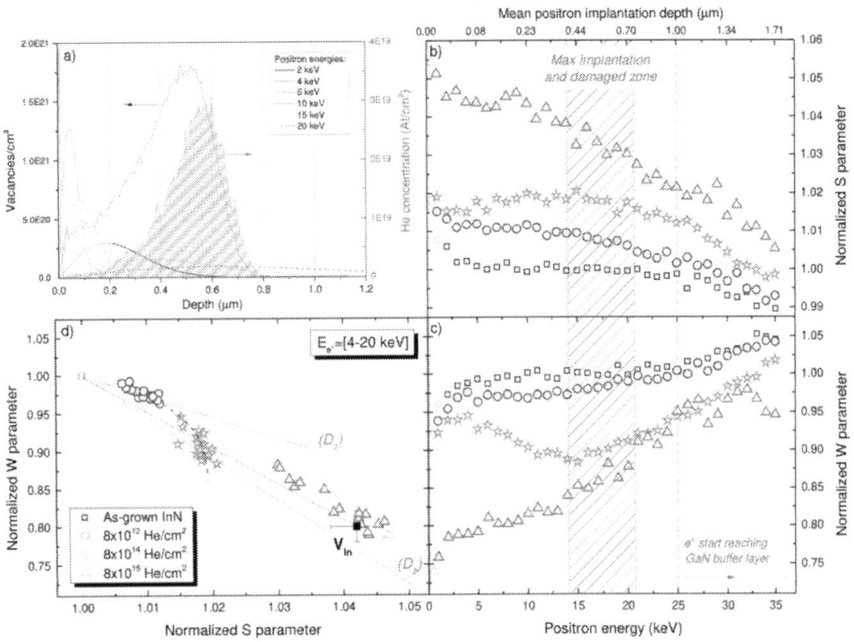

Figure 1. a) Vacancy and He concentration depth profiles as calculated using SRIM 2013. Positron implantation depth profile are plotted for positron energy from 2 to 20 keV. First results: b) and c) Normalized S and W parameter values as a function of positron energy. d) W as a function of S parameter for positron energy from 4 to 20 keV. The full square symbol represents the positron annihilation characteristics in the V_{In} as reported by Rauch et al. [10].

3. Results and discussion

3.1. First measurements

The results of the measurements performed two weeks after implantation are plotted in figure 1. S and W parameters values are normalized with those of the as-grown sample ones that coincide with the InN lattice values [10] ($S_{AG} = 0.460 \pm 0.001$, $W_{AG} = 0.0491 \pm 0.0008$). The figure 1-b. shows the S parameter as a function of the positron energy. The as-grown curve can be shared in three parts: up to 3 keV positrons probe a part of the surface, between 4 and 25 keV the plateau corresponds to the bulk of the InN layer, and after 25 keV, S decreases because positrons start probing the GaN substrate layer. In the case of implanted samples, one can observe that whatever the fluence, the S value is higher than that of the as-grown sample that means defects are detected. At the fluence of 8×10^{12} cm^{-2}, S (resp. W) values form a plateau up to energy of 15 keV where it starts slightly decreasing (resp. increasing). At the fluence of 8×10^{13} cm^{-2}, the S and W values are respectively higher and lower than the previous fluence ones and the curves shapes are different. The S values slightly increase and start decreasing clearly for positron energies higher than 20 keV. The W curve decreases strongly between 5 and 15 keV then increases up to 35 keV. Finally at the fluence 8×10^{15} cm^{-2}, the S values seems to form a plateau up to 12 keV then decrease progressively in the range 15-25 keV.

In the figure 1-d the W parameter is plotted as a function of the S one for the positron energy range 4-20 keV corresponding to the damaged area. The (S, W) points for the three fluences are not aligned with the InN reference point (1, 1) indicating that the defects detected by positrons have different nature. For the lowest fluence, the points form a straight line named *(D_1)*. For the highest fluence, the points are grouped close to the point V_{In} giving the positron annihilation characteristics in the Indium vacancy [10], suggesting that this defect is detected by all the positrons. In the case of the intermediate fluence, the (S, W) points are not aligned with the as-grown sample annihilation characteristics indicating that more than one type of defect is detected. These points being partly on the *(InN-V_{In})* line, one of these defects could be the V_{In} one. The rest of the points form a line *(D_2)* with the InN reference point.

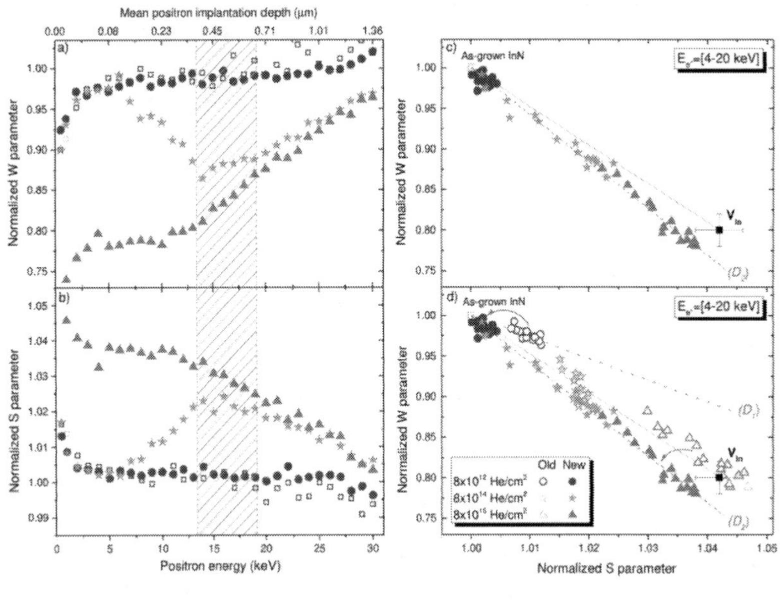

Figure 2. Positron results after 2 years of storage in air. a) and b) Normalized W and S parameter values as a function of positron energy for the three fluences and the as-grown reference sample. c) W as a function of S parameter. d) Comparison of data obtained before and after storage for 2 years.

3.2. Measurements 2 years after

The figure 2 shows the results of the Doppler broadening measurements performed two years after the implantation, with the samples stored in ambient conditions. The S and W curves have significantly changed. For the sample implanted with the lowest fluence, the S and W parameter values are equivalent to those of the as-grown sample meaning that no vacancy-type defect is detected. At the highest fluence, the S and W curves have the same shape than for the previous measurements but the S and W values are both lower in the whole positron energy range. For instance, the plateau between 4 and 10 keV clearly decreases for the S parameter (figure 1-b and figure 2-b). That indicates a change in the defect distribution. For the intermediate fluence, the main change occurs in the first part under the surface, for positron energy up to 10 keV. The S (resp. W) parameter values have decreased (resp. increased) to merge with the InN reference ones. The second parts of the curves are roughly as the previous ones. In the figure 2-c showing the evolution of the W parameter as a function of the S one, one can observe that all the (S, W) points are aligned with the as-grown reference point. The error bars indicated for the Indium vacancy annihilation characteristics suggest than this vacancy might be also on the same line.

3.3. Comparison and discussion

The figure 2-d allows us to compare the positron annihilation states observed in both cases. From three lines definied by the (S, W) points during the first measurements, (D_1),$(InN\text{-}V_{\mathrm{In}})$ and (D_2), just (D_2) is remaining after two years. The annihilation state observed at the lowest fluence disappeared and no other defect are detected. At the intermediate fluence, a part of the points which were on the (D_2) line are still there but the other part which was between the (D_1) and the (D_2) lines are now only on the (D_2) line. It is the same at the highest fluence for which all the points moved from the $(InN\text{-}V_{\mathrm{In}})$ line to the (D_2) line. These observations suggest that at least three annihilation states were detected during the first measurements: the lattice and two vacancy type defects. Indeed in the case of the high fluence, the annihilation states observed could be a mixture of those generating the(D_1) and the (D_2) lines respectively called V_X and V_Y. On the other hand, these results highlight that after 2 years, the defect V_X has disappeared indicating that it is not stable at room temperature.

4. Conclusion

InN samples have been implanted with 158 keV-He ions at three different fluences to study the nature of defects generated and their properties. The damaged layer has been probed by positron annihilation spectroscopy first just after the implantation then two years after. The first results have shown that at least two different types of vacancy defects are created. The second measurement session has highlighted that at least one of the induced defects was not stable at room temperature. Complementary analysis and experiments will be needed to identify the involved defects, but the instabilty of a defect at room temperature is already an interesting observation from ion implantation processing point of view.

Acknowledgments

This work has been partially funded by the Academy of Finland.

References

[1] Inushima T, Mamutin V, Vekshin V, Ivanov S, Sakon T, Motokawa M and Ohoya S 2001 *J. of Cryst. Growth* **227** 481–85
[2] Davydov V Y, Klochikhin A, Seisyan R, Emtsev V, Ivanov S, Bechstedt F, Furthmüller J, Harima H, Mudryi A, Aderhold J *et al.* 2002 *Phys. Status Solidi B* **229** r1–r3
[3] Matsuoka T, Okamoto H, Nakao M, Harima H and Kurimoto E 2002 *Appl. Phys. Lett.* **81** 1246–48
[4] Wu J, Walukiewicz W, Yu K, Ager Iii J, Haller E, Lu H, Schaff W J, Saito Y and Nanishi Y 2002 *Appl. Phys. Lett.* **80** 3967
[5] Nanishi Y, Saito Y and Yamaguchi T 2003 *Jpn J. Appl. Phys.* **42** 2549–59
[6] Millot M, Ubrig N, Poumirol J M, Gherasoiu I, Walukiewicz W, George S, Portugall O, Léotin J, Goiran M and Broto J M 2011 *Phys. Rev. B* **83** 125204
[7] Bellotti E, Doshi B K, Brennan K F, Albrecht J D and Ruden P P 1999 *J. Appl. Phys.* **85** 916–23
[8] Foutz B E, OLeary S K, Shur M S and Eastman L F 1999 *J. Appl. Phys.* **85** 7727–34
[9] Tuomisto F and Makkonen I 2013 Defect identification in semiconductors: Experiment and theory of positron annihilation
[10] Rauch C, Makkonen I and Tuomisto F 2011 *Phys. Rev. B* **84**(12) 125201

Investigation of H^+ implanted Fe-Al alloys

F Lukáč[1], J Čížek[1], I Procházka[1], Y Jirásková[2], D Janičkovič[3], J Šimeg Veterníková[4], W Anwand[5] and M Butterling[5]

[1]Charles University in Prague, Faculty of Mathematics and Physics, V Holešovičkách 2, 18000 Praha 8, Czech Republic
[2]Institute of Physics of Materials, Academy of Science of the Czech Republic, Žižkova 513/22, 61662 Brno, Czech Republic
[3]Institute of Physics, Slovak Science Academy, Dúbravská cesta 9, 84511 Bratislava, Slovakia
[4]Institute of Nuclear and Physical Engineering, Faculty of Electrical Engineering and Inform. Technology, Slovak University of Technology, Ilkovičova 3, 81219 Bratislava, Slovakia
[5]Institut für Strahlenphysik, Helmholtz-Zentrum Dresden-Rossendorf, 01314 Dresden Germany

E-mail: lukac@mbox.troja.mff.cuni.cz

Abstract. In the present work hydrogen interaction with vacancies was investigated in Fe-Al alloys with various concentration of vacancies. The Fe-Al samples were implanted with low energy H^+ ions (100 keV). This procedure introduced high hydrogen concentration into relatively narrow sub-surface region in the depth of ~500 nm. Variable energy positron annihilation spectroscopy (VEPAS) was employed for investigation of hydrogen interaction with vacancies in the sub-surface region. This study revealed formation of vacancy hydrogen complexes in the sub-surface region. Thermal stability of vacancy-hydrogen complexes was investigated as well.

1. Introduction

Iron aluminides are widely considered as prospective materials due to high specific mechanical strength and enhanced corrosion resistance at elevated temperatures. However, work machinability of these alloys suffers from poor ductility at room temperature. Since it has been shown that the ductility of Fe-Al alloys is remarkably improved in the absence of hydrogen and water vapor [1], environmental hydrogen embrittlement was suggested to take place in Fe-Al alloys. In addition to this, Fe-Al alloys are well known for a low vacancy formation enthalpy. As a consequence, the equilibrium concentration of vacancies in Fe-Al alloys is substantially high compared to pure metals. Thermal vacancies formed at elevated temperatures in Fe-Al alloys can be relatively easily quenched to room temperature. Hydrogen interaction with vacancies could play very important role in the embrittlement process. The importance of hydrogen interaction with vacancies is further amplified by high mobility of hydrogen in Fe-Al lattice.

In this work we employed VEPAS for investigation of defects introduced by H^+ implantation and interaction of implanted hydrogen with vacancies in Fe-Al alloys with various composition.

2. Experimental

Fe-Al alloys of Al concentration c_{Al} 27 and 35 at.% were prepared by arc melting from high purity Fe (99.99%) and Al (99.99%) in Ti-gettered Ar atmosphere. As-cast alloys exhibited coarse grains with the mean diameter of a few mm. Specimens were quenched to room temperature after one hour of annealing at 1000 °C. The $Fe_{73}Al_{27}$ sample was subsequently annealed at 520 °C for one hour. The samples were implanted by H^+ ions with the energy of 100 keV up to a fluence of 3×10^{18} at./cm^2. The implantation was performed at room temperature on a cascade accelerator using the accelerating voltage of 100 kV.

Variable energy positron annihilation spectroscopy (VEPAS) measurements were performed using magnetically guided energy variable positron beam "SPONSOR" [2] with positron energy adjustable in the range from 30 eV to 35 keV. Energy spectra of annihilation γ rays were measured by HPGe detector having an energy resolution of (1.06 $\pm$ 0.01) keV (FWHM at 511 keV). The Doppler broadening of annihilation profile was evaluated using the line-shape S-parameter.

3. Results and discussion

The aim of this study was to investigate the interaction of hydrogen with vacancies in Fe-Al alloys. For this purpose the Fe-Al alloys with known vacancy concentrations were prepared, $Fe_{73}Al_{27}$ sample exhibits low concentration of vacancies $c_V \approx 4 \times 10^{-6}$ while $Fe_{65}Al_{35}$ contains high vacancy concentration of $c_V \approx 5 \times 10^{-3}$ [3], consistent with $c_V \approx 3 \times 10^{-3}$ for $Fe_{64}Al_{36}$ [4].

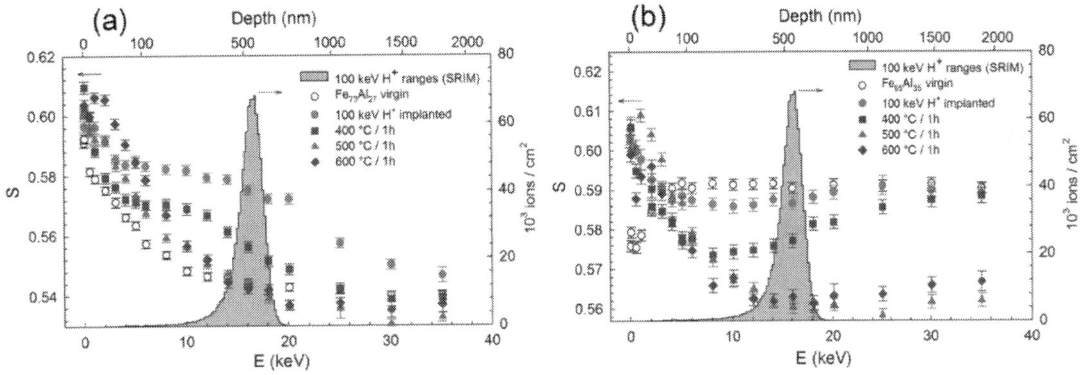

Figure 1. $S(E)$ curves of (a) $Fe_{73}Al_{27}$ and (b) $Fe_{65}Al_{35}$ alloy in the as-prepared state, implanted with 100 keV H^+ ions and subsequently annealed at 400 °C, 500 °C and 600 °C. The hydrogen implantation profile plotted as a gray histogram was calculated by SRIM [5].

The implantation profile of H^+ ions calculated by SRIM code [5] is plotted in Fig. 1 as gray histograms. The mean H^+ penetration depth is $\bar{z}_{H^+} \approx 520$ nm and the width of the implantation profile is characterized by FWHM of ≈ 130 nm.

From the H^+ implantation profile one can calculate that 75% of implanted hydrogen is stopped in a sub-surface layer located at a depth from 470 to 600 nm below the surface. Using the total fluence of H^+ implantation one can easily estimate that hydrogen concentration in this layer is as high as 2.5 H per one host metal atom (Fe, Al). Hence, the hydrogen concentration in the sub-surface region 470-600 nm is much higher than the concentration of vacancies.

The VEPAS results for $Fe_{73}Al_{27}$ alloy are plotted in Fig. 1(a). In the virgin $Fe_{73}Al_{27}$ sample the S-parameter decreases with increasing positron energy from the surface value to the bulk value. Fitting of the S(E) dependence resulted in positron diffusion length $L_+ = (130 \pm 7)$ nm which is in agreement with low concentration of defects in this sample. One can see in Fig. 1(a) that $Fe_{73}Al_{27}$ alloy implanted with H^+ exhibits a bump with significantly enhanced S parameter in the sub-surface region influenced by hydrogen implantation. This is obviously due to new defects created in the sample by H^+ implantation. Protons with energy 100 keV implanted into the sample lose their kinetic energy in collisions with Fe and Al atoms and produce Frenkel pairs in the host lattice. Some of them recombine, i.e. vacancy is filled by interstitial atom, but other ones remain in the sample since the displaced atom was kicked out to sufficient distance from the vacancy. The increase of the S parameter in the sub-surface region in $Fe_{73}Al_{27}$ sample is due to positron trapping at such vacancies created by H^+ implantation. Since it is known that an attractive interaction exists between hydrogen and vacancies [6], vacancies introduced by H^+ implantation are likely associated with hydrogen. However, since the concentration of implanted hydrogen in the sub-surface region exceeds the concentration of vacancies by several orders of magnitude the majority of hydrogen atoms are located in the interstitial sites of Fe-Al lattice.

Fig.1(b) shows S(E) curves for the $Fe_{65}Al_{35}$ alloy. On the contrary to $Fe_{73}Al_{27}$ alloy, the $Fe_{65}Al_{35}$ alloy contains a high concentration of vacancies already in the virgin state, i.e. prior to H^+ implantation. This is demonstrated by a very short positron diffusion length $L_+ = (10 \pm 2)$ nm measured in the virgin sample. From inspection of Figs. 1(a), 1(b) one can deduce that the bulk S parameter in the virgin $Fe_{65}Al_{35}$ sample is higher than S in the H^+ implanted region in $Fe_{73}Al_{27}$. It indicates that $Fe_{65}Al_{35}$ sample in the virgin state contains higher concentration of vacancies than the sub-surface region in H^+ implanted $Fe_{73}Al_{27}$. Protons implanted into $Fe_{65}Al_{35}$ creates vacancies in a similar manner as in $Fe_{73}Al_{27}$ but at the same time hydrogen implanted into the sub-surface region is trapped at vacancies existing already in the sample. The former process increases the S parameter while the latter leads to a decrease of S due to the fact that the electron density in vacancy is enhanced by the presence of hydrogen and therefore reduced positron localization. One can see in Fig. 1(b) that the latter effect prevailed and H^+ implanted $Fe_{73}Al_{27}$ exhibits lowered S parameter in the sub-surface region influenced by implantation. Similar effect of decrease of the S parameter caused by H^+ implantation was recently observed in $Fe_{52}Al_{48}$ [7] which presumably contains very high concentration of vacancies in orders of 10^{-2} due to its high Al content [3].

The implanted samples were subjected to annealing for 1 h in vacuum at various temperatures in order to examine the thermal stability of defects. The S(E) curves for annealed $Fe_{73}Al_{27}$ and $Fe_{65}Al_{35}$ alloys are plotted in Figs. 1(a) and 1(b), respectively.

Annealing at 400 °C activates a diffusion of hydrogen. The implanted hydrogen located in interstitial sites diffuses out of the narrow sub-surface region further into the sample, i.e. the hydrogen concentration profile becomes broader and vacancies located in larger distance from the maximum of the hydrogen implantation profile can be filled with hydrogen. Driving force for this process is a gradient of the hydrogen concentration. This process leads to a decrease of S in a broader sub-surface region which can be seen for $Fe_{73}Al_{27}$ and $Fe_{65}Al_{35}$ alloys annealed at 400 °C in Fig. 1(a) and 1(b), respectively. Positrons with kinetic energy of 16 keV exhibit the mean penetration depth of $\sim$ 510 nm which is close to the maximum of the implantation profile of hydrogen. The implantation profiles for positrons and hydrogen are compared in Fig. 2a. Obviously, the positron implantation profile is significantly wider which testifies that the S parameter value measured at positron energy of 16 keV contains a superposition of a contribution from positrons annihilated inside the region with high hydrogen concentration and also a contribution of positrons annihilated outside this region. The broadening of the hydrogen concentration profile extends the spatial range in which vacancies are filled with hydrogen and thereby decreases the S parameter. An additional mechanism which occurs at higher temperatures is recovery of vacancies which disappear by diffusion into sinks at grain boundaries and on the surface. This process is clearly visible in Fig. 1(b) as a drop of the bulk S parameter of both alloys annealed at 500 °C. Note that enhanced S parameter at low energies in the $Fe_{73}Al_{27}$ alloy annealed 600 °C is due to the formation of a thin oxide layer during annealing.

It is instructive to compare the temperature dependence of the bulk S parameter S(35 keV) determined at the positron energy of 35 keV and the S parameter S(16 keV) determined at energy of 16 keV where the contribution of positrons annihilated in the sub-surface region with high hydrogen concentration was maximal. The behavior of the S parameters S(35 keV) and S(16 keV) for the $Fe_{73}Al_{27}$ and $Fe_{65}Al_{35}$ alloys is plotted in Fig. 3b and 3c, respectively. The S parameter S(16 keV), which characterizes the sub-surface region influenced

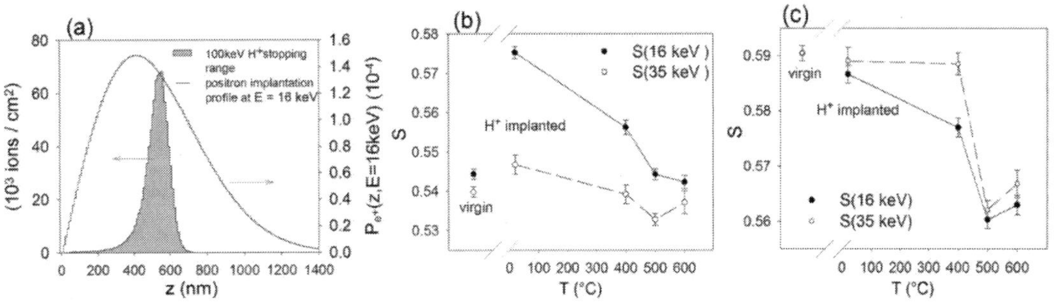

Figure 2. (a) Comparison of the implantation profile of H^+ ions with the energy of 100 keV and the implantation profile of positrons with energy of 16 keV. S-parameter determined at positron energy of E = 16 keV and 35 keV for (b) $Fe_{73}Al_{27}$ and (c) $Fe_{65}Al_{35}$.

by implanted hydrogen, decreases already during annealing at 400 °C due to the spread of hydrogen concentration profile. During further annealing the decrease of S(16 keV) continues due to the recovery of vacancies. The S parameter S(35 keV) for $Fe_{73}Al_{27}$ alloy exhibits a similar but less pronounced decrease since a fraction of positrons implanted into the sample with energy of 35 keV is still able to diffuse back into the sub-surface region containing vacancies created by H^+ implantation. On the other hand, in the $Fe_{65}Al_{35}$ alloy the S parameter S(35 keV) is not affected by the sub-surface region containing implanted hydrogen because of very short positron diffusion length due to a high concentration of vacancies. The S parameter S(35 keV) for the $Fe_{65}Al_{35}$ alloy remains approximately constant during annealing and decreases only after annealing at 500 °C where vacancies in the bulk are annealed out.

4. Conclusions

Fe-Al alloys implanted by H^+ ions with the energy of 100 keV were studied by the VEPAS technique. It was suggested that H^+ implantation creates vacancy-hydrogen complexes in a sub-surface region. Moreover, trapping of hydrogen at vacancies existing already in sample was observed. The investigation of the thermal stability of defects in H^+ implanted alloys revealed that hydrogen diffusion is activated by annealing at 400 °C leading to a spread of the hydrogen concentration profile. Vacancies are annealed out by further annealing at 500 °C.

Acknowledgments

This work was supported by Grant Agency of Charles University (Project No. 348111), the Ministry of Education, Youths and Sports of the Czech Republic (Project 7AMB12SK009), Slovak Research and Development Agency (Project APVV SK-CZ-0096-11), the Czech Science Foundation (Project No.P108/11/1350) and by the grant SVV-2013-267303.

References

[1] Stoloff N and Liu C 1994 *Intermetallics* **2** 75 – 87
[2] Anwand W, Kissener H R and Brauer G 1995 *Acta Phys. Polon. A* **88** 7
[3] Lukáč F, Čížek J, Procházka I, Jirásková Y, Janičkovič D, Anwand W and Brauer G 2013 *Journal of Physics: Conference Series* **443** 012025
[4] Wolff J, Franz M, Broska A, Kerl R, Weinhagen M, Köhler B, Brauer M, Faupel F and Hehenkamp T 1999 *Intermetallics* **7** 289 – 300
[5] http://www.srim.org
[6] Pundt A and Kirchheim R 2006 *Annual Review of Materials Research* **36** 555–608
[7] Komagata S, Kawasuso A, Yabuuchi A, Maekawa M, Batchulun C, Yasuda K, Ishigami R, Kume K, Iwase A and Hori F 2012 *Physics Procedia* **35** 75 – 79

Positron and thermal desorption studies on He ion implanted nuclear graphite

Z Hu[1], Z Li[1,3], Z Zhou[2], C Shi[1], H Schut[2] and K Pappas[2]
[1] State Key Lab of New Ceramic and Fine Processing, School of Materials Science & Engineering, Tsinghua University, Beijing , China.
[2] Faculty of Applied Sciences, Delft University of Technology, Delft, the Netherlands.

E-mail: zcli@tsinghua.edu.cn

Abstract. The positron beam Doppler Broadening (DB) and Thermal Helium Desorption Spectroscopy (THDS) techniques are applied to study the behavior of radiation induced point defects in IG-110 nuclear graphite. The defects are introduced by irradiation at room temperature with 200keV He^+ at doses ranging from 10^{15} to 10^{17} He/cm^2. In the thermal desorption spectroscopy, the release of He^+ is observed between 500K and 800K. With increasing He^+ implantation dose, the fraction of He^+ desorbed decreases from 27% to 3%. In the DB-curves showing the *S* parameter values versus positron implantation depth, the derived vacancy type defect distribution are in accordance with those obtained by SRIM calculations. Subsequent annealing of the implanted samples in steps of 100K for 5 minutes up to 1200K shows a distinct decrease of the *S* parameter value to its reference value between 500K and 700K. This temperature interval corresponds to the literature values for single vacancy migration energy of 1-2eV.

1. Introduction

In the High Temperature Gas-cooled Reactor (HTGR), which is one of the most promising advanced nuclear energy systems due to its inherent safety and economic competitiveness [1,2] nuclear graphite is applied as the moderator, reflector and as structure material because of its excellent performance and mechanical stability. Given the hostile (neutron) radiation conditions, for the safe operation of such a reactor it is therefore important to understand the behavior of radiation induced point defects in this material. To mimic the effects of neutron irradiation IG-110 nuclear graphite is implanted with 200keV He^+ ions. The displacement damage after implantation and annealing treatment is studied by the positron beam Doppler Broadening (DB) technique. In addition, Thermal Helium Desorption Spectroscopy (THDS) is applied to study the thermal behavior of the implanted helium.

[3] To whom correspondence should be addressed.

2. Experimental

IG-110 nuclear graphite produced by Toyo Tanso Co. Ltd. was used in this work. Three specimens (20×20×2 mm^3, density 1.76 g/cm^3) have been implanted with 200keV He^+ ions at the Institute of Semiconductors, Chinese Academy of Science. The doses were 10^{15}, 10^{16} and 10^{17} He/cm^2, respectively at a flux of several $\mu A/cm^2$. The temperature during implantation was about 350K. From the implanted samples smaller pieces for THDS are cut, with masses and dimensions shown in table 1.

Table 1. Masses and dimensions of the THDS samples for different He doses.

Sample NO.	1	2	3	4	5	6	7
dose (He/cm^2)	0	10^{15}	10^{15}	10^{16}	10^{16}	10^{17}	10^{17}
length (mm)	1.45	1.51	1.50	1.44	1.62	1.65	1.48
width (mm)	1.92	1.91	2.01	1.73	1.73	2.20	2.18
thickness (mm)	1.95	1.99	1.92	1.44	1.76	1.98	1.99
mass (mg)	9.55	10.09	10.33	7.74	9.07	12.38	11.41

In the THDS experiments a tuned quadrupole mass spectrometer is used for the detection of the He atoms that are released into the desorption vacuum chamber from the linearly heated sample. The temperature of the sample is measured with a WRe25%-WRe3% thermocouple. The heating rate (controlled by a PID feedback loop) from RT up to 1500K is set at 0.33 K/s. The level of background signal (at a base pressure of 10^{-8} torr) is reduced by overnight baking-out of the desorption vacuum chamber at 375K. Calibration of the quadrupole is performed by admission of a known amount of He gas and measuring the spectrometer response.

The Positron Annihilation Doppler Broadening (PADB) experiments on the three implanted and one *in situ* annealed sample (10×10×2 mm^3) are performed at the Delft Variable Energy Positron beam (VEP) [3]. By selecting positron implantation energies between 0.1 and 25keV the maximum obtained positron mean implantation depth is 4μm. This is well beyond the 1 μm range of the displacement damage distribution (with a 200nm wide maximum at 800nm depth) predicted by SRIM [4]. The measured *S*-parameters are calculated as the ratio of the counts registered in a fixed momentum window, with $|p_{//}| < 3,5 \times 10^{-3} m_0 c$, to the total number of counts in the 511keV photo peak. Here $p_{//}$ is the momentum of the electrons in the direction of gamma emission. The sample implanted with a dose of 10^{17} He/cm^2 has been selected for the PADB annealing study. The annealing was carried out *in situ* of the positron beam for 5 minutes at temperatures from 300K to 1200K in steps of 100K. The PADB experiments were started after the samples had cooled down to RT.

3. Results and Discussion

The desorption spectra of figure 1 show that for the implanted samples the majority of He release occurs in the temperature interval between 400 and 900K. The sharp spike at 310K in all spectra is an artefact caused by switching on the heater filament. In the temperature interval between 1000K and 1300K, an increase in the He counts in both implanted and reference samples is observed. Since in non-graphite samples, this is not reproduced, the origin of this increase may be due to the release of other gases (O, N) adsorbed tightly onto the graphite surface causing an increased background. With these observations in mind the total amount of He

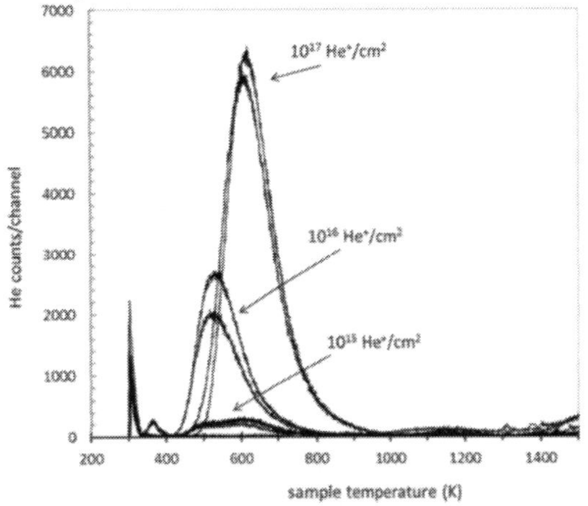

Figure 1. Helium desorption spectra obtained for IG-110 nuclear graphite implanted with 200keV He^+ ions to the doses indicated.

desorbed is determined in the temperature range from 400K to 1000K only. For each implantation dose two samples have been desorbed, the results of which are reproducible. A closer inspection of the spectra shows that for the sample implanted with 10^{15} He/cm^2 two peaks around 500K and 600K are found. For the samples implanted with 10^{16} and 10^{17} He/cm^2 a single broad peak with maxima at 520K and 600K are observed, respectively. Clearly, the release of He is from the traps in the He implantation zone and apparently two types of He traps may be formed or the traps may have different He occupation number. This is in line with hydrogen desorption experiments [5] from which it was concluded that the implantation conditions have an effect on the position of the hydrogen release peaks. The amount of He desorbed (between 400 and 100K) as a function of implantation dose is presented in table 2. It can be seen that the relative amount of desorbed He decreases from 27% to 3% with increasing He^+ implantation dose. This trend is comparable with observed re-emission of He during the implantation of 40 keV He at 300K [6] where complete retention of the implanted He is found at low fluences only. Moreover, the re-emission rate was found to be dependent on the type of graphite and implantation temperature

Table 2. Calculated fraction (%) of He^+ desorbed from IG-110.

Sample NO.	1	2	3	4	5	6	7
dose (He/cm^2)	0	10^{15}	10^{15}	10^{16}	10^{16}	10^{17}	10^{17}
Desorbed He^+ (%)	-	27	24	16	17	3	4

The results of the Positron Annihilation Doppler Broadening experiments are presented in figure 2. The data in the inset shows that in comparison to the reference sample, the implantation at the low dose of 10^{15} He/cm^2 has little effect on the *S* values. For the higher doses the formation of open volume defects is clearly visible by the higher values of *S* which, in the implantation region, increases with He dose. The data has been analysed with VEPFIT [7] using a layered structure model. It is clear that three layers are sufficient to obtain a good fit which is represented by the full lines. The corresponding *S* parameters and their depth ranges are indicated by the dashed lines. For the higher He doses the model thus adequately describes the distribution of vacancies as calculated by SRIM. This model is further used for the analysis of the results obtained after *in situ* annealing of the sample implanted with the highest He dose. The annealing was carried out for 5 minutes at temperatures between 300K and 1200K in steps of 100K. The resulting fitted values of *S* for the three layers are plotted in figure 3. The lines through the data for the *S* parameters in the two near surface layers are obtained by fitting a cumulative normal distribution. The line through the *S* values beyond 1μm represents the average value. From the normal distributions the temperature at which the *S* parameter has decreased to a value halfway between that of the defect and the bulk is found to be 630K with a

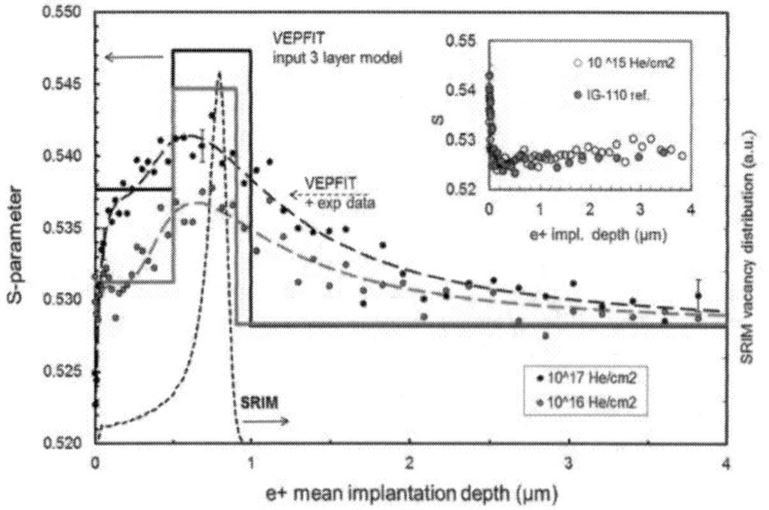

Figure 2. The *S*-parameter as a function of the positron mean implantation depth for the high dose samples. The dashed lines through the data are obtained with VEPFIT using a 3 layer structure. The fitted *S* parameters and layer boundaries are indicated by the profiles together with the distribution of the damage events calculated by SRIM. The inset shows the *S* for the reference and low dose sample.

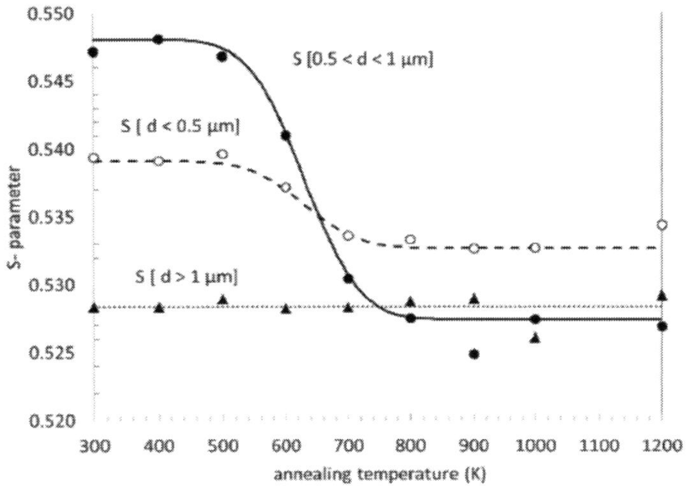

Figure 3. Fitted S values as a function of the annealing temperature for the depth ranges indicated. The lines represent a cumulative normal distribution or average value.

FWHM of approximately 160K. These values correspond well with the maximum temperature and the width of the He desorption peak measured at the sample with the same (high) dose. SRIM calculations show that the He distribution coincides with the peak in the damage profile. We thus conclude that the annealing of the defects is accompanied by the release of He. Recalling that positrons are preferentially trapped at vacancy type defects, the decrease in S between 500 and 700K can be explained by a decrease in the defect concentration. Migration energies for interstitial and vacancy motion are discussed in the paper by Telling [8]. The quoted value of 1.75eV for single vacancy migration corresponds to peak temperature between 500 and 600K for heating rates comparable to that applied in this work. The migration energy for interstitials (of the order of tens of meV) is too low to survive the irradiation temperature. It is noteworthy that the S value nearest the surface does not recover to the reference value. Since in this region the initial defect concentration after implantation is lower compared to the deeper layer, clustering of vacancies to larger, more stable defects is unlikely. Surface erosion caused during the implantation, which may account for the decrease in desorbed He needs further investigation.

4. Conclusion

Thermal desorption and positron annihilation techniques have been applied to study the thermal behaviour of He and implantation defects in IG-110 nuclear graphite. From the observed two desorption temperatures in the THDS spectra two types of He^+ traps are believed to exist which are associated with (single) vacancies which are annealed out between 500 and 700K.

Acknowledgements

I. Carvalho from Delft University of Technology/M2i is gratefully acknowledged for her assistance.

References

[1] Contescu C I, Azad S, Miller D, Lance M J, Baker F S, and Burchell T D 2008 *J. Nucl. Mat.* **381** 15-24
[2] Li Z, Chen D, Fu X, Miao W, and Zhang Z 2011 *Adv. Mat. Sci. Eng.*
[3] Schut H, "A variable energy positron beam facility with application in materials science", Thesis, Delft University of Technology, Delft, The Netherlands, 1990.
[4] Ziegler J, Biersack J, Ziegler M, "SRIM - The Stopping and Range of Ions in Matter", 2008.
[5] Airapetov A A, Begrambekov L B, Vergazov S V, Kuzmin A A, Fadina O C, and Shigin P A 2010 *J. Surf. Inve. X-ray, Sync. Neut. Tech.* **4** 567-571.
[6] Alimov V K, Scherzer B M U, Chernikov V N, and Ullmaier H 1995 *J. Appl. Phys.* **78** 137-148
[7] Veen A van, Schut H, Clement M, Nijs J M M de, Kruseman A, IJpma M R, 1995 *Appl. Surf. Sci.* **85** 216
[8] Telling R H, Heggie M I 2007 *Phil. Mag.* **87**:31 4797-4846.

Journal of Physics: Conference Series **505** (2014) 012015 doi:10.1088/1742-6596/505/1/012015

Structural studies of thin Pd films loaded with hydrogen

O Melikhova[1], J Čížek[1], M Vlček[1], F Lukáč[1], I Procházka[1], W Anwand[2], G Brauer[2]
[1]Charles University in Prague, Faculty of Mathematics and Physics, V Holešovičkách 2, 180 00 Prague 8, Czech Republic
[2]Institut für Strahlenphysik, Helmholtz-Zentrum Dresden-Rossendorf, Postfach 510119, D-01314, Dresden, Germany

E-mail: oksivmel@yahoo.com

Abstract. In this work variable energy positron annihilation spectroscopy was employed for investigation of defects created in Pd films electrochemically charged with hydrogen. The development of hydrogen-induced defects in nanocrystalline, polycrystalline and epitaxial Pd films were compared. It was found that absorbed hydrogen causes plastic deformation and increases defect density in all Pd films studied. Moreover, buckling was observed in nanocrystalline and polycrystalline films loaded above certain critical hydrogen concentration.

1. Introduction

Pd is a very important material for hydrogen technologies and is also used as a model system for study of hydrogen behaviour in metal lattice. In a perfect Pd lattice hydrogen occupies octahedral interstitial sites and causes remarkable volume expansion [1]. Hydrogen strongly interacts with defects: it can be trapped at open volumes defects like vacancies, dislocations and grain boundaries (GBs) and may also introduce new defects [2]. In this work we employed variable energy slow positron spectroscopy (VEPAS) to study the hydrogen interaction with defects in Pd films with various microstructures ranging from epitaxial to nanocrystalline.

2. Experimental Details

Three kinds of Pd films were deposited by cold cathode beam sputtering on (11-20) sapphire substrates: (i) nanocrystalline films deposited at room temperature; (ii) polycrystalline films deposited at room temperature and then annealed in vacuum at 800°C for 1h; (iii) epitaxial films deposited at 800°C. The thickness of the deposited films determined by stylus profilometry was (1080 ± 10) nm and (1040 ± 10) nm for the nanocrystalline and the polycrystalline film, respectively, and (485 ± 5) nm for the epitaxial film. Scanning electron microscopy and X-ray diffraction [3] revealed that films deposited at room temperature exhibit a nanocrystalline structure consisting of column-like crystallites with the mean width of ≈50 nm; subsequent annealing at 800°C led to a pronounced grain growth and the annealed films exhibit polycrystalline structure with the grain size of ≈2.5 μm. The films deposited at 800°C are single crystals with well defined orientation with respect to the sapphire substrate.

Hydrogen loading was performed electrochemically by constant current pulses in a galvanic cell filled with a mixture of H_3PO_4 and glycerine (1:2). The hydrogen concentration x_H introduced into the sample was calculated from the transported charge using the Faraday's law. Hydrogen loading introduces large stresses since the films are fixed on a substrate which prevents in-plane expansion [3].

VEPAS studies were performed on a slow positron beam SPONSOR [4] with positron energy adjustable from 0.03 to 36 keV. Doppler broadening of the annihilation line was measured by a HPGe detector with energy resolution of 1.09 keV at 511 keV and evaluated using the S and W parameters. All S and W parameters were normalized to bulk values S_0, W_0 measured in a well annealed bulk Pd.

Table 1. Dislocation densities estimated from VEPAS results by Eq. (1).

	virgin state		**H-loaded up to $x_H = 1.0$**	
sample	**L_+ (nm)**	**ρ_D (m^{-2})**	**L_+ (nm)**	**ρ_D (m^{-2})**
bulk Pd cold rolled	49 ± 2	$(7.6 \pm 0.6) \times 10^{14}$	-	-
polycrystalline Pd film	106 ± 1	$(9.2 \pm 0.4) \times 10^{13}$	66 ± 3	$(3.8 \pm 0.8) \times 10^{14}$
epitaxial Pd film	16 ± 2	$(8 \pm 1) \times 10^{15}$	2.3 ± 0.8	$(4 \pm 1) \times 10^{16}$

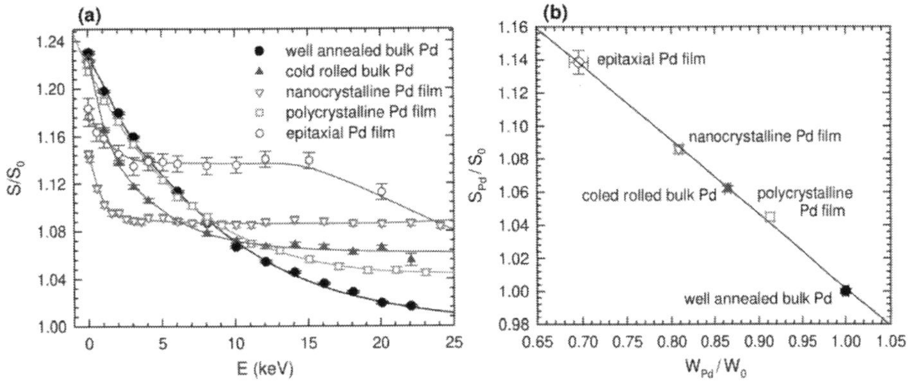

Figure 1. (a) $S(E)$ curves for the virgin films and well annealed and cold rolled bulk Pd. Solid lines show model curves calculated by VEPFIT; (b) S-W plot constructed from bulk values.

3. Results and Discussion

Fig. 1 shows the dependences of S on the positron energy E for the virgin films. At very low energies virtually all positrons are annihilated at the surface. With increasing energy positrons penetrate deeper and deeper into the sample and the fraction of positrons diffusing back to the surface decreases. This is reflected by a decrease of S from a surface value down to a bulk value. Note that in the epitaxial film with lower thickness positrons penetrate into the substrate at $E > 15$ keV which leads to a further decrease of S. The $S(E)$ curves were fitted by VEPFIT [5] and the calculated model curves plotted in Fig. 1 by solid lines describe the experimental points accurately.

The $S(E)$ curves for a well annealed (1000°C/1h) and cold rolled (thickness reduction of 50%) bulk Pd are plotted in Fig. 1 as well. Positron lifetime measurement confirmed that the well annealed bulk Pd exhibits a single component spectrum with the lifetime $\tau_B = (111 \pm 1)$ ps which agrees well with the calculated lifetime of free positrons in a perfect Pd crystal [6]. Hence, the well annealed bulk Pd can be considered as a defect-free material. Indeed, the positron diffusion length $L_{+,B} = (151 \pm 4)$ nm obtained from fitting of the $S(E)$ curve of this sample falls into the range expected for perfect (defect-free) metals. The cold rolled bulk Pd sample exhibits shorter positron diffusion length $L_+ = (49 \pm 2)$ nm and higher bulk S parameter due to positron trapping at dislocations introduced by plastic deformation. The nanocrystalline film exhibits even shorter positron diffusion length $L_+ = (41 \pm 5)$ nm and higher S parameter than the cold rolled bulk Pd. This is not surprising since the average width of nanocrystalline columns is shorter than $L_{+,B}$ and majority of positrons, thereby, diffuse to GBs and are trapped at open volume defects there. Annealing of the film at 800°C led to a decrease of S and an increase of L_+ due to intensive grain growth which substantially reduced the volume fraction of GBs.

The epitaxial film exhibits the highest S and the shortest L_+ among the samples studied. This is likely due to a dense network of misfit dislocations induced by relatively large lattice mismatch (~30%) between the Pd films and the substrate. Note that low angle GBs can be formed in the film by accumulation of misfit dislocations to compensate for the misfit. In VEPAS study low angle GBs cannot be distinguished from isolated misfit dislocations but since XRD characterization of the epitaxial film [3] showed very sharp rocking curve one can conclude that although formation of low angle GBs cannot be excluded they do not represent dominant kind of defects in the sample.

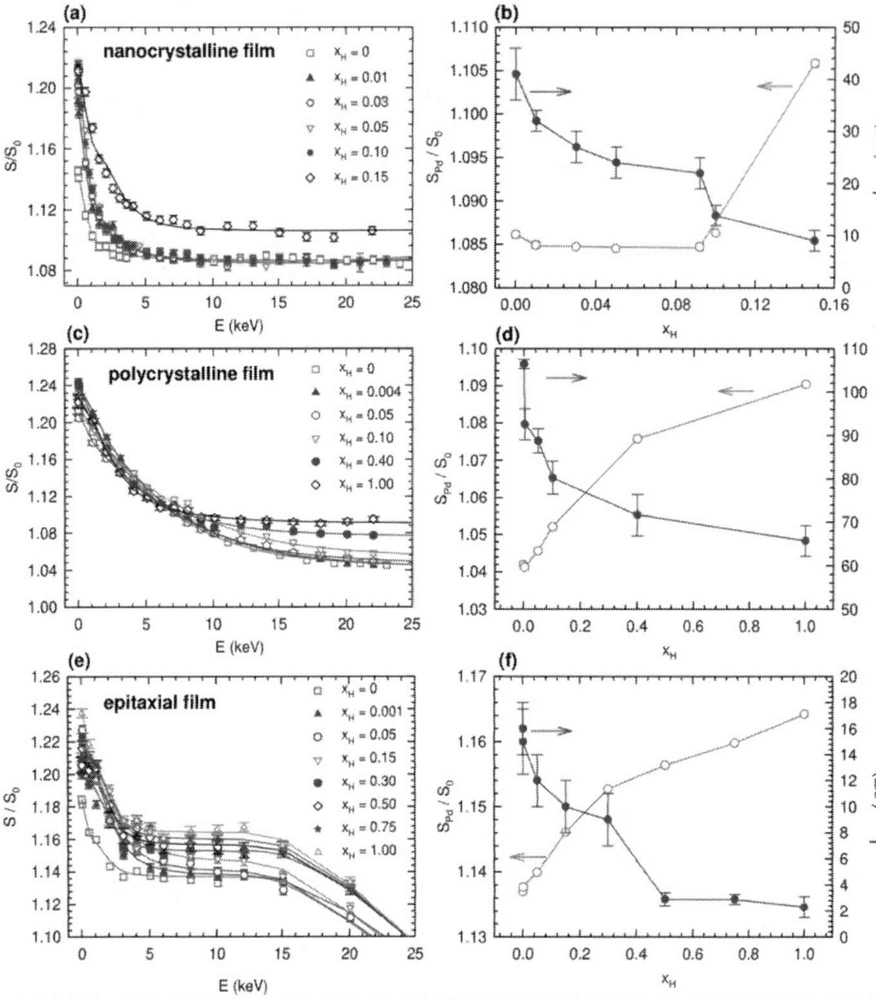

Figure 2. Left panels: dependence of the S parameter on the energy of incident positrons for (a) nanocrystalline, (c) polycrystalline and (e) epitaxial film. Solid lines show model curves calculated by VEPFIT. Right panels: dependence of the bulk S parameter and positron diffusion length on the hydrogen concentration for (b) nanocrystalline, (d) polycrystalline and (f) epitaxial film.

Although a dislocation line itself is a shallow positron trap a positron trapped there quickly diffuses along the dislocation and is finally trapped at vacancies associated with jogs or anchored in the compressive elastic field of dislocation. Hence, positrons trapped at dislocations are finally annihilated at vacancy-like defects. It is known [7] that positrons trapped at GBs exhibit lifetime comparable to a vacancy. Thus, in all Pd samples studied positrons are confined in similar kind of defects characterized by open volume comparable to a monovacancy. This is confirmed by S-W plot constructed from the bulk values and plotted in Fig. 1b. Since all points fall on a straight line connecting the well annealed bulk sample and the epitaxial film positrons are annihilated at similar kind of defects but defect density in the samples differs. Table 1 shows the density of dislocations estimated from the expression

$$\rho_D = \frac{1}{\nu_D \tau_B}\left(\frac{L_+^2}{L_{+,B}^2} - 1\right), \qquad (1)$$

where $\nu_D \approx 10^{-4}$ m^{2}s^{-1} is the specific positron trapping rate for dislocations [8]. Note that Eq. (1) is not applicable for the nanocrystalline film where positrons are trapped predominantly at GBs.

The films were subsequently step-by-step loaded with hydrogen. Left panels in Fig. 2 show the $S(E)$ curves obtained for nanocrystalline, polycrystalline and epitaxial films, respectively. The bulk S parameter and positron diffusion length obtained from fitting are plotted in the right panels as a function of hydrogen concentration x_H. In the polycrystalline and the epitaxial film S increases with x_H while L_+ decreases testifying to formation of new defects introduced by hydrogen loading. It indicates that stresses induced by absorbed hydrogen lead to plastic deformation which creates dislocations in the films. On the other hand, in the nanocrystalline film S firstly decreases because hydrogen fills open volume defects at GBs. At hydrogen concentrations $x_H > 0.02$ all available deep traps at GBs are already filled and S remains approximately constant. When hydrogen concentration exceeds $x_H \approx 0.1$ hydrogen-induced plastic deformation takes place and creates dislocations in the film. Hence dislocation density in hydrogen loaded films increased remarkably. This can be seen in Table 1 containing dislocation densities in the hydrogen loaded films estimated by Eq. (1). Moreover at $x_H >$ 0.1 hydrogen-induced stresses in the nanocystalline film exceeded film adhesion to substrate and buckling of the film occurs. Delamination of the film starts at edges but finally at $x_H > 0.15$ the whole film is detached from the substrate (VEPAS measurement was stopped at that point). Annealing at 800°C improved film bonding to the substrate and the onset of buckling was shifted to $x_H > 0.3$. In the epitaxial film no buckles were formed during the whole loading procedure, i.e. up to $x_H \approx 1$. Buckling is always accompanied by plastic deformation as was confirmed also by acoustic emission [9].

4. Conclusions

Hydrogen interaction with defects in Pd films was studied. It was found that hydrogen is trapped at open volume misfit defects at grain boundaries. Moreover, absorbed hydrogen introduces high stresses which inevitably lead to plastic deformation and introduces a high number of dislocations into the film. When the stresses exceed the film adhesion to the substrate buckling of the film occurs.

Acknowledgements

Financial support from the Ministry of Education, Youths and Sports of the Czech Republic (project LH12173) and the Czech Science Foundation (project P108-13-09436S) are highly acknowledged.

5. References

[1] Flanagan T B, Oates W A 1991 *Annu. Rev. Mater. Sci.* **21**, p. 269
[2] Pundt A, Kirchheim R 2006 *Annu. Rev. Mater. Res.* **36**, p. 555
[3] Čížek J, Melikhova O, Vlček M, Lukáč F, Vlach M, Procházka I, Anwand W, Brauer G, Mücklich A, Wagner S, Uchida H, Pundt A 2013, *Int. J. Hydrogen Energy* **38**, p. 12115
[4] Anwand W, Brauer G, Butterling M, Kissenger H-R, Wagner A 2012 *Defect and Diffusion Forum* **331**, p. 25
[5] van Veen A, Schut H, Clement M, de Nijs J, Kruseman A, Ijpma M 1995 *Appl. Surf. Sci* **85**, 216
[6] Campillo Robles J M, Ogando E, Plazaola F 2007, *J. Phys.: Condens. Matter* **19**, 176222
[7] Dupasquier A, Romero R, Somoza A 1993, *Phys. Rev. B* **48**, p. 9235
[8] Hautojärvi P, Corbel C 1995 *Proceedings of the International School of Physics "Enrico Fermi", Course CXXV* ed A Dupasquier, A P Mills (Varena: IOS Press) pp 491
[9] Čížek J, Procházka I, Vlach M, Žaludová N, Daniš S, Dobroň P, Chmelík F, Brauer G, Anwand W, Mücklich A, Nikitin E, Gemma R, Kirchheim R, Pundt A, 2008 *Appl. Surf. Sci.* **255**, p. 241

Journal of Physics: Conference Series **505** (2014) 012016 doi:10.1088/1742-6596/505/1/012016

PLEPS study of ions implanted RAFM steels

S Sojak[1], V Slugeň[1], W Egger[2], L Ravelli[2], M Petriska[1], J Veterníková[1], M Stacho[1], V Sabelová[1]

[1] Slovak University of Technology, Institute of Nuclear and Physical Engineering, Ilkovičova 3, 812 19 Bratislava, Slovakia
[2] LRT2, Universitaet der Bundeswehr Muenchen, Werner-Heisenberg-Weg 39, D-85577 Neubiberg, Germany

E-mail: stanislav.sojak@stuba.sk

Abstract. Current nuclear power plants (NPP) require radiation, heat and mechanical resistance of their structural materials with the ability to stay operational during NPP planned lifetime. Radiation damage much higher, than in the current NPP, is expected in new generations of nuclear power plants, such as Generation IV and fusion reactors. Investigation of perspective structural materials for new generations of nuclear power plants is among others focused on study of reduced activation ferritic/martensitic (RAFM) steels. These steels have good characteristics as reduced activation, good resistance to volume swelling, good radiation, and heat resistance. Our experiments were focused on the study of microstructural changes of binary Fe-Cr alloys with different chromium content after irradiation, experimentally simulated by ion implantations. Fe-Cr alloys were examined, by Pulsed Low Energy Positron System (PLEPS) at FRM II reactor in Garching (Munich), after helium ion implantations at the dose of 0.1 C/cm^2. The investigation was focused on the chromium effect and the radiation defects resistivity. In particular, the vacancy type defects (monovacancies, vacancy clusters) have been studied. Based on our previous results achieved by conventional lifetime technique, the decrease of the defects size with increasing content of chromium is expected also for PLEPS measurements.

1. Introduction

Structural materials of nuclear power plants (NPP), for example reactor pressure vessel steels, are exposed to high doses of irradiation, heat and mechanical stresses, which may reduce their lifetime during NPP operation [1-3]. Much higher radiation damage is expected in new generations of nuclear power plants, such as Generation IV and fusion reactors. The advanced structural materials are being developed for application in next generations of nuclear power plants. Reduced activation ferritic/martensitic (RAFM) steels are considered as the advanced materials not only for Gen. IV reactors but also as a first wall material in fusion reactors [4,5]. Their microstructures and alloying elements, for example chromium, provide them with high swelling and radiation and thermal (up to 550 °C) resistance [6].

Our work is focused on the study of radiation damage (simulated by ion implantations) evaluation of RAFM steels represented by binary Fe-Cr model alloys. The influence of different content of chromium on the defects size and amount was studied. The experimental analysis of material damage at microstructural level was performed by conventional positron lifetime measurements (PALS) at

Institute of Nuclear and Physical Engineering at Slovak University of Technology and by the Pulsed Low Energy Positron System (PLEPS) [7] at the high intensity positron source NEPOMUC [8] at the Munich research reactor FRM-II.

2. Materials treatment

The detailed chemical composition of studied Fe-Cr alloy can be seen in the table 1. The fabrication processes and treatments of the alloy can be found in [9]. "As-received" material was cut into desired dimensions, ground and polished to mirror-like surface before an exposure to helium implantation.

Table 1. Chemical composition of studied Fe-Cr alloys (wt%) [9].

Alloy	Cr*	O*	N*	C*	Mn	P	Ni	Cu	V
L251	2.36	0.035	0.012	0.008	0.009	0.013	0.044	0.005	0.001
L259	4.62	0.066	0.013	0.02	0.02	0.011	0.06	0.01	0.001
L252	8.39	0.067	0.015	0.021	0.03	0.012	0.07	0.01	0.002
L253	11.62	0.031	0.024	0.028	0.03	0.05	0.09	0.01	0.002

** measured after heat treatment during fabrication*

Accelerated helium ions were used to obtain cascade collisions in the microstructure of the studied material without neutron activation. The helium implantation was performed in two steps with ions energy of 250keV and 100keV, respectively. Implantations at the linear accelerator of the Slovak University of Technology in Bratislava [10] were performed at dose of 0.1 C/cm^2 (6.24x10^{19} cm^{-2}) corresponding to ~ 19 dpa. The maximum temperature during implantation did not exceed 100 °C. The ion energies were chosen to ensure the possibility of application of non-destructive techniques, sensitive in near surface areas (PAS, SEM, etc.). Applicability of these energies was also simulated by SRIM code (Stopping and Range of Ions in Matter) and the maximum damage peaks were calculated for the depths of about 300 nm and 550 nm corresponding to the energies of 100 keV and 250 keV helium ions, respectively.

3. Results

The depth profiling of vacancy type defects was performed by PLEPS on ions implanted Fe-Cr alloys with different chromium content using positron energies, applicable for our materials, in a range from 2 keV to 18 keV corresponding to the positron penetration depth of about 15 – 525 nm. The evaluation of Fe-11.62%Cr measured spectra was performed by PosWin code [11]. Mean positron lifetimes (MLT) were calculated from the three partial components of the lifetime spectra, usually assigned as: τ_1- positron annihilation in bulk, τ_2- positron annihilation in defects (vacancies, vacancy clusters) and τ_3- positron annihilation in large defects (voids).

Results achieved by the PLEPS (figure 1) technique, in case of as-received specimens showed, based on the positron lifetimes reaching levels of about 190 ps, presence of vacancy type defects with the size of about 1-2 vacancies for alloys with 5-12 % of chromium. Specimens with 2.5 % of chromium had the defects slightly bigger and could, according to the literature [12], indicate presence of small clusters with 3 vacancies. These defects, in as-received specimens, can be present in microstructure from the fabrication process or preparation of alloys for measurements. Top layers of microstructure (<100 nm) were not taken into account because of the probable surface annihilation of positrons and because of the thin oxide layer which could also increase the positron lifetimes to the depth of about 100 nm.

The Ion implantation at 0.1 C/cm^2 caused a significant increase of positron lifetimes in comparison to as-received state of specimens. The mean positron lifetimes were of 40-80 ps higher than in the case of as-received specimens. According to the figure 2 the lowest values were observed for the specimens with 11.62% of chromium and with decreasing content of the chromium content increased also the positron lifetime to maximum of 280 ps. This positron lifetime increase corresponds to the increased size of the vacancy type defects to clusters of about 8 vacancies [12].

The lowest values of mean lifetimes (~ 220 ps) correspond to the clusters of about 3-4 vacancies. This is, in comparison to the as-received specimens, a small increase of positron lifetime and positive influence of chromium on the smaller defects formation is observed. The maximum damage peaks from helium ions were not recognized in measured PLEPS spectra probably due to the very high level of damage introduced into the material that was discussed in our previous work [13].

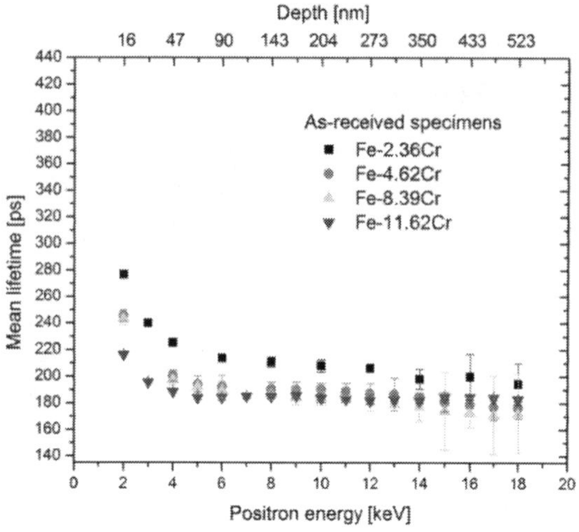

Figure 1. Mean positron lifetimes of as-received Fe-Cr alloys measured by PLEPS.

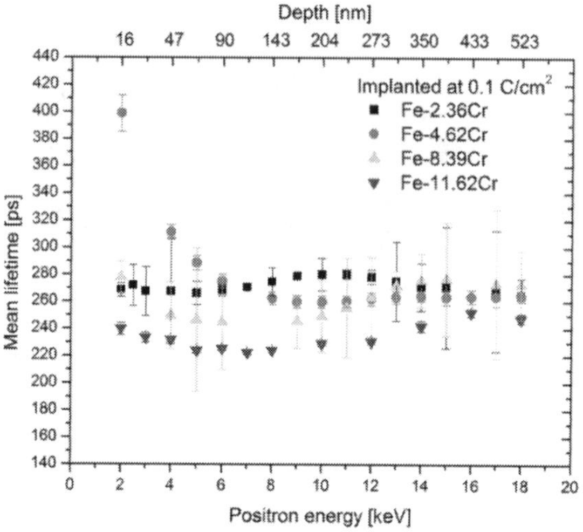

Figure 2. Mean positron lifetimes of ions implanted Fe-Cr alloys measured by PLEPS.

The partial study of these steels was also performed by the conventional positron lifetime measurement (PALS) but the results achieved by this technique were unsatisfactory as the mean positron lifetimes in comparison to PLEPS results were lower of about 60-130 ps (figure 3). Such enormous decrease of MLTs cannot be explained as techniques deviations and it has to be assigned to the ability of PLEPS technique to see the increased amount of defects in specific depths thanks to the monoenergetic positrons. It means that the PALS technique is not so sensitive close to the surface and therefore the final results can be influenced by an undamaged bulk.

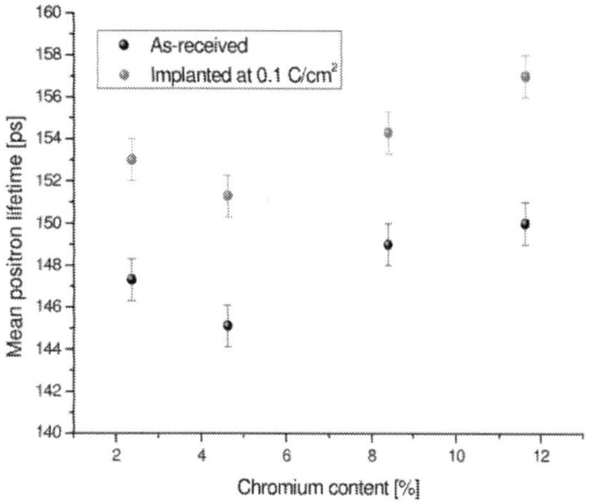

Figure 3. Mean positron lifetimes of as-received and implanted Fe-Cr alloys measured by PALS.

Figure 4. SEM view of Fe-11.62%Cr surface implanted by He ions.

The interference of the bulk structure to the final PALS results could be also indirectly influenced by blistering because of the exposed undamaged structure. Figure 4 shows Scanning Electron Microscopy (SEM) results of the great blisters on the surface of the Fe-11.62%Cr alloy after ions implantation and the undamaged structure under the blisters.

4. Conclusions

The investigation of ions implanted Fe-Cr alloys with a different content of chromium was performed in this work. The Pulsed Low Energy Positron System (PLEPS) and conventional positron lifetime technique (PALS) were applied for Fe-Cr steels investigation. Comparison of these techniques was also performed and the disadvantages of the PALS technique, in form of the inefficiency in the study of material layers close to the surface (up to 1 μm), were revealed. The blisters on the surface also influenced final results of the PALS technique.

Study of the as-received and helium ions implanted Fe-Cr alloys showed expected behavior and the mean positron lifetimes increased after a damage introduction by implantation, therefore the size of the defects increased. The increase in positron lifetime of 40-80 ps for implanted specimens in comparison to as received specimens, where the mono/di-vacancies were present, indicated presence of the vacancy clusters with the size of about 4-8 vacancies based on the chromium content. The implanted alloys with the highest content of chromium contained the smallest defects (4 vacancies). Therefore, the positive influence of the chromium content was observed and the formation of the bigger vacancy clusters, which were present in specimens with low chromium content (~2.5% Cr) was blocked.

Acknowledgments

Financial contributions from the grant VEGA 1/0204/13 and 7RP-Euratom/CU are acknowledged. Support from the Centre of Competence in New Materials, Advanced Technologies and Energy, ITMS: 26240220073 is acknowledged (this is promoted by the Research and Development Operational programme funded by the European Regional Development Fund). The authors gratefully acknowledge the assistance of the staff of the high-intensity positron source (NEPOMUC) and the PLEPS staff at the Munich research Reactor FRM II and the stuff at the Institute of Materials Science at Faculty of Materials Science and Technology, SUT.

References

[1] Slugen V and Magula V 1998 *Nucl. Eng. Design* **186** 323
[2] Slugen V, Segers D, Bakker P M A, Grave E, Magula V, Hoecke T and Waeyenberge B 1999 *J. Nucl. Mat.* **274** 273
[3] De Bakker P.M.A 1997 *Hyp. Inter.* **110** 11
[4] Zeman A, Debarberis L, Slugen V and Acosta B 2006 *Appl. Surf. Sci.* **252** 3290
[5] Zeman A 2007 *J. Nucl. Mat.* **362** 259
[6] Ronald K L and Harries D R 2001 *High-Chromium ferritic and martensitic steels for nuclear applications* (USA: ASTM)
[7] Sperr P, Egger W, Kögel G, Dollinger G, Hugenschmidt Ch, Repper R and Piochacz C 2008 *Appl. Surf. Sci.* **255** 35
[8] Hugenschmidt Ch, Dollinger G, Egger W, Kögel G, Löwe B, Mayer J, Pikart P, Piochacz C, Repper R, Schreckenbach K, Sperr P and Stadlbauer M 2008 *Appl. Surf. Sci.* **255** 29
[9] Matijasevic M and Almazouzi A 2008 *J. Nucl. Mat.* **377** 147
[10] Kovac P Pavlovic M and Dobrovodsky J 1994 *Nucl. Instr. Meth. B* **85** 749
[11] Kirkegaard P and Eldrup M 1972 *Comput. Phys. Commun.* **3** 240
[12] Troev T, Markowski A, Peneva S and Yoshiie T 2006 *J. Nucl. Mat.* **359** 93
[13] Sojak S, Slugen V, Egger W, Ravelli L, Petriska M, Stancek S, Sahul M, Skarba M, Priputen P, Stacho M, Veterníková J, Hinca R and Sabelová V 2012 *J. Phys.: Conf. Ser.* **443** 012036

Study of structural inhomogeneity of commercial oxide-dispersion-strengthened steels

J Veternikova[1], J Degmova[1], M Skraba[2], M Petriska[1], S Sojak[1], V Slugen[1]

[1]Institute of Nuclear and Physical Engineering, Faculty of Electrical and Information Technology, Slovak University of Technology, Ilkovicova 3, 812 19 Bratislava

[2]Institute of Materials Science, Faculty of Material Science and Technology, Slovak University of Technology, Bottova 25, 917 24 Trnava, Slovakia

jana.veternikova@stuba.sk

Abstract. Oxide-dispersion-strengthened steels, candidate materials for construction of new generation of nuclear reactors, can have inhomogeneous structure and anisotropic mechanical properties due to their fabrication, especially by process of mechanical alloying. This paper examines two similar commercial oxide-dispersion-strengthened steels MA 956 and PM 2000 in two different orientations of samples in order to study inhomogeneity of defect presence using positron annihilation techniques. Positron data showed small or negligible differences between samples of individual steels (up to 4 % for average lifetime). Moreover, structural divergence between two orientations (transversal cut and longitudinal cut) of the same steel was also not apparent. However, higher presence of defects and also bigger defects were found for both orientations of MA 956 samples. The structure of PM 2000 probably contains fewer defects with smaller size, though their occurrence in surface and subsurface layers up to 1.6 μm (from Doppler broadening spectroscopy) seem to be less homogeneous than in MA 956.

1. Introduction

Oxide-dispersion-strengthened (ODS) steels belong to perspective candidate materials for new nuclear reactors within the frame of the international program Generation IV and also for fusion reactors. The ODS steels have appropriate mechanical properties and good resistance to corrosion and thermal strains. However, these steels are fabricated by process of mechanical alloying which brings inhomogeneity of their structure and anisotropy of their properties [1]. Methods of homogenization and anisotropy suppression are still studied in many international laboratories, e.g. by addition of variant alloying elements [2, 3] or by different post-manufacture thermo-mechanical treatment (i.e. warm rolling, annealing) [4].

Anisotropy of mechanical properties is not acceptable in construction materials of nuclear facilities. Therefore recent R&D is focused on improvement of technology processes which includes also optimization of milling time and intensity of base material and Y_2O_3 powder or optimization of grain size [5]. As well as laboratory produced steels, also commercial fabricated steels have to be tested in order to eliminate application of inhomogeneous and anisotropic material in construction of the facilities. Mechanical properties are dependent on microstructure, i.e. lattice, structural defects, etc. For that reason, this paper deals with structural study of two commercial ODS steels with almost identical chemical composition and the same process of production but in two different companies.

The emphasis was put on observation of structural defects in samples of two different orientations (transversal and longitudinal cuts) for each investigated material. This is for purpose to compare homogeneity of defect presence in samples depending on their location in extruded steels. Defect presence was observed by positron techniques - Doppler broadening spectroscopy at Aalto University in Finland and positron annihilation lifetime spectroscopy at Institute of Nuclear and Physical Engineering in Bratislava.

2. Experiment

Two following commercial high chromium ferritic ODS steels were measured: MA 956 (products of Incoloy) and PM 2000 (product of Plansee). The chemical composition of investigated steels (See Table 1) was observed by optical emission spectroscopy at Institute of Materials, Slovak University of Technology.

Table 1. Chemical composition of steels (in % wt.).

Steels	C	Mn	Ni	Cr	Mo	Ti	Al	Cu	Si	Nb	N	V	Y_2O_3
MA 956	0.07	0.12	0.07	19.49	0.10	0.33	3.40	0.03	0.04	0.01	0.04	0.02	0.50
PM 2000	0.07	0.07	0.03	19.07	0.13	0.49	4.23	0.01	0.03	0.01	0.01	0.01	0.50

These investigated ODS steels were produced by mechanical alloying, i.e. matrix materials were milled and mixed together with yttria particles to form solid solutions with a uniform dispersion of oxide nano-particles, and the mixtures were then consolidated using HEx at 1150°C under a pressure of 100 MPa. Investigated samples were prepared from as-received material (cylinder shape) by cutting steel sheets one after another into suitable pieces.

Two directions of cuts – longitudinal and transversal ones were prepared (as is shown in Figure 1) for purpose of studying a potential inhomogeneity of defect presence in differently orientated samples, which can support properties anisotropy. Investigation of the longitudinal direction can bring a view of defect presence from inner to outer sides of the received cylinder for the extrusion at the same moment. The transversal direction can show a dependence of defect presence on the extrusion time, as each sample was formed at different moment. After cutting, the sample surfaces were polished in order to remove surface impurities. Mechanical treatment of samples affects surface and subsurface layers, although samples are polished almost into a mirror level.

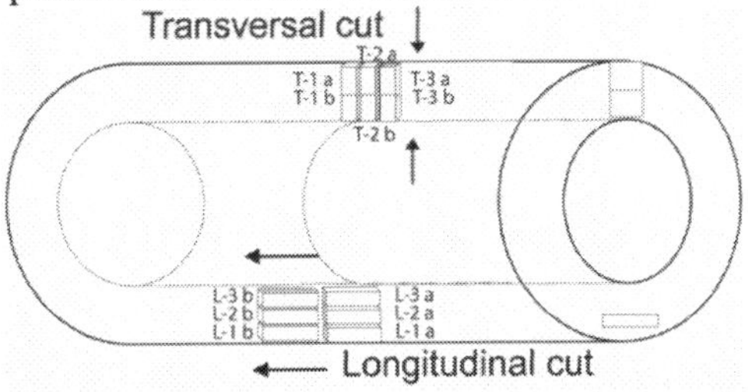

Figure 1. Cutting scheme of two kinds of samples - longitudinal and transversal cuts.

The samples were observed by two different positron annihilation methods describing the presence of vacancy type defects in structure: positron annihilation lifetime spectroscopy (PALS) and positron Doppler broadening spectroscopy (DBS).

The PALS [6] was measured in fast-fast mode [7] with the FWHM parameter up to 240 ps. The variance of fit (reduced chi-square) achieved value in range (0.99; 1.14). The PALS can determine the defect size (lifetime) and defect concentration (intensity) up to the depth ~ 120 μm.

The DBS [8] with a conventional setup of the measurement was recorded by one HPGe detector with Gaussian resolution function of 1.24 keV. The results are shown in S parameters which is calculated from an energy window of $|E\gamma - 511\ \text{keV}| < 0.83$ keV. Positrons used in the DBS measurements are acquired from a slow positron beam [9]. The monoenergetic beam applied energies ranging from 5 to 36 keV during the measurement. This technique and the equipment can be used for a study of the defect profile as a function of a positron implantation depth in samples up to 1.6 μm. The S parameter increases with growth of defect presence (defect size or defect concentration).

3. Results and Discussion

Three samples of each orientation were observed by two positron annihilation techniques. Data of positron annihilation lifetime spectroscopy were treated by spectrum decomposition into two components according to Diffusion trapping model [10] in software LT10 [11]. The shortest lifetime (τ_f) achieved values between 125 and 140 ps which describes positron annihilation in free state - defect free material (bulk) together with small defects, probably dislocations. The bulk values according to the calculation [12] are around 80 ps. Second positron lifetime (τ_D) found in range between 228 and 258 ps describes positrons annihilated in defects with size probably three and four-vacancies in predominance together with annihilation in yttria particles (theoretical lifetime - 240 ps [13]). A paper published by Krsjak et al. in 2010 [12] describes that approximately 10% of τ_D forms annihilation in yttria in our studied materials.

PALS results (Figure 2) show that samples of both investigated steels have small divergences in lifetime of defects (τ_D), although the lifetimes indicate presence of the same defects for the individual steels. The average values of τ_D for MA 956 are 244 ± 6 ps for longitudinal orientations and 256 ± 4 ps for transversal cuts. MA 956 has probably four-vacancies in predominance. The values τ_D for PM 2000 were 230 ± 5 ps for longitudinal cuts and 240 ± 2 ps for transversal ones; therefore PM 2000 contains mostly three-vacancies.

Intensities of defects (I_D), percentage of positrons annihilated with τ_D, have a light dispersion (Figure 2). The average values I_D for MA 956 are 57.7 ± 1.5 % for longitudinal cuts and 53.6 ± 2 % for transversal ones. The values I_D for PM 2000 are 57.1 ± 2 % for longitudinal cuts and 52.1 ± 1.0 % for transversal ones.

The PALS results for individual steels are similar and the data dispersion can be only formed due to the data decomposition. The Average lifetime (τ_{AV}) calculated by LT10 program is presented in Figure 3a. This value is more accurate in comparison to the individual lifetimes, because it is calculated from the measured spectrum without the data decomposition. The average lifetime for MA 956 is 176 ± 2 ps and for PM 2000 163 ± 3 ps. The error of the measurement equipment achieves ± 2 ps in common. Therefore, the average lifetimes lead to assumption that homogeneity of the investigated steels (up to depth of 120 μm) was attained enough during the process of manufacture.

Further, the Average lifetime demonstrated with high probability that steel MA 956 has higher defect presence than PM 2000. It is indicated by all investigated samples in both sample orientations. The minimum difference of τ_{AV} is 6 ps which is sufficient for our statement in term of the measurement error.

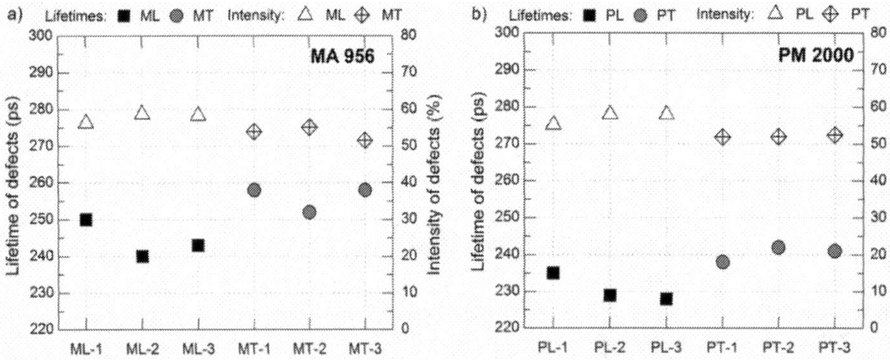

Figure 2. Positron lifetimes and intensities describing defects in MA 956 (a) and in PM 2000 (b).

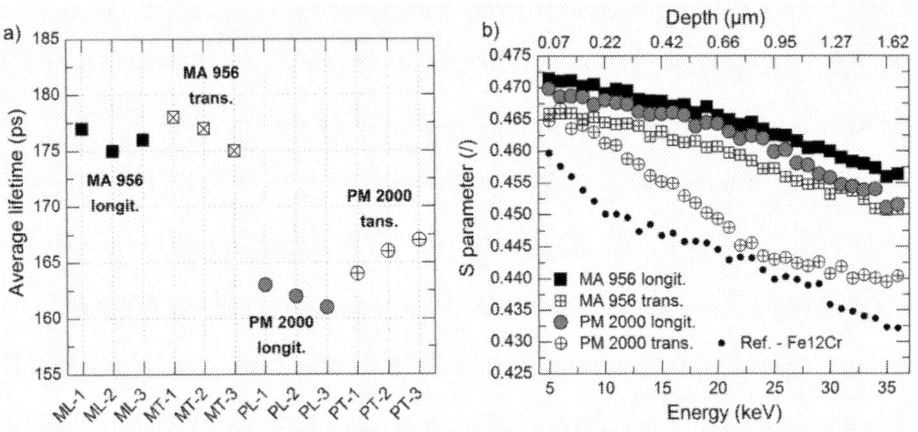

Figure 3. Average lifetime of MA 956 and PM 2000 (a). Dependency of S parameter and positron energy/implantation depth of investigated steels (b).

The investigated samples were later measured by DBS measurement. Results of DBS measurements are shown in Figure 3b - plot of S parameter versus the implantation energy/depth of positrons. The S parameter of PM 2000 in transversal cut demonstrated distinctive difference in comparison to the other samples, while these other samples (both orientations of MA 956 steel and PM 2000 sample in longitudinal cut) present only slight divergences to each other. PM 2000 (transversal cut) has the lowest S parameter and so the defect presence is the smallest there.

DBS measurement was completed by measurement of a reference sample - binary alloy Fe12Cr which was prepared by classical casting and then by the same process of cutting, grinding and polishing as the investigated steels. This sample was annealed after the mechanical treatment at 600°C during 2 hours for elimination of defect presence; therefore it can be considered as reference bulk structure. If the S parameter of the reference sample and the studied samples are confronted, it is found that both orientation of steels MA 956 and PM 2000 (even PM 2000 in transversal cut with minimum defects) contain defects over the equilibrium state described by Fe-12Cr. This is probably influence of mechanical alloying which creates strengthened structure with more lattice defects or it can be also an effect of yttria presence attracting positrons more than bulk as is published in [14].

It is not able to exactly determine the diffusion length (L^+) of positrons from our S parameters; therefore calculations of defect concentration performed according to [12], with using L^+ equal to 40 nm, can be only informative (Table 2). The calculations are based on concentration and size of yttria particles in MA 956 and PM 2000 found in [15], thus the concentration of vacancy clusters can be straight separated from the second component of PALS results (τ_D, I_D).

The average values of four-vacancy concentration for MA 956 are (1.43 ± 0.2) x 10^{23} m^{-3} in longitudinal cuts and (1.14 ± 0.06) x 10^{23} m^{-3} for transversal cuts. For steel PM 2000, the concentrations of thee-vacancies were found as (2.3 ± 0.03) x 10^{23} m^{-3} in longitudinal cuts and (1.47 ± 0.03) x 10^{23} m^{-3} for transversal cuts.

Table 2. Concentration of defects for both orientations of samples - MA 956 and PM 2000.

Sample	ML1	ML2	ML3	MT1	MT2	MT3	PL1	PL2	PL3	PT1	PT2	PT3
Concentration of defects (10^{23} m^{-3})	1.25	1.54	1.50	1.14	1.20	1.08	2.33	2.29	2.30	1.47	1.43	1.43

The difference of S parameters for individual sample orientations (Figure 3b) was visible for both studied steels, while PALS data observed similar presence of defects. The transversal cuts have smaller S parameter which means also lower presence of defects (probably lower concentration). This is in a good accordance with defect concentration calculated from PALS results. Though, it is necessary to realize that these two techniques are not fully comparable due to observed depth of samples. They can rather complete each other.

4. Conclusion

In this paper, the structural inhomogeneity (especially a distribution of vacancy defects) was observed in two similar oxide-dispersion-strengthened steels. Two different orientations of samples (longitudinal and transversal cuts) for steels MA 956 and PM 2000 were studied in term of determining whether properties anisotropy can have connection with defect distribution in a sample. Results from positron annihilation lifetime spectroscopy showed existence of small or almost negligible differences in defect size or defect concentration for the individual samples prepared in both directions of cuts. This demonstrated sufficient homogeneity of the investigated steels concerning to vacancy defect presence up to 120 µm.

Though, the results of Doppler broadening spectroscopy observed visible differences mostly for different cuts of PM 2000. The variance was around 5 % of S parameter for PM 2000 and only around 1 % for MA 956. However, the Doppler broadening spectroscopy was performed only for one sample from each type of material and cut. Further, the measurement also describes shallow depth up to 1.6 µm. Thus, this difference could be formed during the process of sample preparation.

All samples of steel PM 2000 showed smaller presence of defects, probably defect concentration as was calculated form the lifetime spectroscopy, than the samples of MA 956. However, MA 956 introduced greater homogeneity in total.

Acknowledgement: Authors would like to thank to Prof. Filip Tuomisto for management of DBS measurements, to VEGA projects (1/0366/12, 1/0204/13) and EFDA for financial contribution.

References

[1] Klueh R L, Harries D R 2001 High-Chromium Ferritic and Martensitic Steels for Nuclear Applications ASTM Bridgepoint New Jerssy ISBN 0-8031-2090-7.
[2] Coffin L F, Manson S S, Carden A E, Seveud L K, Greenstreet W L 1997 Time-Dependent Fatigue of Structural Alloys ORNL Repot 5073.
[3] Kim S, Weertman J R 1988 Merall. Trans. A19 999.
[4] Baumgaerter W, Bocek M, Petersen C, Rodrian D, Scheibe W, Schmitt R, Schneider H 1989 Pre- and Post Irradiation Fatigue Properties of 1.4914 Martensitic Steel (MANET) Kernforschungzentrum Karlsruhe Report 4774 p. 97.
[5] Bodart M, Baccino R, Moret F 1993 Elaboration and study of mechanically alloyed O.D.S. ferritic stainless steels J de Physique IV 3 p. 709.
[6] Krause-Rehberg R, Leipner S H 1998 Positron annihilation in Semiconductors Springer Berlin Germany ISBN 3-540-64371-0.
[7] Petriska M, Zeman A, Slugen V, Krsjak V, Sojak S 2009 Phys. Stat. Solidi. C 60 2465.
[8] Willardson R K, Weber E R, Stavola M 1998 Identification of Defects in Semiconductors, Academic press, Hardbound, p.376. ISBN-13: 978-0-12-752159-6.
[9] Oila J, Ranki V, Kivioja J, Saarinen K, Hautojaervi P 2002 Appl. Surf. Science 194 38.
[10] Dupasquier A, Romero R, Somoza A 1993 Physical Review B 48 9235.
[11] Giebel D, Kansy J 2010 Material Science Forum 666 138.
[12] Krsjak V, Szaraz Z, Haehner P 2012 J. Nucl. Mater. 428 160.
[13] Alinger M J, Glade S C, Wirth B D, Odette G R, Toyama T, Nagai Y, Hasegawa M 2009 Mat. Sci. Eng. A 518 150.
[14] Simeg Veternikova J, Slugen S, Sojak S, Skarba M, Korhonen E, Stancek S, Degmova J, Sabelova V, Bartosova I 2013 J. Nucl. Mat. In press. Online available from December 8 2013.
[15] Chen J, Hoffelner W 2009 J Nucl Mat 392 360-363,

Vacancy type defect formation in irradiated α-iron investigated by positron beam Doppler broadening technique

C W He[1], K Dawi[1], C Platteau[1], M F Barthe[1], P Desgardin[1] and S Akhmadaliev[2]

[1]CNRS, CEMHTI UPR3079, Univ. Orléans, F-45071 Orléans, France
[2]Institute of Ion Beam Physics and Materials Research, Helmholtz-Zentrum Dresden-Rossendorf, BautznerLandstr. 400, D-01328 Dresden, Germany

E-mail: chenwei.he@cnrs-orleans.fr

Abstract. Vacancy type defects formations have been investigated in virgin and irradiated α-iron samples using slow positron beam Doppler broadening technique. Mono-vacancies and vacancy clusters are observed in 1.5 MeV ^{4}He ions irradiated Fe samples at varying fluences from 1×10^{13} to 1×10^{17} cm^{-2}. In the 1.2 MeV Yttrium ions implantation at low fluence 1×10^{14} cm^{-2}, vacancy clusters with higher concentration and larger size are formed. In this sample, vacancy defects are detected deeper than predicted by SRIM calculation due to channelling.

1. Introduction

Iron based alloys, and for example, Oxide Dispersion Strengthened (ODS) ferritic steels, constitute good candidates for structural components in fission reactors or future fusion reactors which are submitted to extreme conditions of irradiation and temperatures. Changes in microstructure and macroscopic properties of these nuclear materials are governed by the kinetics of defects produced by irradiation, such as vacancy-type clusters or vacancy-impurity complex clusters. In ODS steels, the Yttrium, Titanium oxide nanoparticles are traps for He and pin dislocations allowing to limit the swelling and creep. These materials are fabricated by mechanical alloying and their interesting properties are very dependent of the nanoparticles distribution in size and in space. Fundamental studies are needed to improve the control of this distribution. One of the key-point in the understanding of the formation of these particles is the role of vacancy defects induced during milling. As a key component of these materials, the understanding of defect formation in irradiated bcc iron (α-Fe) is essential. Many efforts, both experimental and computational works, have been put into this study. Some are carried out by using positron annihilation spectroscopy as its sensitivity to the size and density of vacancy clusters in the range from mono-vacancies up to cavities containing 50-100 vacancies [1]. Recently T. Iwai et *al.* [2, 3] have studied the damage induced in iron by various ion species (He, C, O, Fe) by using positron beam Doppler broadening. Mono-vacancies and vacancy clusters have been observed but characteristics S/W of positron annihilation in these defects have not been defined which are important to discriminate between the different vacancy types.

In this study, the vacancy-type defects induced by 1.5 MeV ^{4}He ions at varying fluences in α-Fe are investigated with slow positron beam Doppler broadening (SPBDB) technique at the CEMHTI laboratory. It allowed defining the annihilation characteristics S and W for mono-vacancy and vacancy clusters. Then the defects produced by 1.2 MeV Yttrium implantation in α-Fe have also been studied. This will be the basis for our study on the interaction of vacancy clusters with Y.

2. Experimental

Samples are 99.99% pure iron 7 mm×7 mm×0.5 mm sheets. After polishing, the samples were annealed at 800°C for one hour in vacuum. The grain size is about 20-100 μm. 1.5 MeV ^{4}He irradiations were carried out with the 3.5 MeV Van der Graaff accelerator at CEMHTI Orleans. Eight different fluences have been performed in the range from 1×10^{13} to 1×10^{17} cm^{-2}. 1.2 MeV Yttrium implantations were done with the 3 MeV Tandetron accelerator at laboratory HZDR Dresden at the fluence of 1×10^{14} cm^{-2} producing a similar order of magnitude of damage dose compared to 1.5 MeV ^{4}He irradiation at fluence 1×10^{17} cm^{-2}. Both these two types of irradiations were carried out at room temperature.

The momentum distribution of electron- positron pairs has been measured at 300 K by recording the Doppler broadening of the 511 keV annihilation line characterized by the low (S) and the high (W) momentum annihilation fraction in the momentum range (-2.49, 2.49)× 10^{-3} and (-24.88, -9.64), (9.64, 24.88) × $10^{-3}m_0c$, respectively. To investigate the depth dependence of S and W, the S(E) and W(E) were measured as a function of the positron energy E changed in 500 eV steps in the 0.5 - 25 keV range using the slow positron beam. For this energy range, the maximum positron penetration depth in pure Fe is at most ~2 μm.

Figure 1 shows the depth profiles of Y and He concentrations and the corresponding damage calculated by using SRIM 2008 [4] for pure Fe irradiated by 1.5 MeV ^{4}He and 1.2 MeV Y. In these calculations, the Fe displacement threshold energy has been fixed at a value of 40eV [5]. In the same figure, the slow positron implantation profile is also illustrated for different energies [6]. In the track region of ^{4}He ions located between surface and 2μm which is probed with the slow positron beam, the damage dose is 0.1 dpa (displacement per atom) and the helium concentration is 4 atomic ppm for the highest fluence 10^{17} cm^{-2}. This concentration is quite low, thus the effect of helium on vacancy defects could be ignored in the following. 1.2 MeV Y ions are implanted between surface and 500 nm in Fe. The projected range (Rp) is 265 nm. Thus, the positrons could detect the entire implanted region. At fluence 1×10^{14} cm^{-2}, the damage dose is 0.14 dpa at the surface and increases up to reach 0.24 dpa at the depth of 170 nm and the Y concentration reaches the maximum value of 58.9 appm at the depth of 270 nm. This concentration is lower than yttrium solubility in iron (290 at. ppm in iron at 800°C [7]). The profile of Y concentration has also been calculated using Marlowe code considering that sample is a polycrystal. It suggests that Y ions stop deeper than Rp up to ~1500 nm.

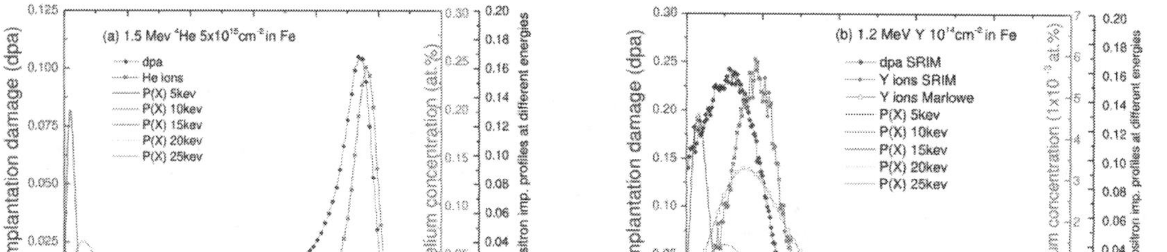

Figure 1. Depth profile of defect production and atomic concentration (a) 1.5 MeV ^{4}He in Fe (b) 1.2 MeV Y in Fe.

3. Results and discussion

3.1. Virgin iron samples

All as-received polished and 800°C/1 h/vacuum annealed Fe samples have given SPBDB reproducible results and a typical representation is illustrated for two samples in Fig 2. S (respectively, W) decreases slowly (increases) as a function of positron energy. This slow evolution indicates that a large fraction of positrons diffuses back to the sample surface at intermediate positron energy and annihilation occurs both at surface and in the bulk of sample. The adjustment of experimental results with VEPFIT [8] program using a simple assumption of one homogeneous layer reveals an effective positron diffusion length $L^+ = 160 \pm 2$ nm in the bulk. The positron diffusion lengths reported for defect-free metals are ~ 100 nm [9]. Higher L^+ value of 184 nm, has been obtained by Iwai and *al.*, for pure Ni (99,997 wt.%) annealed at 800°C in vacuum for one hour [10]. All these results reveal that the quantity of open-volume defects in the bulk of the virgin samples is low. The annihilation characteristics of the Fe lattice considered as the reference values in this work are: $S_{ref} = 0.357(2)$ and $W_{ref} = 0.111(3)$.

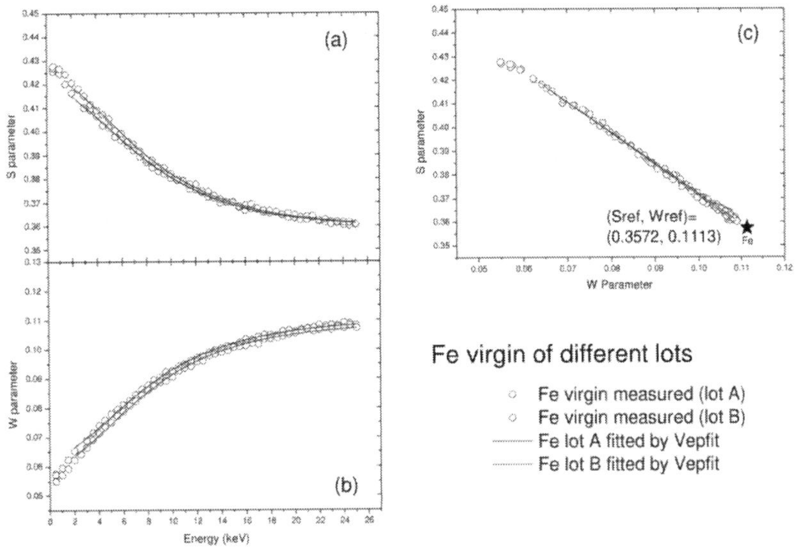

Figure 2. Low (S) (a) and high (W) (b) momentum fraction as a function of positron energy and S as a function of W (c) for the virgin pure Fe samples. VEPFIT is used to fit the (S, W) curves with a model of one homogeneous layer.

3.2. 1.5 MeV ^{4}He irradiated iron samples

Figure 3(a) shows low momentum fraction S as a function of positron energy for pure Fe samples irradiated with 1.5 MeV ^{4}He at fluences from 1×10^{13} to 1×10^{17}cm^{-2} (10^{-5} to 10^{-1} dpa) and for Fe virgin sample. The S and W positron annihilation characteristics in the ^{4}He irradiated Fe samples differ from the ones measured in the virgin Fe samples. Whatever the fluence, all the S(E) values are higher in the irradiated samples than in the virgin ones when all the W(E) values are, respectively, lower. This indicates that ^{4}He ions have induced vacancy-related defects in their track region. The general look of W(E) curves is reverse to that of S(E). S(E) curves depend on the fluence. For low fluencies from 1×10^{13} to 5×10^{14}cm^{-2}, the curves suggest the diffusion back of positrons to the surface and a large fraction of annihilation in the surface states. For higher fluences, diffusion back to the surface becomes negligible. From approximately 22.5 keV, the S(E) increases as a function of E for low fluences and decreases for the highest ones. In Figure 1.a, we can see that 25 keV positrons probe a region where the damage dose and the He concentration begin to increase because of the accumulation of cascades. S(E) increases with the damage dose. But when the fluence becomes equal or higher than $1\text{x}10^{16}$ cm^{-2}, it decreases. At this fluence the He concentration becomes non negligible in the region probed by high energy positrons. He can be trapped in vacancies and it leads to a change of electron density and momentum at the trap [11]. This effect of helium is negligible at low fluences when the He concentration is very low. In the following we choose to consider the S and W values measured in the energy range between 20 and 22 keV as the annihilation characteristics in the track region of the ions. In this range S(E) changes very slightly, the diffusion back of positron to surface and the annihilation fraction in cascade region are negligible. In this energy range S increases with the fluence (see arrow in Fig 3.a). It indicates that damage increases when the irradiation fluence increases in the pure Fe samples.

The mean values of S and W estimated in this energy range will be called S and W respectively to lighten the text. The evolution of S as a function of W is plotted on Figure 3(b). At low fluence from 1×10^{13} to 2×10^{14} cm^{-2} (from 1×10^{-5} to 2×10^{-4}dpa), the S,W points follow the same straight line D1 which goes through the (S_{ref}, W_{ref}) point. It indicates that the same nature of vacancy defects has been created for these fluences. In these conditions of irradiation, SRIM predicts that the energy of the Primary Knocked-on Atoms (PKA) is low (91.8eV) so the recoiled atoms collisions should lead to the formation of small cascades with essentially single vacancies as molecular dynamics modelling predicts [12]. Thus, we will propose that this D1 line is more likely characteristic of annihilation in single vacancy in Fe.

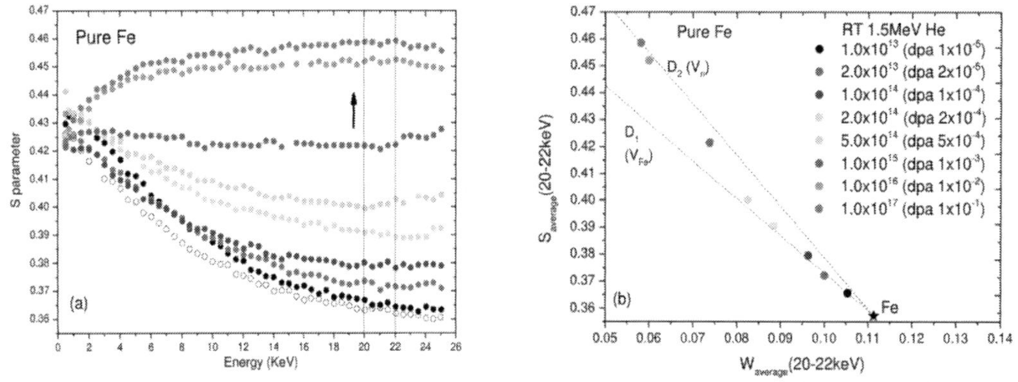

Figure 3. Low (S) momentum fraction as a function of positron energy (a) and S as a function of W obtained by calculating the average value between positron energy 20 and 22 keV (b) in Fe samples before irradiation and after 1.5 MeV ^{4}He irradiation at different fluences in the range from 1.0×10^{13} to 1.0×10^{17} cm^{-2}.

From the fluence of 5×10^{14} cm^{-2} (5×10^{-4} dpa), the S,W points begin to leave the single vacancy line. This indicates that the nature of vacancy defects has changed and larger vacancy-type defects are detected. When fluence increases, the shift from D1 characteristic line increases indicating that the size of vacancy clusters grows with the damage dose. Such clusters formation could be explained by the migration of single vacancies generated by irradiation that could agglomerate leading to the formation of larger clusters when their concentration increases that means with the fluence. Indeed, migration temperature of V in Fe is below the room temperature (-53 to 5°C) as suggested by Takaki et *al.* [14]. T. Iwai et *al.* [2] investigate accumulation of vacancy-type defects in iron produced by 1 MeV He irradiations at room temperature by positron beam both in-situ during irradiations and after irradiations. Vacancy clusters have been observed from $\sim1\times10^{-2}$ to 0.1 dpa and the results suggest the S values are not likely to change under and after irradiation. Thus, the vacancy clusters in Fe by irradiation with ^{4}He 1.5 MeV (this study) could be considered to form during the irradiation by migration and agglomeration. Between 1×10^{16} cm^{-2} and the highest fluence of 1×10^{17} cm^{-2} the S and W values don't change too much indicating that saturation occurs. This saturation could correspond to a maximum size of clusters that can be formed in these irradiation conditions. We will define the straight line going from lattice S,W point and S,W measured in sample implanted at the highest fluence as D2 characteristic of annihilation of positrons trapped in vacancy clusters.

Being given the migration energy of the single vacancies it is therefore unlikely that they survive in the irradiated sample at room temperature. They should either disappear by diffusion into sinks or agglomerate with other vacancies into clusters or form pairs with some impurities which stabilizes them. It has to be noted, the carbon concentration in these iron samples is about 56 appm. They could be bound to single vacancies and stabilize them as it has been shown in [13]. It suggests that D1 could be characteristic of the detection of single vacancies as proposed above.

3.3. 1.2 MeV Yttrium implanted iron samples

As shown in Figure 4(a), S(E) increases fast from 0.5 to 4 keV and reaches a plateau value from 4 to 15 keV. From 15 keV, S(E) begins to decrease. Figure 4(b) shows, from 0.5 to 4 keV, the corresponding S,W points go towards the D2 line which indicates vacancy clusters are formed in this region. S,W points remain superposed from 4 to 15 keV which means the size and the concentration of vacancy clusters do not change or that positrons are not able to detect this change. In the last positron energy range from 15 keV, S,W points evolve along the D2 line towards the (S_{ref}, W_{ref}) point. It indicates that the size of defects does not change but their concentration decreases in the probed depth range.

Compared to the result of 1.5 MeV ^4He irradiated at highest fluence 1.0×10^{17} cm^{-2} Fe sample, the S(E) values of 1.2 MeV Yttrium implanted Fe are much higher and S,W points are further from the (S_{ref}, W_{ref}) point and slightly above the D2 line. It indicates that the vacancy cluster concentration is much higher in Y implanted sample. It could be explained by the higher damage dose accumulated with Y irradiation 0.242 dpa compared to 0.1 dpa for sample irradiated with ^4He at the highest fluence. Moreover their size is slightly larger. It could be due to the difference in PKA (Primary Knock-on Atom) energies for both projectile ions. Heavier ions produce more PKAs with high energy than lighter ions. PKAs which can induce collision cascades in which the displacement local density depends on their energy. Half of PKAs produced by 1.5 MeV ^4He in Fe have energies lower than 91.8 eV whereas this energy is 232 eV for 1.2 MeV Y. Molecular dynamics simulations show PKA with higher energy produce larger vacancy clusters [15].

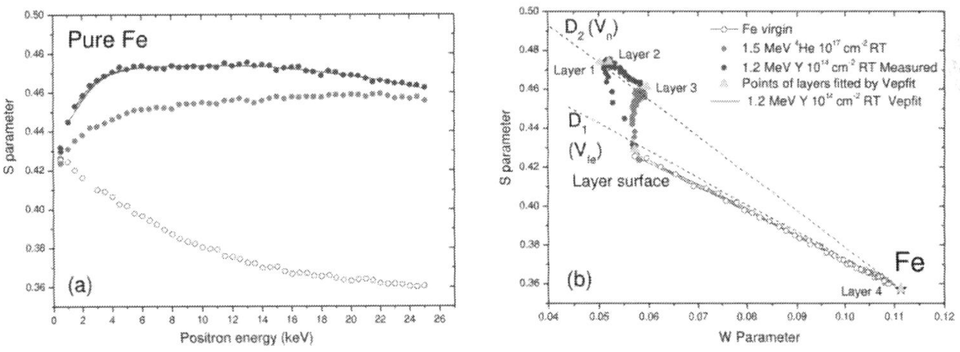

Figure 4. Low (S) momentum fraction as a function of positron energy (a) and S as a function of W (b) in Fe samples before implantation and after 1.2 MeV ^+Y implantation at fluence 1.0×10^{14} cm^{-2}.

Another observation has attracted our attention. The S(E) curve of 1.2 MeV ^+Y implantation begins to decrease from 15 keV instead of 10 keV energy from which a fraction of positron should annihilate in the defect-free Fe lattice as suggested by the depth damage profile calculated by SRIM (Figure 1.b). It indicates that positrons detect vacancy clusters deeper than expected by SRIM. A structure of four layers allows to fit the S(E) and W(E) experimental curves with Vepfit. The S and W values of each of these layers are plotted in Figure 4 and indicate that vacancy clusters are detected in the first three layers. The first layer spreads from surface to 57.5±6.4 nm, the second spreads up to 609.4±11.3 nm and the third stops at the depth of 1661.80±25.5 nm. For these three layers L^+ changes from 10 to 15 nm. In the fourth layer the S, W and L^+ correspond to the one of Fe lattice. So vacancy clusters are detected up to 1660 nm. Two phenomena which are not taken into account in SRIM code could explain this observation. First the migration of vacancies or vacancy clusters toward the bulk could lead to agglomeration of these defects deeper than the cascade region and the stopping depth of ions. Secondly channeling of ions in crystalline solids can occur. The Fe samples irradiated in this study are polycrystalline with mean size of the grains of 20 - 100 μm dramatically larger than the range of the Y ions. Calculations with Marlowe code in which the sample has been modeled as a polycrystal show that this effect is not negligible in these irradiation conditions. Kinomura et *al.* have observed this channeling phenomenon for 150 keV Ar implantation in polycrystalline Fe with grain sizes of the order of 10μm [16].

4. Summary

Vacancy type defect formations in 1.5 MeV ^4He and 1.2 MeV Y irradiated bcc Fe samples are investigated by slow positron beam Doppler- broadening measurements. The positron diffusion length L^+ in Fe virgin sample and the reference value (S_{ref}, W_{ref}) for Fe lattice are determined to be 160±2 nm and (0.357, 0.111). Single vacancies are observed in 1.5 MeV ^4He irradiated Fe samples at low fluences from 1×10^{13} to 2×10^{14} cm^{-2} and vacancy clusters are detected at high fluences from 5×10^{14} to 1×10^{17} cm^{-2}. In 1.2 MeV Y implanted Fe sample, vacancy clusters are also formed even at low fluence 1×10^{14} cm^{-2}. Their concentration and size are larger than that of 1.5 MeV ^4He irradiated sample at the highest fluence 1×10^{17} cm^{-2}. In Y implanted Fe sample, defects profiles obtained by positron measurements are deeper than those calculated by SRIM calculation which could be explained by ion channelling predicted by Marlowe modelling.

Acknowledgments
This work was supported by PERFORM60 and EFDA.

References

[1] Eldrup M and Singh B N 2003 Accumulation of point defects and their complexes in irradiated metals as studied by the use of positron annihilation spectroscopy- a brief review *Journal of Nuclear Materials* **323** 346-53
[2] Iwai T, Ito Y and Koshimizu M 2004 Vacancy-type defect production in iron under ion beam irradiation investigated with positron beam doppler broadening technique *Journal of Nuclear Materials* **329-333** 963-6
[3] Iwai T and Tsuchida H 2012 In situ positron beam Doppler broadening measurement of ion-irradiated metals - Current status and potential *Nuclear Instruments and Methods in Physics Research B* **285** 18- 23
[4] Ziegler J F, Ziegler M D and Biersack J P 2010 The stopping and range of Ions in matter. *Nucl. Instr. and Meth. in Phys. Res. B* **268** 1818- 23
[5] Nordlund K 2006 *Nucl. Instr. and Meth. in Phys. Res. B* **246** 322- 32
[6] Krause-Rehberg R and Leipner H S 1999 *Positron Annihilation in Semiconductors*
[7] Li L and Xing Z 1993 *Acta Metall. Sin.* **29** A136
[8] Veen A V, Schut H, Clement M, Nijs J D, Kruseman A and Ijpma M 1995 VEPFIT applied to depth profiling problems *Applied Surface Science* **85** 216- 24
[9] Paulin R and Ripon R 1974 *Appl. Phys.* **4** 343- 7
[10] Iwai T, Tsuchida H and Awano M 2007 Application of positron beam Doppler broadening technique to ion beam irradiation in nickel *Journal of Nuclear Materials* **367-370** 372-6
[11] Maekawa M and Kawasuso A 2009 Characterization of helium bubbles in Si by slow positron beam *Journal of Physics: Conference Series* **225**

[12] Stoller R E 2000 The role of cascade energy and temperature in primary defect formation in iron *Journal of Nuclear Materials* **276** 22- 32
[13] Fu C C and Willaime F 2007 First priniciples calculations in iron: structure and mobility of defect clusters and defect complexes for kinetic modelling *C.R. physique*
[14] Takaki S, Fuss J, Kugler H, Dedek U and Schultz H 1983 The resistivity recovery of high purity and carbon doped iron following low temperature electron irradiation *Radiat. Eff* **79** 87-122
[15] Bacon D J, Osetsky Y N, Stoller R and Voskoboinikov R E 2003 MD description of damage production in displacement cascades in copper and α -iron *Journal of Nuclear Materials* **323** 152-62
[16] Kinomura A, Suzuki R, Ohdaira T, Oshima N, Ito K, Kobayashi Y and Iwai T 2011 Defect profiles in ion-irradiated metal samples by slow positron beams in comparison with simulation profiles *Journal of Physics: Conference Series* **262**

Dose effects in He implanted Eurofer97 steel

I Carvalho,[1,2,3] H Schut[2], A Fedorov[3], N Luzginova[3], P Desgardin[4], J Sietsma[5]

[1]Materials innovation institute (M2i), Mekelweg 2, 2628 CD, Delft, The Netherlands
[2]Delft University of Technology, Faculty of Applied Sciences, Mekelweg 15, 2629 JB, Delft, The Netherlands
[3]Nuclear Research and consultancy Group (NRG), Westerduinweg 3,1755 LE, Petten, The Netherlands
[4]CEMHTI-CNRS, 3A rue de la Férolerie, 45071 Orléans Cedex, France.
[5]Delft University of Technology, Faculty of Mechanical, Maritime and Materials Engineering, Mekelweg 2, 2628 CD, Delft, The Netherlands

Email: i.carvalho@m2i.nl

Abstract. Reduced Activation Ferritic/Martensitic steels are being extensively studied because of their foreseen application in fusion reactors. To reproduce neutron irradiation conditions, Eurofer97 samples were implanted at room temperature with helium ions at energies of 500 keV and 2 MeV and doses of $1x10^{15}$-10^{17} He /cm^2. The implantation induced defects were characterized by positron beam Doppler Broadening (DB). The samples were annealed in the range 300 – 1500 K, in 100 K steps. As the temperature increases, the annealing of vacancies and vacancy clusters is noticed and followed by the coalescence of He_nV_m clusters. At temperatures around 1200 K HeV pairs dissociate and bubbles are formed. Above 1300 K the helium release from bubbles is observed. The *S-W* graphs reveals that the samples have similar positron traps up to 1200 K. At 1200 K helium bubbles are noticed and the *S,W* pair shows a clearly distinct behaviour from the *S,W* values of vacancy type defects. As the temperature increases and helium is release, the *S,W* pairs shift towards the *S,W*'s of vacancies.

1. Introduction

Eurofer97 is a Reduced Activation Ferritic/Martensitic (RAFM) steel used as reference structural material for future fusion reactors. Eurofer97 steel is known to possess a high resistance against swelling caused by gaseous transmutation products (hydrogen, helium), has attractive mechanical properties and shows reduced activation behavior in a fusion neutron spectrum [1]. The effects of neutron irradiation and the accumulation of defects in this material are extensively studied as Eurofer97 will be used for constructing test blanket modules in the ITER fusion reactor [1, 2]. As a consequence of the neutron irradiation, displacement damage will be created together with the production of hydrogen and helium [1, 2]. This leads to changes in the microstructure and ultimately to alterations in the material's mechanical properties. The work described here is part of a larger project in which Thermal Desorption Spectroscopy (TDS) and electron microscopy studies will be performed on Eurofer97 neutron irradiated in the High Flux Reactor at NRG [3]. To mimic these neutron irradiations, in this work Eurofer97 samples are implanted with 500 keV and 2 MeV He^+ ions. The defect structures after implantation and annealing treatments are characterized with the positron beam Doppler Broadening technique (DB). The present work is the continuation of a previous study

[4] in which the behaviour of He in tempered Eurofer97 steel was studied and helium release mechanism was suggested.

1. Experimental

The samples used in this study were cut from the Eurofer97 batch, produced by Böhler, Austria, with a nominal composition of Fe-9Cr-1W-0.2V-0.1Ta-0.1C (wt.%). The dimensions were 1.2x1.2x0.05 cm^3. In order to anneal the defects introduced by cutting and to obtain a tempered martensitic microstructure, the samples were first annealed for 30 min at 1253 K, and then tempered for 90 min at 1033 K, both followed by cooling in argon ambient.

Low dose ($1x10^{15}$ – $1x10^{16}$ He/cm^2) 500 keV and 2 MeV room temperature helium implantations were performed at CEMHTI-CNRS Orléans, France. In the case of the 2 MeV energy implanted samples, the dose range was extended to $1x10^{17}$ He/cm^2. Details regarding the implantations can be found elsewhere [4]. The beam DB measurements were performed up to positron implantation energies of 25 keV, corresponding to a maximum probing depth in steel of approximately 1 μm.

2. Results and discussion

Figure 1 shows the DB *S*-parameter as a function of the positron implantation energy (bottom axis) and positron mean implantation depth (top axis) for the reference and the implanted samples. For all implantations a higher *S*-parameter reflects a higher implantation dose. The lines through the data are the fits obtained with the VEPFIT program [5]. In case of the 2 MeV implantation, one layer defining the *S*-parameter plus a layer representing the bulk was sufficient to obtain a good fit. For the 500 keV implantation two layers (with thicknesses of 250 and 800 nm) plus a layer for the bulk were needed. This is consistent with the TRIM results which show that for the 2 MeV implantation the defect production profile is flat over the positron probing range while for the 500 keV implantation the defect profile peaks at the end of the positron range. The He/V (He per vacancy) ratio in the first layer (250 nm thick) of the 500 keV sample is comparable to that of the 2 MeV sample over the whole region probed by the positrons.

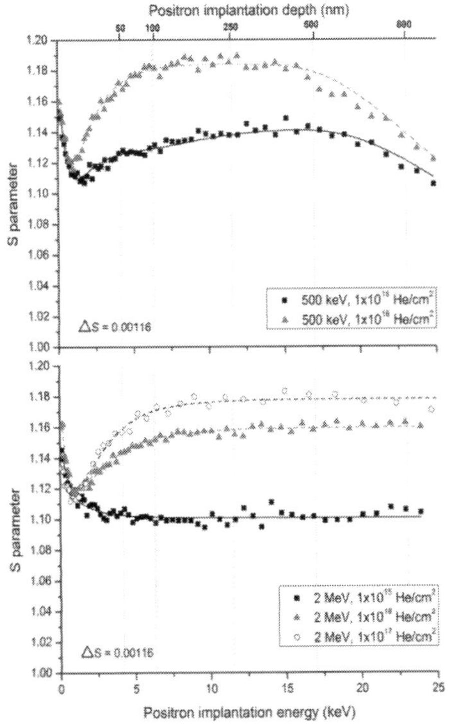

Figure 1. Normalized *S* parameter vs. positron implantation energy (bottom axis) and depth (top axis). Data fitted with VEPFIT.

The fitted positron diffusion lengths ranges between 60 nm for the reference sample and less than 15 nm after helium implantation, indicative for high positron trapping efficiency.

Figure 2 shows the fitted *S*-parameter (measured at room temperature) as a function of annealing temperature. The annealing was performed *in situ*, inside the DB setup. With increasing annealing temperature the thickness of the second layer of the 500 keV samples was decreased in order to obtain a good fit. Nonetheless, the sum of the thicknesses of the first and second layers was kept constant. Above 1300 K, one layer describing the implantation effects and a layer associated with the bulk was sufficient to achieve satisfactory results.

Below 600 K a drop in the *S*-parameter is observed that is associated with the increase in the mobility of the vacancies and dissociation of vacancies from vacancy clusters [6]. This effect is more pronounced in the first layer of the 500 keV samples in comparison with the second layer for each implantation dose, respectively.

As the temperature increases to 800 K, a further slower decrease in the *S* values is observed in the first layer of the 500 keV implanted sample whereas a stable *S* parameter is noticed in the second layer. This decrease can be interpreted as a decrease of the number of positron traps (e.g. continuation of vacancy annealing) in the first layer.

In the range 800 – 1300 K multiple helium release mechanisms take place [4]. Between 800 – 1000 K the *S* parameter increases significantly both in the second fitted layer of the 500 keV samples and in the high dose ($1x10^{17}$ He/cm^2) 2 MeV implanted samples. This increase can be explained by growth of vacancy-helium clusters through coalescence or by Ostwald ripening. In the case of the latter process ultimately the more stable He_nV_m (n, m = 1) pairs are created. For the first layer of the 500 keV as well as for the lower dose (10^{15}-10^{16} He/cm^2) 2 MeV samples, the relatively stable *S* shows that the initial defect density in combination with the presence of Helium is crucial for the above growth processes to take place. Also the smaller distance to the surface which acts as a sink for defects is believed to play a role.

As the annealing temperature increases from 1000K to 1300 K the *S*-parameter drops because of the dissociation of HeV pairs. Retrapping of helium in the larger clusters occurs, forming He bubbles [4]. In this temperature interval the phase transition of Eurofer97 at 1160 K takes place [7].

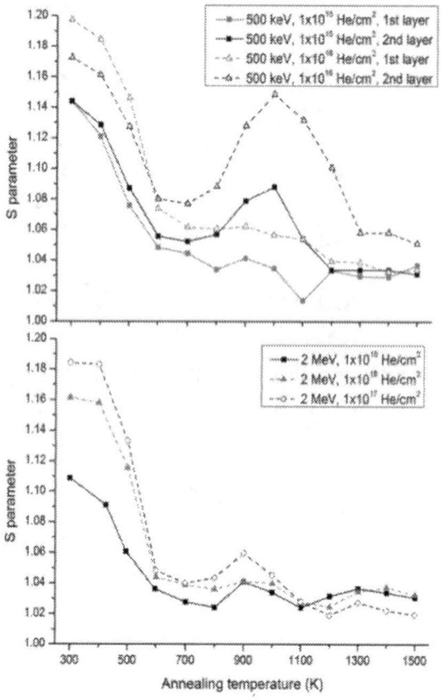

Figure 2. *S* parameter vs. annealing temperature. Data fitted with VEPFIT.

For temperatures above 1300 K the release of He bubbles is expected [8, 9] and this is noticed as a decrease in the *S* value of the 500 keV , $1x10^{16}$ He/cm^2 implanted sample.

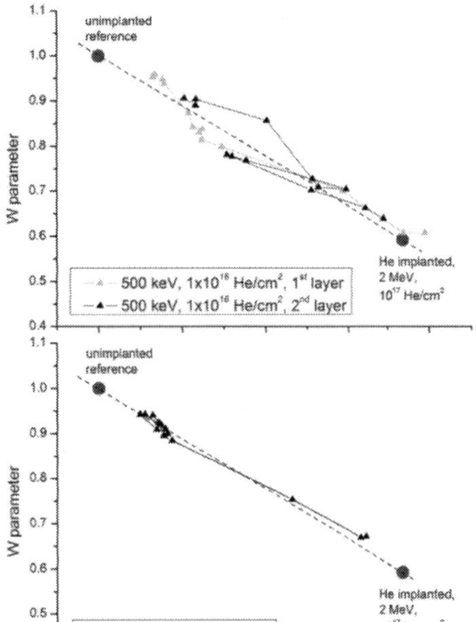

Figure 3. *S* vs. *W* parameter. Data fitted with VEPFIT.

Figure 3 shows the *S*-*W* values of the near surface layers in the $1x10^{16}$ He/cm^2 samples. The dashed line connects the *S*-*W* values of an unimplanted Eurofer97 reference sample and the sample with the highest concentration of defects (2 MeV energy and a dose of $1x10^{17}$ He/cm^2) and is therefore associated with Eurofer97 vacancy type defects. For temperatures up to 500 K the points of the second layer of the 500 keV sample are located very close to this line, revealing the vacancy type character of these defects. As the temperature increases to 800 K, the *S*-*W* points move down from this line due to an increase in the He/V ratio in the positron traps (over-pressurized He_nV_m clusters). In the temperature range 800 – 1200 K the *S*-*W* pairs shift back towards the vacancy type defects line. This behavior is in accordance with the idea of the Ostwald ripening as it produces He_nV_m (n, m = 1) pairs, sensed as vacancies. By coalescence bigger clusters may be formed as well but apparently with less probability. At 1200 K the *S*-*W* pairs appear above the line. This hints at the formation of larger helium filled voids, where due to the relatively low amount of He per vacancy the helium is not sensed by positrons. At temperatures above 1300 K the bubbles may start to migrate and the remaining low concentration traps are again similar to vacancies.

Up to the temperature of 800 K, the *S-W* pairs the 2 MeV and the first layer of the 500 keV samples show a similar behavior as the ones of deeper layer of the 500 keV sample with *S-W* pairs ending up close to the line. However, the tendency to return into the direction of *S-W* characteristic for higher density of He_nV_m (n, m = 1) defects is not observed. This is explained by a lower overall initial defect concentration and a smaller He/V ratio. As the temperature further increases to 1500 K the location of the *S-W* pairs o slightly moves into the direction of the reference defect free material, indicating that no major changes take place in this region.

3. Conclusions

He implantations have been performed in Eurofer97 steel with energies of 500 keV and 2 MeV and doses in the range of $1x10^{15}$ -10^{17} He/cm^2. By annealing the samples up to 1500 K it is possible to observe an initial decrease of the *S*-parameter, associated with the annealing of vacancies and vacancy clusters, followed by a significant increase of this value due to growth of the helium-vacancy clusters. At temperatures in the range 1100 – 1200 K HeV pairs dissociate and retrapping by the bigger clusters occurs and helium filled bubbles are formed. Above 1300 K these bubbles anneal out by dissociation or migration, causing a drop in the *S*-parameter.

The *S-W* graph reveals that the positron traps of the samples implanted with a dose of $1x10^{16}$ He/cm^2 are of similar character up to a temperature of 800 K. The different behavior of *S-W* pairs above this temperature show that implantation conditions, such as dose and energy, are determining the thermal evolution of implantation (and radiation) induced defects.

Acknowledgements

This research was carried out under project number M74.10.10393 in the framework of the Research Program of the Materials innovation institute M2i (www.m2i.nl).

References

[1] Lindau R, Moslang A, Rieth M, Klimiankou M, Materna-Morris E, Alamo A, Tavassoli F, Cayron C, Lancha AM, Fernandez P *et al.* 2005 *Proc. 23rd Symp. of Fusion Technology (Venice)* **75-79** 989-96

[2] van der Schaaf B, Tavassoli F, Fazio C, Rigal E, Diegele E, Lindau R and LeMarois G 2003 *Fusion Eng. and Design* **69** 197-203

[3] Rensman J, Lucon E, Boskeljon J, van Hoepen J, den Boef R and ten Pierick P 2004 *Journal of Nuclear Materials* **329 - 333** 1113-1116

[4] Carvalho I, Schut H, Fedorov AV, Luzginova N, Desgardin P and Sietsma J 2013 *Journal of Nuclear Materials: in press*

[5] van Veen A, Schut H, Vries J, Hakvoort RA and IJpma MR 1991 *Proc. 4th Int. workshop on Slow positron beam techniques for solids and surfaces* **218** 171-98

[6] Morishita K, Sugano R, Wirth BD and de la Rubia TD 2003 *Nuclear Instr. & Methods in Physics Research Section B: Beam Interactions with Materials and Atoms* **202** 76-81

[7] Rieth M, Schirra M, Falkenstein A, Graf P, Heger S, Kempe H, Lindau R and Zimmermman H 2003 *Forschungszentrum Karlsruche GmbH* report number 6911

[8] Oo KZ, Chernov I, Staltsov M, Kalin B, Kalashnikov A, Binyukova SY, Zaw AK, Ageev V and Nikitina A 2010 *Atomic Energy* **110** 151-159

[9] Morishita K, Sugano R, Iwakiri H, Yoshida N and Kimura A 2001 *Proc. 4th Pacific Rim Int. Conf. on Advanced Materials and Processing (Hawaii)* **1-2** 1395-98

Sintering of zirconia-based nanomaterials studied by variable-energy slow-positron beam

I Prochazka[1], J Cizek[1], O Melikhova[1], W Anwand[2], G Brauer[2], T E Konstantinova[3] and I A Danilenko[3]

[1] Charles University in Prague, Faculty of Mathematics and Physics, Department of Low Temperature Physics,V Holesovickach 2, CZ-180 00 raha 8, Czech Republic
[2] Helmholtz-Zentrum Dresden-Rossendorf, Institute of Radiation Physics, POB 51 01 19, D 01314 Dresden, Germany
[3] National Academy of Sciences of Ukraine, Donetsk Institute for Physics and Engineering Named after O O Galkin, Luxemburg Street 72, 83114 Donetsk, Ukraine

E mail: ivan.prochazka@mff.cuni.cz

Abstract. A variable-energy slow-positron beam was applied to the investigations of the tetragonal yttria-stabilised zirconia (YSZ), the YSZ co-doped with small amount of Cr_2O_3. The initial nanopowders exhibiting the mean particle size of ≈ 20 nm were prepared by co-precipitation technique. Prior the sintering, the nanopowders were calcined and compacted using a pressure of 500 MPa. The ordinary shape parameters of the Doppler-broadened annihilation peak and the relative positronium 3γ-fractions were determined as functions of positron energy. The results are consistent with a remarkable sintering-induced grain growth and disappearance of porosity which is driven out from the sample interior toward a thin sub-surface layer.

1. Introduction

Zirconium dioxide (zirconia) is a substance widely used in practice, e.g. in ceramic industry, due to its advantageous thermal, electronic, mechanical and chemical properties [1]: a high melting point (2750 °C), a low thermal conductivity, a good high-κ dielectricity, a low electronic conductivity, a good oxygen ionic conductivity at increased temperatures, a high hardness combined with a reasonable fracture toughness and a good resistance to corrosion and wear.

Pure zirconia is monoclinic (*m*) at room temperature, but it exhibits transitions to the denser tetragonal (*t*) and cubic (*c*) phases at temperatures of ≈ 1100 and ≈ 2350 °C, respectively. In order to avoid a potential mechanical deterioration of zirconia in high-temperature applications, therefore, it is necessary to stabilize the *t*- and *c*-phases down to room temperature. Required phase stability is fulfilled e.g. by the yttria-stabilised zirconia (YSZ) which is a binary oxide system with a small amount of Y_2O_3 (yttria) immersed as a solid solution in zirconia lattice (≈ 3 mol. % of yttria is sufficient for a stabilization of the *t*-phase). Due to the different valences of Zr (4^+) and Y (3^+) cations, a large amount of oxygen vacancy-like defects is brought into the lattice.

Ceramic materials are commonly manufactured from fine particle powders by sintering [2]. The sintering process activates a mass transfer resulting in a grain growth and a disappearance of pores.

Important driving parameters of sintering are the sintering temperature, the mean particle size of the initial powder and the diffusivity of the constituents.

The ternary solid solution, YSZ plus Cr_2O_3 (YSZC), is also an interesting material [3]. A smaller ionic radius of Cr^{3+} compared to Zr^{4+} and a possible multiple Cr valence may influence e.g. the diffusivity of Cr ions and/or grain size of YSZC.

In the present work, the tetragonal (*t*) binary YSZ and ternary YSZC nanopowders were prepared and subjected to a sequential sintering at various temperatures. The potentially depth-dependent sintering-induced diffusion processes were studied using variable-energy slow-positron annihilation spectroscopy which is a non-destructive technique with high sensitivity to open-volume defects.

2. Experimental

The binary ZrO_2+3 mol. % Y_2O_3 (Z3Y) and the ternary ZrO_2+3 mol. % Y_2O_3+1 mol. % Cr_2O_3 (Z3Y1C) nanopowders and sintered ceramics were studied in the present work. The initial nanopowders were prepared by the co-precipitation technique similar to that of Ref. [3] from the water solutions of respective salts taken in stoichiometric proportions. The nanopowders were then calcined at 500 °C/2 h in air. The calcined nanopowders were uniaxially pressed under 500 MPa into disk-shaped pellets of ≈ 10 mm diameter and ≈ 3 mm thickness. The pressure-compacted specimens were then sintered at selected temperatures in an increasing order from 600 toward 1500 °C in air. Each sintering step lasted for 1 h with a subsequent controlled slow cooling down to room temperature and positron annihilation measurement. The *t*-phases of nanopowders as well as sintered ceramics were confirmed by means of X-ray diffraction (XRD). A mean crystallite size was estimated by XRD as 17.0±0.9 and 13±1 nm for the Z3Y and Z3Y1C, respectively. The bulk densities of specimens were also measured after each sintering step.

A magnetically guided variable-energy slow-positron beam SPONSOR [4] was utilized in the present work. The energies of incident monoenergetic positrons covered the range from 0.03 to 35 keV. The Doppler broadening (DB) of the annihilation peak was measured with a HPGe γ-spectrometer exhibiting a resolution of 1.03 keV (fwhm) at 511 keV γ-ray energy. About 5×10^5 counts were typically accumulated in the peak area. The ordinary peak-shape parameters, S and W, were evaluated to characterise the DB effects. In case of Z3Y when positronium (Ps) formation was observed in our recent lifetime measurements [5], the relative ortho-Ps 3γ-fractions, F, were also deduced from measured spectra of annihilation photons following to the method described in [6]. A flat valley between the 480 and 500 keV γ-ray energy was considered for this purpose. Since no Ps formation was detected in the Z3Y1C bulk [5] the measurements on this specimen for positron energies above 10 keV have provided a reference level for the evaluation of F-values.

3. Results and discussion

In figure 1, shape parameters S were plotted against positron energy E_+ for various sintering temperatures T_S. For E_+ below 2 keV, the observed S-parameters fell from their surface value close to the bulk one, S_{bulk}. In addition, a contribution of para-Ps formation in Z3Y specimen can be seen in figure 1 as (i) enhanced S-parameters, compared to the Z3Y1C specimen, and (ii) a bulge with a wide maximum at ≈ 4 keV. The contribution of Ps vanished for $T_S > 900$ °C. The E_+-dependences of W-parameters displayed a mirror patterns compared to those for S-parameters, as expected. Present findings are consistent with our earlier positron lifetime studies [5] in which all positrons in the compacted YSZ nanopowders and the YSZ ceramics sintered up to 1200 °C were shown either to annihilate from trapped states in defects (saturated trapping) or to form Ps. Ps formation was found, moreover, to disappear after sintering of the YSZ nanopowders at 1000°C and Ps formation was even not observed at all in YSZC. The latter phenomenon is a consequence of an efficient suppression of Ps formation by the Cr cations segregated along grain boundaries (GBs) [5].

In the energy region $E_+ > 20$ keV, the back-diffusion of positrons toward the specimen surface becomes negligible. Hence the S_{bulk}-parameters can be well-approximated as the S-values of figure 1 averaged over $E_+ > 20$ keV. The S_{bulk}-values for Z3Y and Z3Y1C were plotted against T_S in figure 2. Overall decrease in S_{bulk}-values with increasing T_S is seen from the figure. For $T_S < 900$ °C, the S_{bulk}-values for Z3Y exceed significantly those for Z3Y1C. Above this temperature, both specimens exhibit mutually the same values of S_{bulk}. In addition, the measured bulk densities, ρ, were plotted in figure 2, exhibiting a sharp increase between 900 and 1100 °C sintering temperatures which coincides roughly with a region of disappearance of pores in the bulk [5]. This sharp increase is then followed by a slow approaching to a ρ-value for the *t*-YSZ single crystal (6.1 g·cm^{-3}).

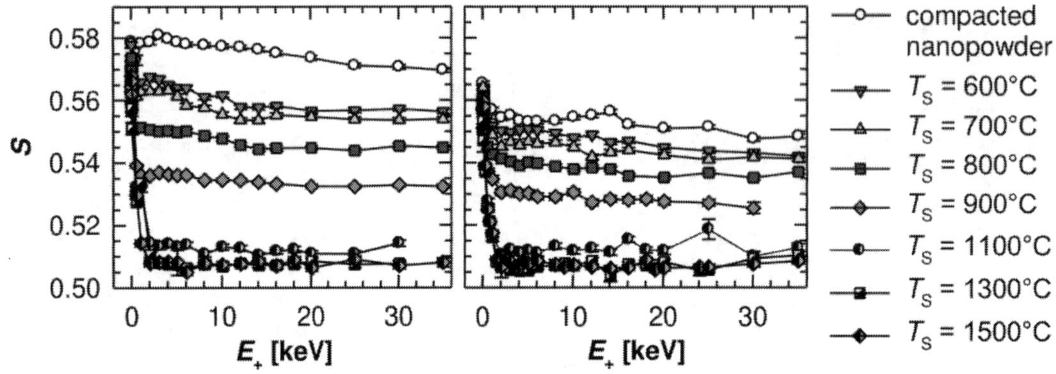

Figure 1. Measured S-parameters plotted as functions of positron energies E_+ for different sintering temperatures T_S : the left panel – Z3Y, the right panel – Z3Y1C.

Sintering leads to a decrease of the S_{bulk}-parameters due to two processes: (1) the grain growth which causes a decrease in the concentration of defects associated to GBs and (2) the disappearance of pores leading to a diminishing of the Ps contribution. These processes were observed and treated in more details in our previous LT and CDB study of bulk zirconia [5]. In this work, both processes were obviously observed in the Z3Y sample, while only the process (1) can be seen for the Z3Y1C specimen due to an absence of the Ps contribution. Hence, the difference of the S_{bulk}-parameters for Z3Y and Z3Y1C can be regarded as an indicator of the porosity in Z3Y. Inspection of figure 2 clearly reveals that the porosity is essentially removed by sintering at 1000 °C. A decrease of the S_{bulk}-values after further sintering at higher temperatures is caused solely by the sintering-induced grain growth.

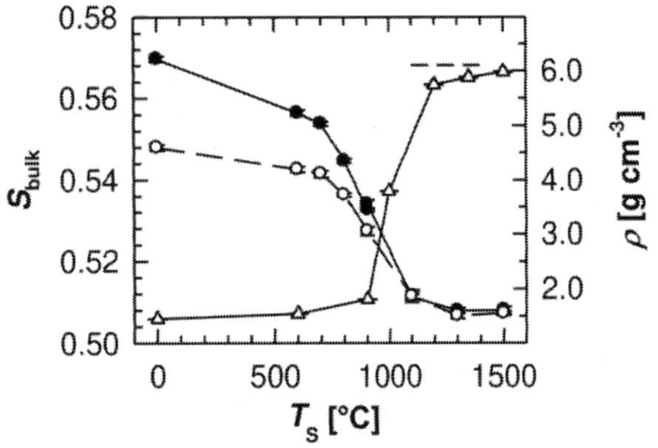

Figure 2. S_{bulk}-values (averaged S-parameters for $E_+ > 20$ keV) and bulk densities ρ plotted as functions of T_S : ● – S_{bulk} for Z3Y, ○ – S_{bulk} for Z3Y1C, Δ – ρ for Z3Y. Dashed horizontal line stands for a density of the *t*-YSZ single crystal.

The above picture is further supported by the observed relative ortho-Ps 3γ-fractions, F, which were plotted against E_+ for various sintering temperatures T_S in figure 3. Note that a bulge similar to that seen in S-values for Z3Y nanopowder around $E_+ \approx 6$ keV corresponding to a mean penetration depth of ≈ 200 nm is shown also by F-values. Moreover, F-fractions vanished for $T_S > 900$ °C, similarly like the para-Ps contributions to S-parameters did (see figure 1). In this way, an existence of a subsurface layer with enhanced porosity is independently testified.

4. Concluding remarks

Zirconia-based nanograined materials exhibit

significant structure non-homogeneities on a scale of ≈ 100 nm. Behaviour of positrons and Ps in such materials is thus given a rather complicated pattern. Positrons thermalised in grain interiors with a relatively low defect concentration may easily diffuse to the nearest GB where they will get trapped in GB-associated open-volume defects (misfit defects of a mono-vacancy size or triple points [5]). Such positrons, therefore, can hardly diffuse to a distance exceeding the mean grain size (≈ 20 nm). On the other hand, GBs and nanopores may facilitate diffusion of Ps and those positrons which get thermalised there.

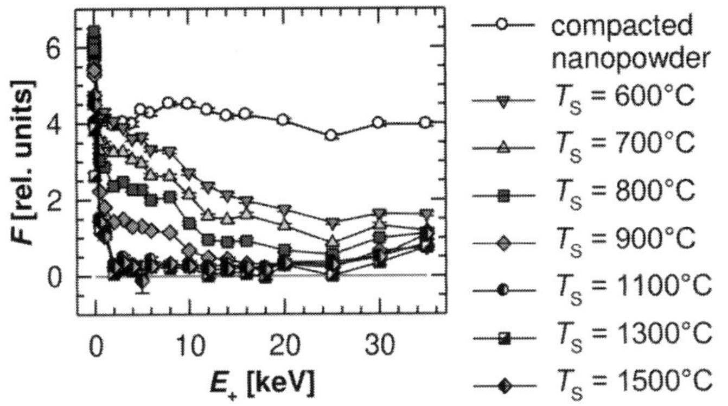

Figure 3. Relative ortho-Ps 3γ-fractions, F, in Z3Y plotted as functions of positron energies E_+ for different sintering temperatures T_S.

Experimental data obtained in the present work, illustrate that the depth profiling of sintering-induced microstructure changes by means of the variable-energy slow-positron beam reveal pronounced effects which are complementary to the bulk studies of zirconia-based nanomaterials by positron annihilation techniques. The two near-surface regions could be distinguished according to the observed depth profiles of S- and F-parameters in Z3Y and Z3Y1C specimens. Below $E_+ \approx 2$ keV, positrons and Ps may diffuse back to the surface which becomes likely a dominating process responsible for a sharp decrease in S- and F-values in this region. Then, a subsurface region of residual porosity occurs, ranging up to $E_+ \approx 6$ keV. This region is likely a result of sintering-induced mass transfer accompanied with the grain growth and disappearance of pores. The remnant gas contained in the pores is gradually diffused from the specimen interior toward its surface during sintering and may thus assist to more stable pore structures in the subsurface region. This subsurface residual porosity vanishes above $T_S \approx 1000$ °C. Taking into account large variations of densities illustrated by figure 2, the estimated depths of the two near-surface regions could be set decreasing from 80 to 20 nm and 500 to 130 nm, respectively, when T_S is risen from its initial value to 1500 °C.

Acknowledgements
This work was supported by The Czech Science Foundation (contract GA P108/11/1396) and The National Academy of Science of Ukraine (project 89/12-H). OM is grateful for funding from Charles University in Prague (program UNCE "Physics of Condensed Matter and Functional Materials").

References
[1] Badwal S P, Bannister M J and Hannink R H J 1993 *Science and technology of zirconia V* (Lancaster, Pennsylvania: Technomic Pub. Co.)
[2] Castro R H R 2013 *Sintering (Engineering Materials* vol 35*)* ed R H R Castro and K van Benthem (Berlin Heidelberg: Springer Verlag) pp 1 – 16
[3] Korduban A M, Yashchishyn I A, Konstantinova T K, Danilenko I A, Volkova G K and Glazunova V A 2007 *Functional Materials* **14** 454
[4] Anwand W, Kissener H R and Brauer G 1995 *Acta Phys. Polonica* A **88** 7
[5] Cizek J, Melikhova O, Prochazka I, Kuriplach J, Kuzel R, Brauer G, Anwand W, Konstantinova T E and Danilenko I A 2010 *Phys. Rev.* B **81** 024116
Melikhova O, Cizek J, Prochazka I, Konstantinova T E and Danilenko I A 2012 *Phys. Procedia* **35** 134
[6] Wu Y C, Jiang J, Wang S J, Kallis A and Coleman P G 2011 *Phys. Rev.* B **84** 064123

Journal of Physics: Conference Series **505** (2014) 012021 doi:10.1088/1742-6596/505/1/012021

Defect studies of thin ZnO films prepared by pulsed laser deposition

M Vlček[1], J Čížek[1], I Procházka[1], M Novotný[2], J Bulíř[2], J Lančok[2], W Anwand[3], G Brauer[3], J-P Mosnier[4]

[1]Faculty of Mathematics and Physics, Charles University in Prague, V Holešovičkách 2, 180 00 Praha 8, Czech Republic

[2]Institute of Physics, Academy of Sciences of the Czech Republic, Na Slovance 2, 182 21 Praha, Czech Republic

[3]Institut für Strahlenphysik, Forszungszentrum Dresden-Rossendorf, 01314 Dresden, Germany

[4]National Centre for Plasma Science and Technology, School of Physical Sciences, Glasnevin, Dublin 9, Ireland

E-mail: `vlcek@mbox.troja.mff.cuni.cz`

Abstract. Thin ZnO films were grown by pulsed laser deposition on four different substrates: sapphire (0 0 0 1), MgO (1 0 0), fused silica and nanocrystalline synthetic diamond. Defect studies by slow positron implantation spectroscopy (SPIS) revealed significantly higher concentration of defects in the studied films when compared to a bulk ZnO single crystal. The concentration of defects in the films deposited on single crystal sapphire and MgO substrates is higher than in the films deposited on amorphous fused silica substrate and nanocrystalline synthetic diamond. Furthermore, the effect of deposition temperature on film quality was investigated in ZnO films deposited on synthetic diamond substrates. Defect studies performed by SPIS revealed that the concentration of defects firstly decreases with increasing deposition temperature, but at too high deposition temperatures it increases again. The lowest concentration of defects was found in the film deposited at 450°C.

1. Introduction

Preparation of high-quality ZnO films is of great interest for efficient excitonic UV lasers exploiting externally pumped lasing observed in epitaxial ZnO films at room temperature [1]. Furthermore, ZnO films have favorable properties for application in flat panel displays and optical coatings for sollar cells [2]. Several techniques for production of ZnO films have been developed so far. In this work we employed pulsed laser deposition (PLD) [3] which enables the production of high-quality ZnO films at lower temperatures than other methods.

2. Experimental details

Samples were prepared in stainless-steel vacuum chamber with base pressure of 2×10^{-4} Pa. Deposition process was carried out using 266 nm frequency-quadrupled Nd:YAG laser providing 90 mJ pulses of 6 ns duration. High-purity ceramic ZnO target was ablated by laser light with fluence of 2.8 $\mathrm{Jcm^{-2}}$ at distance of 55 mm from the substrate. Repetition rate of laser was set to 10 Hz and each sample was grown as a result of 5200 shots. Ambient atmosphere of 5N-pure oxygen at pressure of 10 Pa and 40 sccm flow rate was present during the film growth. These parameters are favorable for ZnO p-type doping in order to avoid compensation by oxygen vacancies [4]. Four different substrates were used: sapphire (0 0 0 1), MgO (1 0 0) and fused silica (FS) supplied by MaTecK GmbH and nanocrystalline diamond (NCD) film on silicon (1 0 0) wafer. NCD films were prepared using plasma-enhanced linear antennas microwave CVD system [5]. Prior to deposition, the substrates except of NCD were baked out in vacuum at 950°C for 5 min to remove remaining surface contamination. ZnO thin films were deposited on sapphire, MgO and FS at 300°C and subsequently annealed in situ at 750°C in 10 Pa oxygen temperature. Another set of films was deposited on NCD at various temperatures. Bulk ZnO (0 0 0 1) single crystal, supplied by MaTecK GmbH, with O-terminated surface grown by hydrothermal method was used as reference material for slow positron implantation spectroscopy (SPIS) measurements.

SPIS studies were carried out using slow positron beam SPONSOR [6] with positron energy adjustable in the range from 30 eV to 36 keV. Doppler broadening (DB) of annihilation line was measured by high-purity germanium detector with resolution of 1.09(1) keV at 511 keV. DB was evaluated using the line shape S parameter. Central energy interval used for the calculation of the value of S parameter was selected so that S parameter for the reference ZnO bulk single crystal S_0 was close to 0.5, where it is most sensitive to changes in line shape of the annihilation peak. Actual value $S_0 = 0.5068(5)$ was calculated from the energy region 510.07-511.93 keV.

3. Results

XRD investigations revealed that ZnO films deposited on all substrates exhibit wurtzite structure. ZnO films deposited on single crystal substrates (MgO, sapphire) exhibit local epitaxy while films deposited on amorphous FS or NCD show (0 0 0 1) fiber texture with random lateral orientation of crystallites in the plane of substrate [7, 8]. Comparison of $S(E)$ curves, i.e. dependence of S parameter on positron implantation energy E, for ZnO films deposited on various substrates at 300°C is shown in figure 1(a).

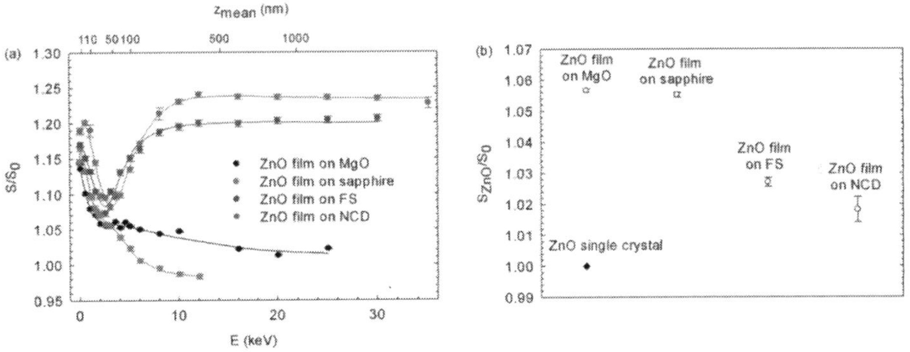

Figure 1. (a) $S(E)$ curves measured on ZnO films deposited on various substrates at 300°C. (b) Fitted S parameters for ZnO layers on various substrates and the reference ZnO crystal.

At the lowest energy $E = 0.03$ keV almost all positrons annihilate on the surface of ZnO films. With increasing energy positrons penetrate into the ZnO films and subsequently into the substrate. The largest fraction of positron annihilates in ZnO films at $E \sim 3$ keV. At energies $E > 20$ keV virtually all positrons annihilate in the substrate and measured S parameters approach bulk values for corresponding substrates. FS and NCD substrates exhibit significantly higher S parameters than MgO or sapphire due to lower electron density in inter-atomic regions. Measured data were analyzed by VEPFIT software package [9] using two-layer model system (ZnO film and substrate). Fitted values for the S parameter and mean positron diffusion length L_+ for the ZnO model layer are shown in figures 1(b) and 2(a), respectively. Trapping rate K for positrons trapping at defects in ZnO films were calculated as $K = 1/\tau_B[(L_{+0}/L_+)^2 - 1]$, where L_+ is the positron diffusion length measured in the ZnO film, $L_{+0} = 280$ nm is the positron diffusion length in a perfect ZnO crystal obtained by theoretical calculations [10] and

$\tau_B = 154$ ps is the bulk positron lifetime in ZnO [11].

Trapping rates for films grown on different substrates and the reference ZnO crystal are compared in figure 2(b). The reference ZnO crystal exhibits positron diffusion length $L_+ = 59$ nm which is similar to the value reported by Uedono *et al* [12]. The fact that $L_+ < L_{+0}$ indicates that the hydrothermally grown crystal contains positron trapping defects. This was confirmed by positron lifetime spectroscopy [13, 14] and the defects were identified as Zn vacancies associated with H atoms [11].

S parameters and trapping rates for all ZnO films are higher than those in the ZnO crystal while the positron diffusion lengths are shorter. Thus, all ZnO films contain higher concentration of defects than the reference ZnO crystal. While ZnO films likely contain point defects similar to the reference ZnO crystal, additional types of defects may be present in polycrystalline films: (i) misfit dislocations compensating the lattice mismatch between the film and the substrate and (ii) open-volume defects at grain boundaries or crystallite interfaces. Moreover the concentration of defects in the ZnO films deposited on single crystal MgO and sapphire substrates is higher than in the films deposited on amorphous FS and NCD diamond substrates which is demonstrated by considerably higher S parameters, see figure 1(b). This is supported also by lower L_+ and higher K values for the films deposited on single crystalline substrates, see figure 2.

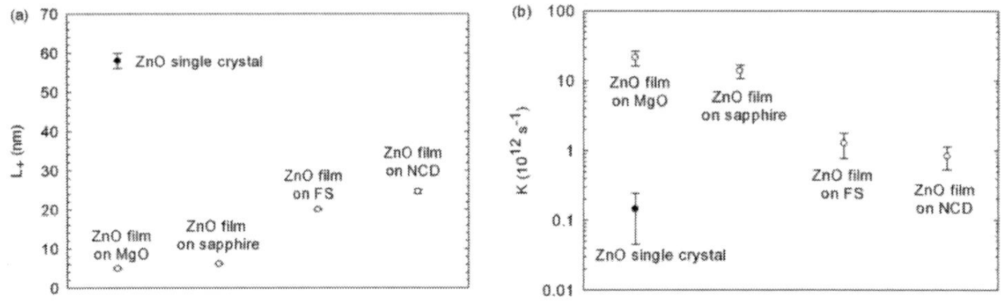

Figure 2. Comparison of parameters of ZnO films deposited on various substrates and reference ZnO single crystal: (a) mean positron diffusion length, (b) positron trapping rate.

Furthermore, the influence of deposition temperature on ZnO film properties was studied on films deposited on NCD substrate. Films at four different temperatures were deposited: room temperature, 300°C, 450°C and 600°C. Measured $S(E)$ curves are plotted in figure 3(a).

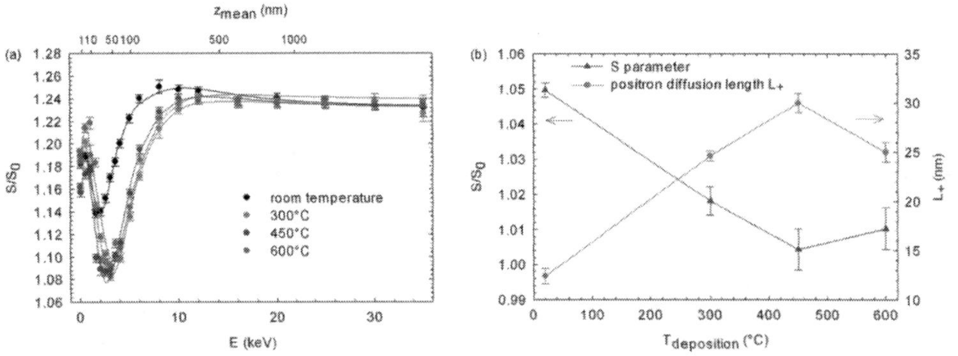

Figure 3. (a) $S(E)$ curves for ZnO films deposited at different temperatures on NCD. (b) Temperature dependence of the S parameter (blue triangles) and positron diffusion length (red circles) of ZnO layer deposited on NCD.

There is evident difference between the $S(E)$ curve of film deposited at room temperature and that of films deposited at elevated temperatures. Measured data were again analyzed by VEPFIT software and results are summarized in figure 3(b). ZnO film deposited at room temperature exhibits the lowest diffusion length and therefore contains the highest concentration of defects. On the other hand, the highest positron diffusion length and lowest value of S parameter, indicating the lowest concentration of defects, were observed in film deposited at 450°C.

4. Discussion

XRD investigations [8] suggested that ZnO films grown on single crystal substrates (MgO, sapphire) exhibiting local epitaxy are of higher quality. Surprisingly, SPIS measurements revealed that these films have significantly higher concentration of defects than ZnO films deposited on FS and NCD. ZnO films deposited on single crystalline substrates exhibit well defined orientation with respect to the substrates. This inevitably results in formation of misfit dislocations compensating lattice mismatch between the substrate and ZnO layer while in the films deposited on FS and NCD the lattice mismatch is to some extent released by various orientation and tilting of the crystallites. Hence, SPIS measurements detect higher concentration of defects in film deposited on MgO and sapphire due to higher density of misfit dislocations.

Studies of ZnO films deposited on NC at various temperatures revealed that the film deposited at 450°C exhibits the lowest concentration of defects. Deposition at elevated temperatures increases chance of atoms to reach proper position in crystal lattice corresponding to the lowest interface energy. XRD investigations revealed that the mean size of crystallites increases with increasing temperature of deposition and therefore concentration of misfit defects at film-substrate interface increases as well. Minimum in temperature dependence of concentration of defects observed at 450°C is created as result of these antagonistic effects.

5. Conclusions

ZnO films deposited on single crystal MgO and sapphire substrates exhibit significantly higher concentration of defects than films deposited on amorphous FS and NCD substrates due to presence of misfit dislocations at film-substrate interface.

Influence of deposition temperature on properties of ZnO films deposited on NCD were investigated. The lowest concentration of defects was found in the film deposited at 450°C.

Acknowledgements

This work was supported by the Czech Science Agency (project P108/11/0958).

References

[1] Bagnall D M, Chen Y F, Zhu Z, Yao T, Koyama S, Shen M and Goto T 1997 *Appl. Phys. Lett.* **70** 2230–2232
[2] Hu J H and Gordon R G 1992 *J. Appl. Phys.* **71** 880–890
[3] Vispute R D, Talyansky V, Trajanovic Z, Choopun S, Downes M, Sharma R P, Venkatesan T, Woods M C, Lareau R T, Jones K A and Iliadis A A 1997 *Appl. Phys. Lett.* **70** 2735–2737
[4] O'Haire R, Meaney A, McGlynn E, Henry M O, Duclère J R and Mosnier J P 2006 *Superlattices Microstruct.* **39** 153 – 161
[5] Fendrych F, Taylor A, Peksa L, Kratochvilova I, Vlcek J, Rezacova V, Petrak V, Kluiber Z, Fekete L, Liehr M and Nesladek M 2010 *J. Phys. D: Appl. Phys.* **43** 374018
[6] Anwand W, Brauer G, Butterling M, Kissenger H R and Wagner A 2012 *Def. Diff. Forum* **331** 25
[7] Novotný M, Čížek J, Kužel R, Bulíř J, Lančok J, Connolly J, McCarthy E, Krishnamurthy S, Mosnier J P, Anwand W and Brauer G 2012 *J. Phys. D: Appl. Phys.* **45** 225101
[8] Kužel R, Čížek J and Novotný M 2013 *Metall. Mater. Trans. A* **44A** 45–57
[9] van Veen A, Schut H, Clement M, de Nijs J, Kruseman A and IJpma M 1995 *Appl. Surf. Sci.* **85** 216 – 224
[10] Lukáč F, Čížek J, Procházka I, Melikhova O, Anwand W and Brauer G 2014 *Acta Phys. Pol., A* In press
[11] Brauer G, Anwand W, Skorupa W, Kuriplach J, Melikhova O, Moisson C, von Wenckstern H, Schmidt H, Lorenz M and Grundmann M 2006 *Phys. Rev. B* **74** 045208
[12] Uedono A, Koida T, Tsukazaki A, Kawasaki M, Chen Z Q, Chichibu S and Koinuma H 2003 *J. Appl. Phys.* **93** 2481
[13] de la Cruz R M, Pareja R, González R, Boatner L A and Chen Y 1992 *Phys. Rev. B* **45**(12) 6581–6586
[14] Tuomisto F, Ranki V, Saarinen K and Look D 2003 *Phys. Rev. Lett.* **91** 205502

Nanohole formation in TEOS-HMDSO hybrid CVD films elucidated by positron beams

K Ito, N Oshima, A Yabuuchi and B E O'Rourke

National Institute of Advanced Industrial Science and Technology (AIST), Tsukuba, Ibaraki 305–8565, Japan

E-mail: k-ito@aist.go.jp

Abstract. Silicon-oxide-backboned hybrid thin films with thicknesses around 600 nm were fabricated through plasma enhanced chemical vapor deposition from tetraethyl orthosilicate mixed with hexamethyldisiloxane as precursors, and their subnano-scaled holes, generated by annealing at 560 °C, were investigated by means of the positron lifetime technique with a pulsed, low-energy positron beam. The hole dimension was quantified from the *ortho*-positronium lifetime for the as-prepared and annealed films with different compositions. The effect of the heat treatment and the precursor composition on the subnano holes was discussed.

1. Introduction

Manufacturing silicon oxide thin films with engineered nanoscaled holes is a key issue for developing various functional materials such as gas sensors [1] and separation membranes [2]. In this work, we fabricated silicon-oxide-backboned thin films with thicknesses of around 600 nm through plasma enhanced chemical vapor deposition (PECVD) from tetraethyl orthosilicate (TEOS) mixed with hexamethyldisiloxane (HMDSO) as precursor of the film, and investigated their nanoscaled holes generated by annealing. By using a pulsed, low-energy positron beam, the hole dimension was quantified from the longest-lived lifetimes of *ortho*-positronium *o*-Ps for the as-prepared and the annealed films with different compositions of the precursors. The effect of the heat treatment and the precursor composition on the nanohole structure is examined in comparison of the positron results with those from Fourier transform infrared spectroscopy and spectroscopic ellipsometry.

2. Experimental

Films were deposited on 4-inch silicon wafers at a substrate temperature of 100 °C using a capacitive coupled PECVD reactor with parallel plate electrodes (Fig. 1) [3, 4]. Mixtures of TEOS and HMDSO diluted in argon and oxygen gas were used as precursors. The deposition was carried out under a pressure of 150 Pa with 200 W RF power. The fabricated films were then annealed at 560 °C under dry nitrogen gas flow. The deposition condition, thickness,

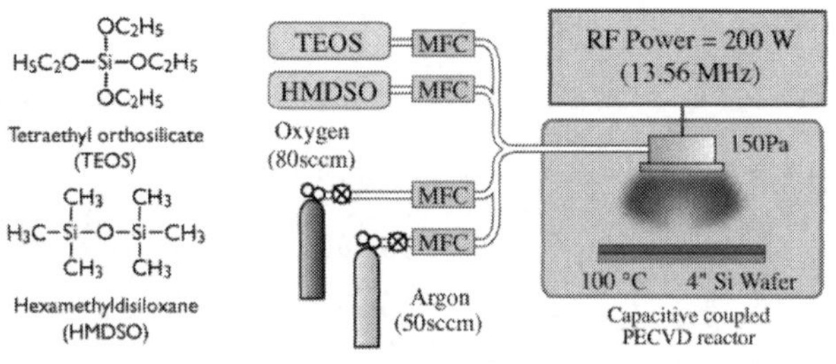

Figure 1: Monomers and PECVD system.

Table 1: Deposition condition (TEOS and HMDSO flow rates, HMDSO feed fraction ϕ_H), thickness L, and refractive index n at 630 nm for the as-deposited and annealed films.

Sample No.	TEOS [sccm]	HMDSO [sccm]	ϕ_H	As deposited		Annealed	
				L [nm]	n	L [nm]	n
1	15	0	0	533	1.412	459	1.412
2	9	6	0.4	587	1.417	539	1.406
3	3	12	0.8	578	1.419	585	1.359
4	0	15	1.0	632	1.421	617	1.362

*L and n were determined by means of spectroscopic ellipsometry at 26 °C.

and refractive index at 630 nm for the as-deposited and annealed films are listed in Table 1. As in the table, by annealing the respective refractive index n decreased, especially the higher the HMDSO feed fraction ϕ_H, the lower the refractive index, indicative of the enhanced porosity.

The chemical structure of films was analyzed by Fourier transform infrared (FT-IR) spectroscopy.

Positron annihilation γ-ray energy spectra were measured by using a ^{22}Na-source-based magnetically-guided beam [5] at positron incident energies E from 0.01 keV to 25 keV. The line-shape S parameter, a measure of Doppler broadening of annihilation radiation, was determined as the ratio of the counts appearing in the central region (510.3 keV–511.7 keV) to the total counts around the 511-keV annihilation photo peak (506.8 keV–515.2 keV).

Positron lifetime measurements were carried out by utilizing a ^{22}Na-based pulsed-positron beam at AIST. Multi-exponential analysis was applied to the recorded lifetime data to deduce the hole radius r [nm] from the longest-lived o-Ps lifetime τ [ns] based on the following equation [6, 7],

$$\tau = 0.5\left[1 - \frac{r}{r+0.1656} + \frac{1}{2\pi}\sin\left(\frac{2\pi r}{r+0.1656}\right)\right]^{-1} \quad . \tag{1}$$

3. Results and discussion

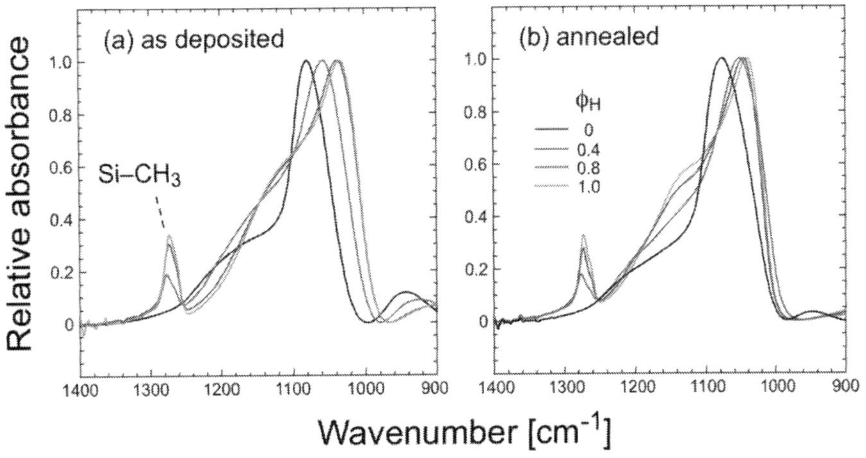

Figure 2: FT–IR spectra for the present PECVD films; (a) as-deposited films, (b) annealed films.

Figure 2 shows the FT–IR spectra for the PECVD films before and after annealing. The distinct absorption peak around 1100 cm^{-1} for the film with ϕ_H = 0 is due to the Si–O–Si stretching mode [8].

Journal of Physics: Conference Series **505** (2014) 012022 doi:10.1088/1742-6596/505/1/012022

With increasing $\phi_H = 0$ the peak for both the as-deposited and annealed films is shifted to the lower wavenumber, indicative of the enhanced absorption relevant to C–Si–O bonds in the silica network [9]. The absorption intensity for the peak around 1280 cm^{-1}, ascribable to Si–CH_3 stretching, increases with increasing ϕ_H, signifying the increased number of the terminal CH_3 in the network structure.

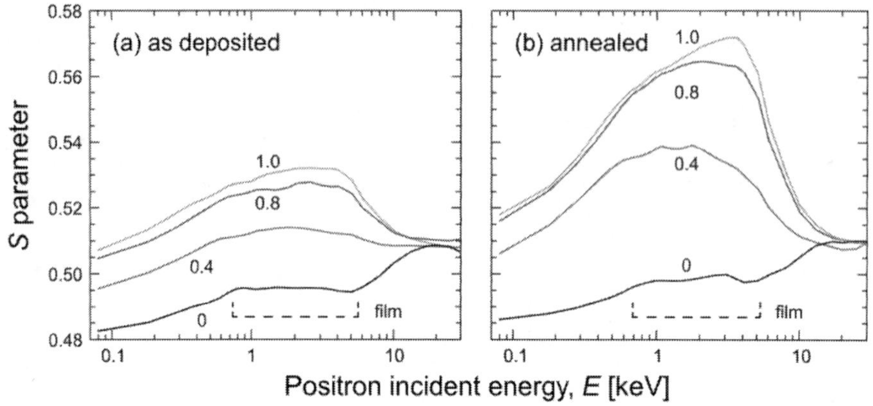

Figure 3: S–E curve for the as-deposited films (a) and the annealed films at 560 °C (b). The values represent the HMDSO feed fraction ϕ_H.

Figure 3 shows the variation of the line-shape S parameter for the present samples as a function of positron incident energy E (S–E curve). The S–E curve is divided roughly into four stages, namely, (1) $E < 0.7$ keV, (2) 0.7 keV$\lesssim E <$ 5.5 keV, (3) 5.5 keV$\lesssim E <$ 15 keV, and (4) 15 keV$\lesssim E$, and those are reasonably ascribed to the annihilation of positrons (1) in the film subsurface region, (2) in the film, (3) in the interface region between the film and the silicon substrate, and (4) in the silicon substrate, respectively. For the as-deposited films the S value at 0.7 keV$\lesssim E <$ 5.5 keV increases with increasing ϕ_H as in Fig. 3(a). On the other hand, for the annealed films (Fig. 3(b)) although the tendency of S on ϕ_H is similar, the variation with ϕ_H is more enhanced, probably due to the increased Ps formation with the increased film porosity after annealing at 560 °C and/or due to the specific annihilation of positrons with the terminal CH_3 groups on the hole surface, leading to higher S value [10].

Figure 4 shows the positron lifetime data obtained at $E = 3.0$ keV for the as-deposited and annealed films. The lifetime data for the films annealed at 560 °C have the longer lifetimes than for the respective as-deposited films. This suggests that larger holes are introduced, possibly as a result of the decomposition of 'unreacted' components in the as-deposited films by annealing. Table 2 lists the analyzed longest-lived o-Ps lifetimes τ, their relative intensities I, and hole radii r, obtained from the respective τ, for the as-deposited and annealed films. As in the table, τ for the as-deposited films ranges from 2.6 ns to 3.5 ns, slightly increases with increasing ϕ_H. By annealing each τ increases, as a result, it ranges from 2.9 ns to 8.6 ns, indicating that the larger holes, ranging from 0.36 nm to 0.61 nm in radius, are formed in the films with the higher ϕ_H by annealing. This may be relevant to the enhanced S by the annihilation of positrons with the terminal hydrocarbon groups at the hole surface in addition to the slight increase of I.

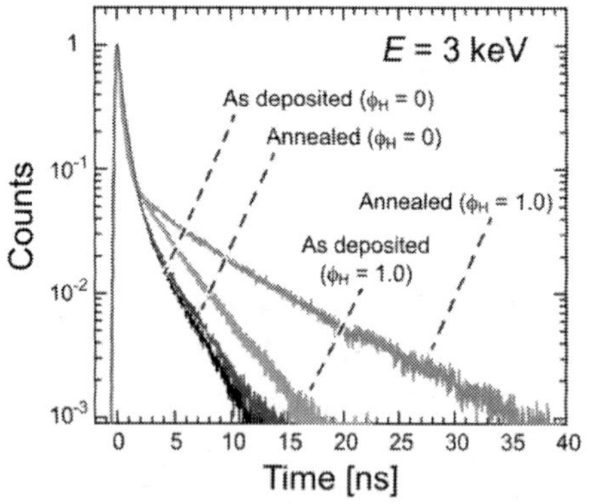

Figure 4: Typical positron annihilation lifetime data (corrected for the background).

Table 2: Longest-lived *o*-Ps lifetimes τ, their relative intensities I and hole radii r obtained from τ for the as-deposited and annealed films. The values in the parenthesis represent the respective standard deviation on the analysis.

Sample No.	ϕ_H	As deposited			Annealed		
		τ [ns]	r [nm]	I [%]	τ [ns]	r [nm]	I [%]
1	0	2.67 (0.009)	0.344	22.5 (0.1)	2.91 (0.009)	0.360	24.5 (0.1)
2	0.4	3.34 (0.007)	0.388	35.9 (0.1)	4.44 (0.009)	0.448	40.5 (0.1)
3	0.8	3.43 (0.007)	0.393	39.4 (0.1)	7.57 (0.220)	0.576	46.8 (0.2)
4	1.0	3.48 (0.007)	0.396	41.1 (0.1)	8.62 (0.004)	0.612	45.2 (0.3)

4. Conclusion

We have investigated nanopore formation upon heat treatment for TEOS–HMDSO hybrid PECVD films by means of the variable-energy positron annihilation γ-ray and lifetime techniques. The IR and ellipsometry results clearly indicated that the absorption intensity for the Si–CH_3 bond is enhanced with increased HMDSO feed fraction ϕ_H, while the refractive index after annealing is reduced, suggesting increased porosity. The positron lifetime results demonstrated that the larger holes are formed in the annealed films, signifying that subnanoscopic hole size up to 0.6 nm in radius for the TEOS–HMDSO hybrid system is tunable with the precursor composition at deposition in combination with pyrolysis.

Acknowledgments

This work was supported by a Grant-in-Aid for Scientific Research (24246126) from JSPS, Japan. KI thanks Dr. Hideaki Hagihara of AIST for his assistance to the lifetime measurements.

References

[1] Guizard C, Bac A, Barboiu M, and Hovnanian N 2001 *Sep. Purif. Technol.* **25** 167–180
[2] Wang J, Gong G, Kanezashi M, Yoshioka T, Ito K, Tsuru T 2013 *J. Membrane Sci.* **441** 120–128
[3] Oka T, Ito K, Muramatsu M, Ohdaira T, Suzuki R, and Kobayashi Y 2006 *J. Phys. Chem. B* **110** 20172
[4] Ito K, Oka T, Kobayashi Y, Suzuki R, and Ohdaira T 2007 *Radiat. Phys. Chem.* **76** 213–216
[5] Kobayashi Y, Kojima I, Hishita S, Suzuki T, Asari E, and Kitajima M 1995 *Phys. Rev. B* **52** 823–828
[6] Tao S J 1972 *J. Chem. Phys.* **56** 5499–5510
[7] Eldrup M, Lightbody D, and Sherwood J N 1981 *Chem. Phys.* **63** 51–58
[8] Tripp C P and Hair M L 1991 *Langmuir* **7** 923–927
[9] Morent R, Geyter N D, Vlierberghe S V, Dubruel P, Leys C, Schacht E 2009 *Surf. Coat. Tech.* **203** 1366–1372
[10] Sato K, Ito K, Hirata K, Yu R S, and Kobayashi Y 2005 *Phys. Rev. B* **71** 012201

Journal of Physics: Conference Series **505** (2014) 012023 doi:10.1088/1742-6596/505/1/012023

Structural and defect characterization of Gd-doped GaN films by X-ray diffraction and positron annihilation

A Yabuuchi[1], N Oshima[1], B E O'Rourke[1], R Suzuki[1], K Ito[1], S Sano[2], K Higashi[2], Y-K Zhou[2] and S Hasegawa[2]

[1] National Institute of Advanced Industrial Science and Technology (AIST), 1-1-1 Umezono, Tsukuba, Ibaraki 305-8568, Japan

[2] The Institute of Scientific and Industrial Research (ISIR), Osaka University, 8-1 Mihogaoka, Ibaraki, Osaka 567-0047, Japan

E-mail: atsushi.yabuuchi@aist.go.jp

Abstract. Molecular-beam-epitaxy-grown $Ga_{1-x}Gd_xN$ films were investigated by X-ray diffraction and slow positron beams. From the positron lifetime results, N-vacancy-related defects may be expected in the $Ga_{0.9}Gd_{0.1}N$ film grown under Ga-rich conditions which exhibits a lattice expansion in the *c*-axis direction. In contrast, Ga vacancies more than 10^{19} cm^{-3} were detected in the $Ga_{0.9}Gd_{0.1}N$ film grown under N-rich conditions which does not exhibit the lattice expansion, implying that the highly-concentrated Ga vacancies contribute to a relaxation of the lattice distortion caused by incorporating oversized Gd atoms.

1. Introduction

Gadolinium-doped GaN ($Ga_{1-x}Gd_xN$) is promising material for semiconductor spintronics, operable at room temperature [1]. Saturation magnetization of GaGdN films, a key factor of the device capability, is known to be enhanced by the GaGdN/GaN short-period superlattice (SL) structure [2]. GaGdN films grown through molecular beam epitaxy (MBE) are classified into two types; one with a natural GaGdN/GaN short-period SL structure, and the other without [3]. The former type is reported to have a diffraction peak from the GaGdN(0002) plane, observed at a scattering angle lower than that from the GaN(0002) plane [4–6]. This indicates that introducing excess amounts of Gd (covalent radius: 1.96 Å [7]), which is larger than Ga (1.22 Å [7]), induces an expansion of the *c*-axis lattice constant. On the contrary, a film grown at certain condition without SL structure, i.e., a single GaGdN layer with a homogeneous Gd distribution has no diffraction peak from the GaGdN(0002) plane. As reported in an earlier positron study [8], vacancy-type defects are incorporated in a Gd-doped GaN system. We expect that those defects allow a lattice relaxation, leading to a reduction of the *c*-axis expansion during crystal growth. In this study, MBE-grown GaGdN films were investigated by X-ray diffraction (XRD) and slow positron beams in order to elucidate the relationship between vacancy formation and lattice relaxation.

2. Experiments

Two wurtzite types of $Ga_{1-x}Gd_xN$ were prepared through a radio-frequency plasma-assisted MBE (PA-MBE) method. (The crystal structure of wurtzite GaN is illustrated in Fig. 1.) GaN-cap (3 nm thick) /$Ga_{1-x}Gd_xN$ (200 nm thick) /GaN-buffer (15 nm thick) layers were grown at 700 °C with a growth

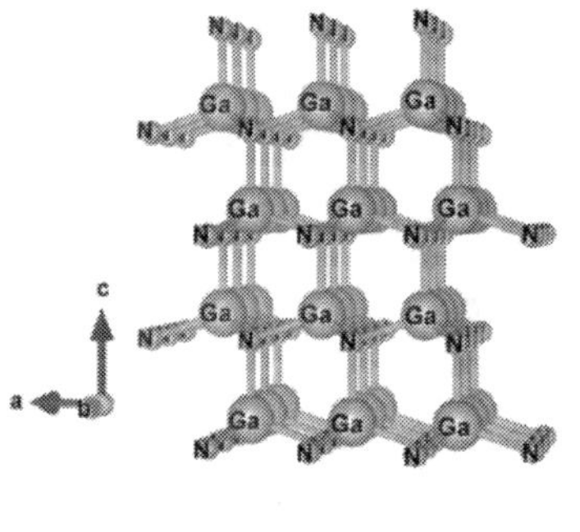

Figure 1. Crystal structure of wurtzite GaN.

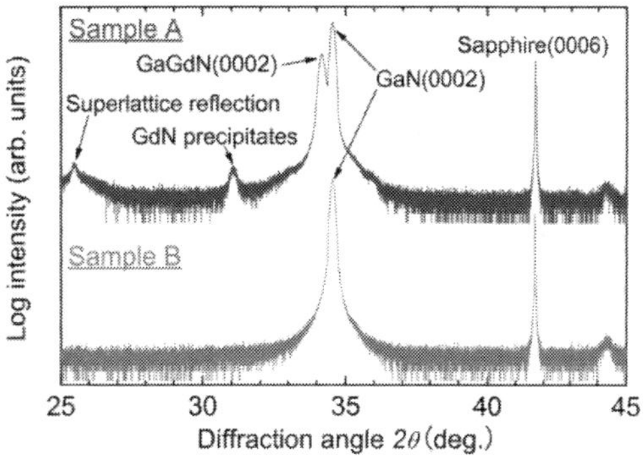

Figure 2. XRD $\omega/2\theta$ scan profiles for Samples A and B.

rate of 200 nm/h on commercial GaN (2 μm thick, fabricated by metal-oxide chemical vapor deposition (MOCVD)) /Al_2O_3(0001) templates. The deposition conditions were 6.8×10^{-7} Pa of Gd flux, 1.5 sccm of N_2 flow, and 260 W (230 W during the buffer-layer growth) of RF power. Ga fluxes during growth for the two films were 2.02×10^{-5} Pa (Sample A) and 1.46×10^{-5} Pa (Sample B).

The crystalline structures of the samples were characterized by means of XRD with a Cu Kα source (λ = 1.54 Å).

Positron annihilation γ-ray spectra as a function of positron incident energy E from 0 to 20 keV were measured using a ^{22}Na-based magnetically-guided beam. The line-shape S parameter was determined as the ratio of the counts appearing in the central region to the total counts of the 511-keV annihilation photo peak. For the measurements, a moderator using six stacks of tungsten meshes with a diameter of 30 μm and a mesh density of 100 wires/inch was prepared by sandwiching it by two sheets of tungsten foil with a 8-mm window in diameter. Before installing to the system, the moderator was annealed in vacuum by electron-bombardment heating with a maximum electric power of 1 kW. The annealing procedure was as follows: 1) evacuating the chamber to get 2×10^{-5} Pa, 2) introducing air into the chamber with a pressure of 1×10^{-4} Pa, 3) annealed for 1 min at 1400 °C to remove carbon impurities from the tungsten surface as carbon monoxide gas [9, 10], 4) evacuating the chamber again, and annealed at 1700 °C for a few seconds at 3×10^{-5} Pa to eliminate tungsten oxide on the mesh surface [9], 5) repeating Step 1–4 for 3 cycles to eliminate carbon impurities in the tungsten mesh, 6) finally, annealed at 2300 °C for 1 min at 7×10^{-5} Pa, and allowed to cool in vacuum with a cooling rate of roughly 50 °C/s.

Positron annihilation lifetime measurements at a positron incident energy of 5 keV were performed by using a ^{22}Na-RI-based pulsed positron beam system with a time resolution of about 270 ps (full width at half-maximum).

3. Results and discussion

Figure 2 shows XRD $\omega/2\theta$ profiles for Samples A and B. Both profiles show a sharp diffraction peak at 41.7° from sapphire(0006) in the template. As for Sample A it exhibits a diffraction peak attributable to GaGdN/GaN superlattice reflection around 25.5°, and a small diffraction peak due to NaCl-type GdN precipitates is observed around 31.0°. Note that a peak at 34.2° from the GaGdN(0002) plane is observed at a lower scattering angle side than from the GaN(0002) plane (34.6°), indicative of a lattice expansion in the c-axis direction. From the change in the c-axis lattice constant, the Gd concentration for Sample A is estimated [5] to be about 10% (x = 0.1). Since the growth conditions of both samples were similar to each other, the Gd concentration in Sample B was also estimated to be about 10%. Nevertheless, in the profile for Sample B only a GaN(0002) diffraction peak is observed, meaning that the c-axis lattice constant of the GaGdN layers in Sample B is comparable with GaN crystal. This signifies that Ga vacancies allow the relaxation of the structural distortion due to incorporating oversized Gd atoms, and

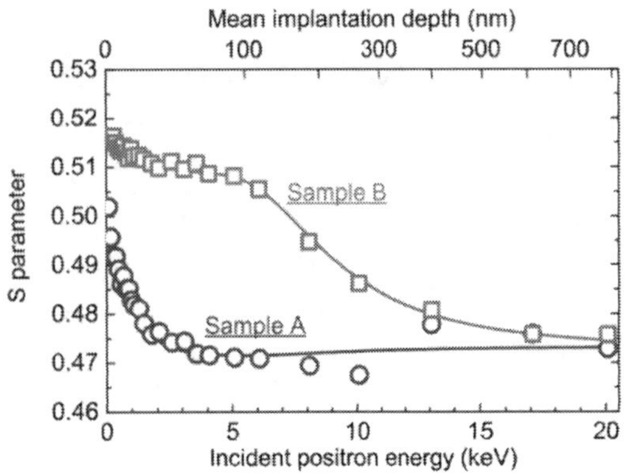

Figure 3. S parameters as a function of incident positron energy obtained for Samples A and B.

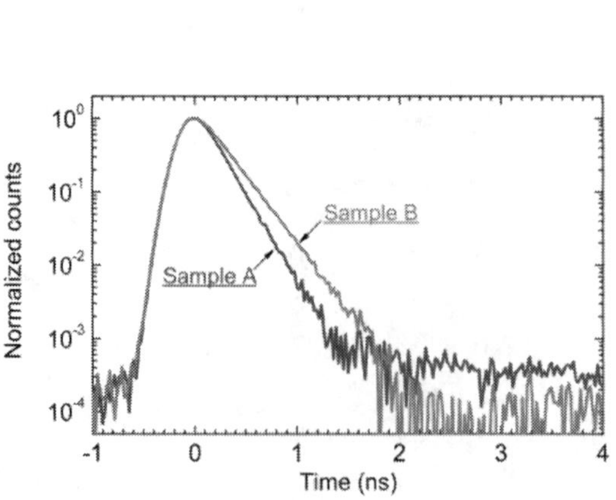

Figure 4. Positron annihilation lifetime spectra for Samples A and B obtained at $E = 5$ keV. The higher background for the spectrum of Sample A is due to the longer positron diffusion length, expected from the S-E curves, as well as to the relatively high positron re-emission efficiency for GaN system [9].

as a result, lattice expansion in the c-axis direction may be inhibited.

Figure 3 shows the S parameters for Samples A and B as a function of incident positron energy E. By taking into account of the positron mean implantation depth, S at 2 keV< E <5 keV is due to the positron annihilation in the 200-nm-thick GaGdN layers. The S value for Sample B at each E is significantly higher than that of Sample A, suggesting the existence of vacancy defects. For Sample A the value of S is almost the same as that of the template GaN layer (E = 20 keV). In order to specify the defect species contained in these samples, positron annihilation lifetime measurements were then conducted.

Figure 4 shows positron annihilation lifetime spectra obtained at E = 5 keV for Samples A and B. Both spectra show a single component with mean lifetimes τ of 179 ps and 234 ps for A and B, respectively, significantly longer than for GaN bulk (154 ps) reported by Hautakangas *et al.* [11]. This indicates that positrons are fully trapped at defect sites in both samples, suggesting defect concentration is around 10^{19} cm^{-3} or higher, that is, over the upper quantification limit. Although τ for Sample A (179 ps) agrees with a prediction for nitrogen vacancy V_N (178 ps [11]), it is conceivable that positrons are repulsed by positively charged V_N [12]. Hautakangas *et al.* previously reported a positron lifetime for V_N-Mg_{Ga} in Mg-doped GaN crystal as 180 ps, similar to that for V_N [11]. While the ratio of the covalent radii Gd/Ga (1.61) is much larger than that of Mg/Ga (1.16) [7], we expect the obtained τ for Sample A may be attributed to a V_N complexed with Gd for our system. To clarify this, a further study of the coincidence Doppler broadening of positron annihilation radiation technique in parallel with theoretical calculations is ongoing. On the other hand, τ for Sample B is in good agreement with a reported τ for a gallium vacancy V_{Ga} (235 ps [11]). The origin of τ = 234 ps can be specified as V_{Ga}. A large amount of Ga vacancies contribute to a relaxation of the lattice distortion caused by incorporating Gd atoms.

4. Conclusion

$Ga_{1-x}Gd_xN$ films, grown with Ga-rich and N-rich conditions, have been investigated by XRD and slow positron beams. It was found that a $Ga_{0.9}Gd_{0.1}N$ film with Ga-rich condition exhibits a lattice expansion in the c-axis direction, while that with N-rich condition does not. From the positron results, N-vacancy-related defects such as V_N-Gd_{Ga} complexes may be expected as candidate defect species in the *c*-axis-expanded film. On the other hand, the *c*-axis-non-expanded film contained Ga vacancies more than 10^{19} cm^{-3}. These highly-concentrated Ga vacancies signified the suppression of the lattice expansion even in such a heavy Gd-doping.

Acknowledgments
The crystal structure shown in Fig. 1 is illustrated using the computer program VESTA [13].

References

[1] Teraguchi N, Suzuki A, Nanishi Y, Zhou Y-K, Hashimoto M and Asahi H 2002 *Solid State Commun.* **122** 651
[2] Choi S W, Zhou Y K, Kim M S, Kimura S, Emura S, Hasegawa S and Asahi H 2006 *Phys. Status Solidi A* **203** 2774
[3] Ishimaru M, Higashi K, Hasegawa S, Asahi H, Sato K and Konno T J 2012 *Appl. Phys. Lett.* **101** 101912
[4] Hasegawa S, Komori S, Higashi K, Abe D, Zhou Y-K and Asahi H 2012 *Phys. Status Solidi C* **9** 741
[5] Higashi K, Hasegawa S, Abe D, Mitsuno Y, Komori S, Ishikawa F, Ishimaru M and Asahi H 2012 *Appl. Phys. Lett.* **101** 221902
[6] Sano S, Hasegawa S, Mitsuno Y, Higashi K, Ishimaru M, Sakurai T, Ohta H and Asahi H 2013 *J. Cryst. Growth* **378** 314
[7] Cordero B, Gómez V, Platero-Prats A E, Revés M, Echeverría J, Cremades E, Barragán F and Alvarez S 2008 *Dalton Trans.* 2832
[8] Roever M, Malindretos J, Bedoya-Pinto A, Rizzi A, Rauch C and Tuomisto F 2011 *Phys. Rev. B* **84** 081201(R)
[9] Suzuki R, Ohdaira T, Uedono A, Cho Y K, Yoshida S, Ishida Y, Ohshima T, Itoh H, Chiwaki M, Mikado T, Yamazaki T and Tanigawa S 1998 *Jpn. J. Appl. Phys.* **37** 4636
[10] Suzuki R, Amarendra G, Ohdaira T and Mikado T 1999 *Appl. Surf. Sci.* **149** 66
[11] Hautakangas S, Oila J, Alatalo M and Saarinen K 2003 *Phys. Rev. Lett.* **90** 137402
[12] Walle C G V and Neugebauer J 2004 *J. Appl. Phys.* **95** 3851
[13] Momma K and Izumi F 2011 *J. Appl. Crystallogr.* **44** 1272

Structural studies of thin Mg films

P Hruška[1], J Čížek[1], W Anwand[2], J Bulíř[3], J Lančok[3], J Stráská[1], O Melikhova [1], I Procházka[1]

[1] Charles University in Prague, Faculty of Mathematics and Physics, V Holešovičkách 2, 18000 Praha 8, Czech Republic

[2] Institute of Radiation Physics, Helmholtz-Zentrum Dresden-Rossendorf, PO Box 510119, D 01314 Dresden, Germany

[3] Institute of Physics, Academy of Sciences of the Czech Republic, Na Slovance 2, 182 21 Praha, Czech Republic

E-mail: `peta.hruska.1@gmail.com`

Abstract. In the present work variable energy positron annihilation spectroscopy (VEPAS) was employed for investigation of defects in Mg films. VEPAS characterization was combined with scanning electron microscopy and X-ray diffraction in order to determine grain size and texture respectively. The aim of this study was to examine the effect of deposition temperature and various substrates on structure and defects in Mg films prepared by RF magnetron sputtering. SEM observations revealed that films deposited on sapphire (0001) substrate exhibit always smaller grains than films deposited on amorphous fused silica and silicon (100) substrates, which have comparable grain size. Defect studies by VEPAS showed that positrons in Mg films studied are trapped at misfit dislocations and at vacancy-like defects in grain boundaries. Moreover, the films deposited on a substrate heated at 300°C exhibited lower concentration of defects and larger grain size compared to the films deposited at room temperature. Subsequent annealing at 300°C for 1 h of the films deposited at room temperature causes a slight decrease of defect density due to coarsening of grains.

1. Introduction

Magnesium hydride (MgH_2) is an attractive material suitable for on-board hydrogen storage in mobile applications due to its safety and high hydrogen absorption capacity up to 7.6 wt. %[1]. However slow sorption rate in the MgH_2 is the major obstacle for its use in practical applications. Structure refinement down to nanoscale could remarkably speed up the hydrogen sorption kinetics since a high volume fraction of grain boundaries provides a network for easier hydrogen distribution into grains. The effect of microstructure on the hydrogen sorption kinetics can be investigated on thin films since samples with nanocrystalline to epitaxial structure can be relatively easily prepared by varying the deposition conditions. Obviously structure characterization of Mg thin films is necessary as a first step of this investigation.

Variable energy positron annihilation spectroscopy (VEPAS) provides information about defect structure in various depths in the studied material, hence it is a very useful tool for studies of the microstructure of Mg thin films [2, 3, 4, 5].

2. Experimental

A set of 4 different Mg films was prepared by RF (13.6 MHz) magnetron sputtering with a deposition rate of 1.5 - 2.1 nm/s and using a pressure of 3 Pa (Ar atmosphere): (i) films with thickness of $\sim 1\ \mu$m deposited at room temperature (RT), (ii) thin films ($\sim$ 100 nm) deposited at RT, (iii) thin films ($\sim$ 100 nm) deposited on a substrate heated at 300°C and thin films ($\sim$ 100 nm) deposited at RT and then annealed in UHV deposition chamber at 300°C for 1 h. Subsequently 20 nm thick Pd cap was deposited on each film. The Pd cap prevents oxidation of the films on the air and facilitates hydrogen absorption during electrochemical charging [6]. Three different substrates were used: sapphire (0001), silicon (100) and amorphous fused silica. Structural studies of Mg films were performed using FEI Quanta 200 SEM.

VEPAS measurements were performed using magnetically guided variable energy positron beam "SPONSOR" [7] with slow positrons of energies from 27 eV to 35 keV. Energy spectra of annihilation gamma rays were measured by HPGe detectors having a relative efficiency of $\sim$ 30% and an energy resolution of (1.06 ± 0.01) keV (FWHM at 511 keV). The Doppler broadening of annihilation profile was evaluated using the S parameter. The dependence of the S parameter on the positron energy was analyzed using the VEPFIT code [8].

3. Results and discussion

Fig. 1 shows SEM micrographs of the thin Mg films deposited on fused silica. The mean grain sizes estimated by SEM are listed in Table 1. The film deposited at 300°C (Fig. 1a) exhibits the largest grains with the mean grain size of $\sim$ 253(6) nm. The mean grain size of the film deposited at RT is $\sim$ 71(2) nm (Fig. 1b), subsequent annealing of the film at 300°C led to increase of the mean grain size up to $\sim$ 89(6) nm (Fig. 1c). The thicker film (thickness $\sim 1\ \mu$m) deposited at RT on fused silica exhibits the mean grain size of $\sim$ 375(6) nm as determined by SEM. The increase of the mean grain size with the thickness of the film is likely due to the fact that the first generation crystallites attached directly to the substrate are smaller in size than the next generations growing on the top the previous generation. The films deposited on Si substrate exhibit comparable grain sizes as those on fused silica, but films deposited on sapphire are characterized by lower grain sizes as a rule, see Table 1.

Figure 1. SEM micrographs of thin Mg films ($\sim$ 100 nm) on fused silica deposited at 300° C (a), at RT (b) and at RT and subsequently annealed at 300° C for 1 h (c).

XRD revealed that Mg films are always strongly textured independently on the type of substrate, the deposition temperature or the thickness. The XRD patterns measured in the Bragg-Brentano geometry contains Mg(0002), Mg(0004), Pd(111) and Pd(222) peaks only, i.e. Mg films grew with preferred orientation of the most densely populated basal plane (0001) parallel to the surface of the substrate while Pd layer grew on the top of Mg film with (111) plane parallel with the surface.

The dependences of the S parameter on the energy of incident positrons for Mg films deposited on sapphire and fused silica substrate are plotted in Fig. 2a and 2b, respectively. The $S(E)$ curves for the bare substrates are plotted in the figure as well. In addition, the $S(E)$ curve for a bulk Mg sample annealed (250°C) in UHV chamber and covered by 20 nm Pd over-layer prior

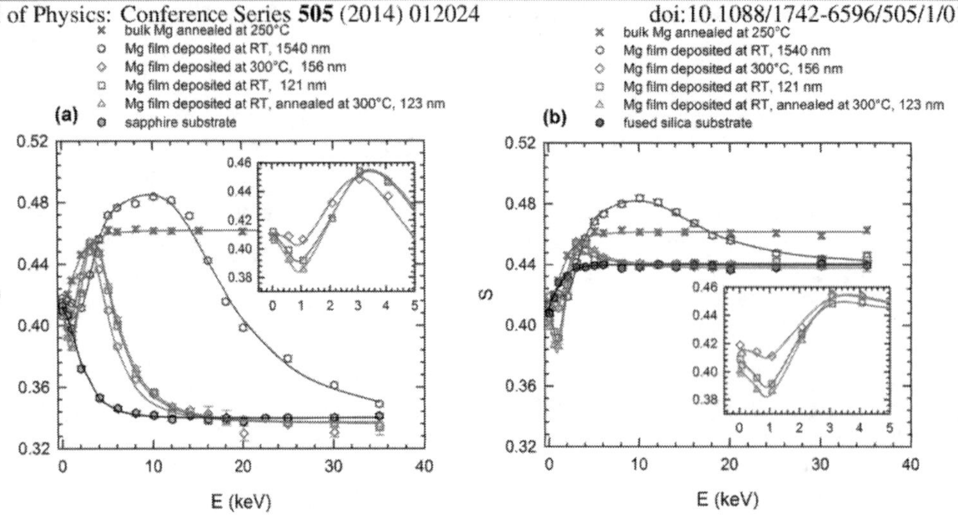

Figure 2. Dependence of the S parameter on positron energy for Mg films deposited on sapphire (a) and fused silica substrate (b). The $S(E)$ curves for the bare substrates and annealed (250°C) bulk Mg reference are plotted in the figure as well. All data were fitted by VEPFIT code (solid lines). Insets show zoomed low energy parts of the $S(E)$ curves for thinner films ($\sim$ 100 nm).

to its removal from the chamber to prevent oxidation is shown in Fig. 2 for comparison. The mean positron diffusion length $L_+ = (195 \pm 8)$ nm obtained from fitting the S vs E values for the annealed bulk Mg agrees well with the value reported in Ref. [3].

The $S(E)$ curves for Mg films exhibit firstly a drop of S due to positron annihilations in Pd cap. With increasing energy positrons penetrate into the Mg layer and the S parameter increases. In the Mg films with thickness of $\sim$ 1.5 μm the S parameter becomes approximately constant in the range 8 - 12 keV (see Fig. 2) indicating that practically all positrons are annihilated inside the Mg film. The S parameter for the Mg films is higher than the S parameter measured in the well annealed bulk Mg. It indicates that Mg films contain defects which trap positrons. At higher energies ($E > 12$ keV) positrons implanted into the Mg film become to penetrate into the substrate leading to a decrease of S which gradually approaches the bulk S parameter for the bare substrate. The $S(E)$ curves for the thiner films ($\sim$ 100 nm) exhibit similar behavior but the decrease of S starts at lower energies because of lower thickness of the Mg layer.

The $S(E)$ curves for Mg films were fitted by VEPFIT assuming 3 layers: (i) Pd cap with thickness of 20 nm, (ii) Mg film with thickness of $\sim$ 1.5 μm or $\sim$ 100 nm and (iii) substrate. The model curves calculated by VEPFIT are plotted in Fig. 2 by solid lines and are in a good agreement with the experimental points. Each layer is characterized by its positron diffusion length L_+ and the S parameter. In order to reduce the number of fitting parameters L_+ and S for the substrates were fixed at the values obtained for the bare substrates. The positron diffusion length and the S parameter obtained for the Mg films are listed in Table 1. All Mg films studied exhibit shorter L_+ and higher S than the annealed bulk Mg. It testifies that the Mg films contain open volume defects which trap positrons. These defects are misfit dislocations necessary for accommodation of the lattice mismatch between the film and the substrate and open volume point defects (vacancies and vacancy clusters) located at grain boundaries which were identified by a pulsed low-energy positron beam [4].

The insets in Fig. 2 show detail of the low energy parts of $S(E)$ curves for the thinner Mg films ($\sim$ 100 nm). The $S(E)$ curves for the films deposited at RT and the films deposited at RT and subsequently annealed at 300°C are almost the same, but in the films deposited at 300°C the rise of S has considerably lower slope indicating longer positron diffusion length. Indeed from inspection of Table 1 one can conclude that films deposited at 300°C exhibit higher L_+ than the films deposited at RT. This is in agreement with SEM observations which revealed that films

Table 1. Results of fitting of the $S(E)$ curves for Mg films, bulk Mg and bare substrates by the VEPFIT code and the mean grain size determined by SEM; n.a. = not analyzed.

thickness (nm)	T_{dep}	T_{ann}	substrate	L_+ (nm)	S	grain size (nm)
1540	RT	–	sapphire	45(2)	0.4878(4)	77(2)
1540	RT	–	fused silica	46(1)	0.4843(4)	375(6)
1540	RT	–	silicon	n.a.	n.a.	325(8)
156	300°C	–	sapphire	60(5)	0.483(2)	133(4)
156	300°C	–	fused silica	77(7)	0.480(2)	253(6)
156	300°C	–	silicon	n.a.	n.a.	195(7)
121	RT	–	sapphire	35(4)	0.485(2)	59(2)
121	RT	–	fused silica	39(5)	0.478(2)	71(2)
121	RT	–	silicon	n.a.	n.a.	77(3)
123	RT	300°C	sapphire	50(5)	0.483(2)	81(3)
123	RT	300°C	fused silica	52(6)	0.482(2)	89(3)
123	RT	300°C	silicon	n.a.	n.a.	92(2)
–	–	250°C	bulk Mg	196(9)	0.4634(5)	–
–	–	–	sapphire subst.	46(5)	0.3400(2)	–
–	–	–	fused silica subst.	24(1)	0.4367(3)	–

deposited at 300°C are characterized by larger grains leading to reduced volume fraction of grain boundaries containing open volume defects. The fitted parameters for the films deposited at RT are very similar taking into account the experimental uncertainties. But as a common trend it can be noticed that post-growth annealing at 300°C led to a slight increase of L_+ which is again likely connected with grain growth which decreases the density of defects at grain boundaries.

4. Conclusions

A set of Mg films deposited under various conditions on 3 different substrates was prepared by RF magnetron sputtering. The films exhibit (0001) texture and the mean grain size increases with the thickness of the film. Defect studies by positron annihilation spectroscopy revealed that positrons in the Mg films are trapped at misfit dislocations and in vacancy-like defects at grain boundaries. Films deposited at 300°C exhibit larger grains and lower concentration of defects than films deposited at RT due to reduced volume fraction of grain boundaries.

Acknowledgments

Financial support by the Czech Science Agency (project P108/12/G043) and the grant SVV-2013-267303 is gratefully acknowledged.

References

[1] Sandrock G 1999 *J. Alloys. Comp.* **293-295** 877

[2] Checchetto R, Bazzanella N, Miotello A, Brusa R S, Zecca A and Mengucci A 2004 *J. Appl. Phys.* **95** 1989

[3] Eijt S W H, Kind R, Singh S, Schut H, Legerstee W J, Hendrikx R W A, Svetchnikov V L, Westerwaal R J and Dam B 2009 *J. Appl. Phys.* **105** 043514

[4] Checchetto R, Bazzanella N, Kale A, Miotello A, Mariazzi S, Brusa R S, Mengucci P, Macchi C, Somoza A, Egger W and Ravelli L 2011 *Phys. Rev. B* **84** 054115

[5] Macchi C, Maurizio C, Checchetto R, Mariazzi S, Ravelli L, Egger W, Mengucci P, Bazzanella N, Miotello A, Somoza A and Brusa R S 2012 *Phys. Rev. B* **85** 214117

[6] Kirchheim R 1988 *Prog. Mater. Sci.* **32** 261

[7] Anwand W, Brauer G, Butterling M, Kissenger H R and Wagner A 2012 *Def. Diff. Forum* **331** 25

[8] van Veen A, Schut H, Clement M, de Nijs J M M, Kruseman A and Ijpma M R 1995 *Appl. Surf. Sci.* **85** 216

Journal of Physics: Conference Series **505** (2014) 012025 doi:10.1088/1742-6596/505/1/012025

Ab-initio calculation of CDB spectra - a case study on transition metals

M Reiner[1,2], T Gigl[1,2] and C Hugenschmidt[1,2]

[1] Technische Universität München, Physik-Department, Lehrstuhl E21, James-Franck-Straße, D-85748 Garching Germany

[2] Technische Universität München, Forschungs-Neutronenquelle Heinz Maier-Leibnitz, Lichtenbergstraße 1, D-85748 Garching, Germany

E-mail: markus.reiner@frm2.tum.de

Abstract. Experimental results of Coincident Doppler Broadening Spectroscopy (CDBS) in Si, Cr, Ni, Cu, Pt and Au references were compared with ab-initio calculations. These calculations were performed by use of the MIKA Doppler software package, which is provided for the community by Aalto University. This software calculates electron wavefunctions by an atomic superposition method. It is shown that despite this simple approach, the element specific information obtained by CDBS can be calculated with high reliability for the examined elements.

1. Introduction

Coincident Doppler broadening spectroscopy (CDBS) enables the measurement of the longitudinal electron momentum p_L even for high Doppler shifts, which stem from core electrons [1]. Due to this it can be used for an element specific analysis of trapping sites in multicomponent systems or depth profiling in layered structures. For such an analysis, experimental data can be compared with measured or calculated reference spectra.

A theoretical description of the complete electron momentum spectra requires band structure calculations for the valence electrons. Calculations based on an atomic superposition method for electron wave functions [2] can be expected to well describe the annihilation with core electrons. Software packages based on such calculations are offered for the community, e.g. the MIKA-Doppler program [3] that superposes orbital wave functions from free atoms. Within this study, it was systematically examined how calculations performed by use of MIKA-Doppler describe the element specific part of measured CDB spectra taken in various transition metals and Si.

2. Methods

2.1. Calculation of CDB spectra

The Doppler broadening of the annihilation line is given by the momentum distribution $\rho(\mathbf{p})$ of the annihilating electron-positron pair. This is given as a sum over all orbital momentum distributions $\rho_j(\mathbf{p})$ by equation (1), in which r_0 denotes the classical electron radius and c is the speed of light:

$$\rho(\mathbf{p}) = \sum_j \rho_j(\mathbf{p}) = \pi r_0^2 c \sum_j u_j(0)^2 \left| \int \mathrm{d}\mathbf{r} \exp\left(-i\mathbf{p}\cdot\mathbf{r}\right) \psi_+(\mathbf{r}) \psi_j(\mathbf{r}) \right|^2 \tag{1}$$

The state-dependent enhancement factor $u_j(0)$ is used in order to model the electron-positron correlation for each electronic state. In this study, for the enhancement a generalized gradient approximation [4] with a parametrization based on data by Arponen and Pajanne was used [5].

In MIKA Doppler, the calculation of $\rho(\mathbf{p})$ is based on the two-component density functional theory [6] in the limit of a vanishing positron density. The orbital electron wave functions $\psi_j(\mathbf{r})$ are calculated by the code employed considering the Dirac equation for atoms. The Schrödinger equation for the positron wave function $\psi_+(\mathbf{r})$ is solved by a three-dimensional real-space solver. For comparison with the experimental data, the calculated results were convolved with a Gaussian function in order to mimic the experimental energy resolution.

2.2. Measurement of CDB spectra

The measurements were performed at the CDB-spectrometer located at the high-intensity positron beam NEPOMUC [7, 8], which enables quasi background free coincident photopeak detection without any source component. In a collinear setup of High Purity Ge (HPGe) detectors, more than 10^7 counts per CDB spectrum were accumulated in the coincident photopeak with an effective energy resolution of 1.0 keV FWHM at 511 keV. CDB spectra were measured in Si, Cr, Ni, Cu, Pt and Au references and evaluated as described in [9]. For the Si reference, a Si(100)-wafer was taken; due to the measured positron diffusion length of 228 nm it can be assumed to be defect free. The metal specimens were annealed before performing the measurements.

3. Results and Discussion

3.1. Raw spectra

Fig. 1 shows the experimental and calculated spectra, which were multiplied by a scaling factor. This scaling factor was obtained by evaluating the error weighted mean ratio of measured and calculated CDB spectra in the momentum range between $30{\cdot}10^{-3}\,m_0c$ and $50{\cdot}10^{-3}\,m_0c$, where the annihilation with core electrons is dominant.[1] The same method also has been applied on CDB spectra of vapor deposited Cu films, in which saturated trapping of positrons in open volume defect occurs [10]. Due to the decreased core annihilation probability in these defects, one obtains a smaller scaling factor compared to the spectra of the annealed references used here.

Generally, for all elements good quantitative agreement can be seen for electron momenta $p_L > 30{\cdot}10^{-3}\,m_0c$, where the annihilation with core electrons dominates [11–14]. The agree -

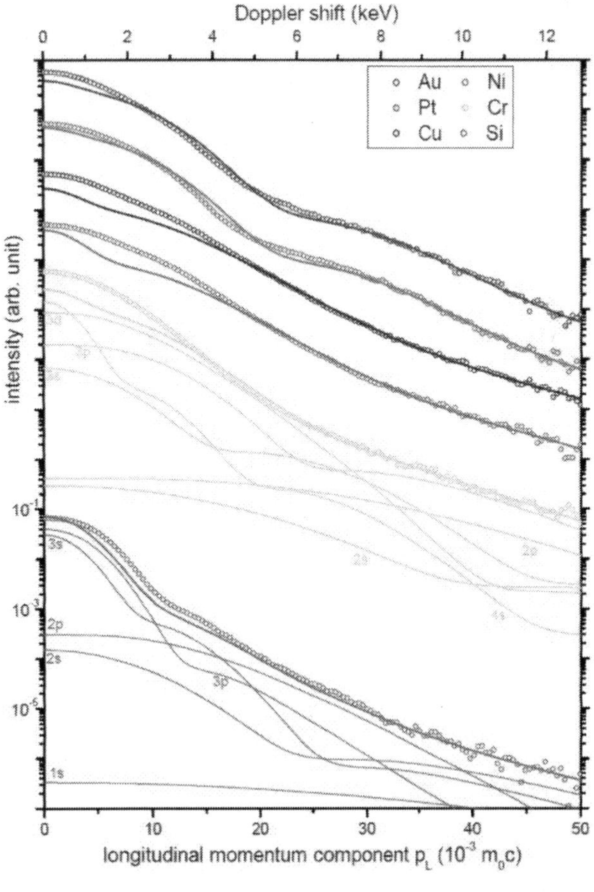

Figure 1: Measured (o) and calculated (–) CDB spectra.

[1] The spectra of each element (except Si) were multiplied by an additional factor for a better visibility in fig. 1.

ment is not that good for smaller momenta, where the annihilation with valence electrons gets dominant and hence, the applied atomic superposition method cannot be expected to give full agreement.

In the case of Si, the calculations exactly fit the data for $p_L > 30 \cdot 10^{-3}\, m_0c$. The small discrepancy for smaller momenta is attributed to the rough calculational treatment of the annihilation with valence electrons (from the 3s- and 3p-shell), which also cannot describe the well known anisotropy of the valence electron momentum distribution in Si [15].

For the 3d transition metals, i. e. Cr, Ni and Cu, calculations and measurements compare in the same way. The applied calculation method clearly underestimates the contribution of low momentum valence electrons to the annihilation, which leads to a discrepancy in the small momentum area. Here, the calculated curves also exhibit a shape different from that of the measured ones. But already above $20 \cdot 10^{-3}\, m_0c$, and in the case of Cr even above $15 \cdot 10^{-3}\, m_0c$, the ab-initio calculations well agree with the experimental data. For Ni and Cu, the contribution of the various electronic states are very similar as in Cr. Details will be discussed in the following (sec. 3.2).

In the case of the 5d transition elements Au and Pt, in addition to the quantitative agreement in the high momentum area, theoretical and experimental curves show the same qualitative behavior for small momenta. The dip seen in the experimental data at $20 \cdot 10^{-3}\, m_0c$ is also exhibited by the calculated curves, but slightly shifted to higher momenta Furthermore, it seems that the annihilation with low momentum valence electrons ($p_L < 20 \cdot 10^{-3}\, m_0c$) is slightly underestimated. The characteristic features of the spectra cannot be attributed to a single electronic state, because for a wide range of momenta three or four different electronic states contribute in a similar extent to the annihilation.

3.2. Ratio curves

In order to reveal to what extent the calculations can be used for the element specific analysis of CDB spectra, the ratio curves to the calculated Si spectrum were investigated (fig. 2) and are shown in the high momentum area ($20 \cdot 10^{-3}\, m_0c < p_L < 50 \cdot 10^{-3}\, m_0c$). The measured Si spectrum

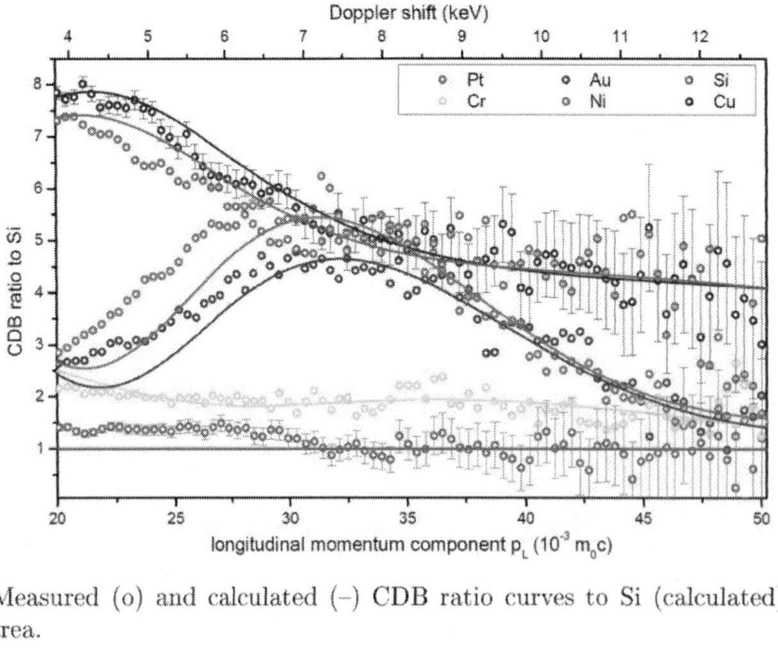

Figure 2: Measured (o) and calculated (–) CDB ratio curves to Si (calculated) in the high momentum area.

could serve as reference for the other measured spectra, which leads to the same results for momenta smaller than $30{\cdot}10^{-3}\,m_0c$. For higher momenta the increasing statistical error of the measured Si spectrum complicates a comparison between theory and experiment and hence, the calculated Si spectrum is used as reference for all measured spectra.

It is noteworthy that both the theoretical and measured Cu ratio curve show a nearly identical decrease between $20{\cdot}10^{-3}\,m_0c$ and $50{\cdot}10^{-3}\,m_0c$. In former studies based on band structure calculations, where the 3d band of Cu was treated as semicore orbital and attributed to the valence electrons, it was found that for Cu the annihilation with core electrons is dominant for momenta higher than $30{\cdot}10^{-3}\,m_0c$ [16]. The astonishingly good agreement already above $20{\cdot}10^{-3}\,m_0c$, despite the rough treatment of the 3d electrons within the atomic superposition method, was also reported in previous studies [12, 17] and is very similar for Ni and Cr (fig. 1).

Furthermore, the element specific CDB signatures of Au and Pt with their maximum at $31{\cdot}10^{-3}\,m_0c$ are well described by the calculations. Although the agreement is not that good for $p_L < 30{\cdot}10^{-3}\,m_0c$, calculations and measurements show the same tendency. Hence, a distinction of both elements is possible due to a qualitative comparison.

4. Conclusion

The MIKA Doppler software package with its simple atomic superposition method has been applied for various transition metals and Si. It was shown that the element specific high momentum part of the CDB spectra can be calculated with high reliability. For 3d transition metals, a very good quantitative agreement was found for even smaller momenta, where the annihilation with valence electrons plays a significant role. In the case of 5d transition metals, for these momenta at least a qualitative agreement was found.

Acknowledgments

The authors gratefully thank I. Makkonen from Aalto University for the support in using the MIKA Doppler software package. Financial support within the project no. 05KI0WOB by the BMBF is gratefully acknowledged.

References

[1] Asoka-Kumar P, Alatalo M, Ghosh V J, Kruseman A C, Nielsen B and Lynn K G 1996 *Phys. Rev. Lett.* **77** 2097–100
[2] Puska M J and Nieminen R M 1983 *J. Phys. F. Met. Phys.* **13** 333
[3] Torsti T, Eirola T, Enkovaara J, Hakala T, Havu P, Havu V, Höynälänmaa T, Ignatius J, Lyly M, Makkonen I, Rantala T T, Ruokolainen J, Ruotsalainen K, Räsänen E, Saarikoski H and Puska M J 2006 *Phys. Status Solidi B* **243** 1016–53
[4] Barbiellini B, Puska M J, Torsti T and Nieminen R M 1995 *Phys. Rev. B* **51**(11) 7341–4
[5] Arponen J and Pajanne E 1979 *Ann. Phys.* **121** 343–89
[6] Schultz P J and Lynn K G 1988 *Rev. Mod. Phys.* **60** 701–79
[7] Reiner M, Pikart P and Hugenschmidt C 2013 *J. Phys. Conf. Ser.* **443** 012071
[8] Hugenschmidt C, Piochacz C, Reiner M and Schreckenbach K 2012 *New J. Phys.* **14** 055027
[9] Pikart P and Hugenschmidt C 2014 *Nucl. Inst. Meth. A* accepted
[10] Reiner M, Pikart P and Hugenschmidt C 2012 *Phys. Procedia* **35** 104–10
[11] Calloni A, Dupasquier A, Ferragut R, Folegati P, Iglesias M M, Makkonen I and Puska M J 2005 *Phys. Rev. B* **72**(5) 054112
[12] Makkonen I, Hakala M and Puska M 2006 *Physica B: Condensed Matter* **376–77** 971–4
[13] Ghosh V J, Alatalo M, Asoka-Kumar P, Nielsen B, Lynn K G, Kruseman A C and Mijnarends P E 2000 *Phys. Rev. B* **61**(15) 10092–9
[14] Hakala M, Puska M J and Nieminen R M 1998 *Phys. Rev. B* **57**(13) 7621–7
[15] Erskine J C and McGervey J D 1966 *Phys. Rev.* **151**(2) 615–20
[16] Tang Z, Hasegawa M, Nagai Y and Saito M 2002 *Phys. Rev. B* **65**(19) 195108
[17] Makkonen I, Hakala M and Puska M J 2006 *Phys. Rev. B* **73**(3) 035103

Journal of Physics: Conference Series **505** (2014) 012026 doi:10.1088/1742-6596/505/1/012026

Design of the Slow POsitron faciliTy (SPOT) in Israel

S May-Tal Beck[1], D Cohen[2], E Cohen[3], A Kelleher[2], O Hen[3], J Dumas[2], E Piasetzky[3], N Pilip[3], G Ron[2], I Sabo-Napadensky[4] and R Weiss-Babai[41]

[1]Physics department, NRCN, P.O.Box 9001, Beer-Sheva, Israel 84190
[2]Racah Institute of Physics, Hebrew University of Jerusalem, Jerusalem, Israel 91904
[3]School of Physics and Astronomy, Tel-Aviv University, Tel-Aviv, Israel 69978
[4]NDT department, Soreq NRC, Yavne, Israel 81800

E-mail: smtbeck@gmail.com

Abstract. A slow positron beam is being built in the Hebrew University, the Slow POsitron faciliTy (SPOT). In Israel, the beam will introduce a new tool for both fundamental and applied research. Here we present the design process of the beam, where the leading goals are safety and high efficiency, with a flexible choice of the positron source. The challenges in the design of a moderator unit, based on frozen Neon, the pre-accelerator section and the full beam-line were addressed by simulation programs using various packages: COMSOL, SIMION and GEANT4. First measurements in SPOT are expected within a year.

1. Introduction

Positron Annihilation Spectroscopy (PAS) is a well-established method used in the fields of solid-state physics, chemistry, materials science and materials engineering [1]. The two most common PAS measurement systems are based on table-top gamma spectrometers, and slow positron beams. The operational concept of most slow positron beams is similar: high-energy positrons emitted from a β^+ radioactive source are moderated, low energy positrons (few eV) are then selected and re-accelerated to create a mono-energetic positron beam with controllable energy that varies from several tens of eV up to several tens of keV [2]. The physical implementation of this concept varies from beam-to-beam mainly in the choice of moderator material (frozen gas vs. metal with negative e^+ work function), in its geometry and source coupling, and in the details of the slow positron selection and re-acceleration technique. The final beam efficiency, defined as the ratio of the number of mono-energetic positrons impacting the target to the source activity, strongly depends on the exact beam design and varies from $\sim10^{-5}$ to $\sim10^{-2}$ [3].

We are currently building the first slow positron beam in Israel. The beam will be the center of the newly formed **S**low **PO**sitron facili**T**y (SPOT) at the Hebrew University. This beam, together with the PAL spectrometer available in Israel [4], will bring new material research capabilities to the Israeli community. The SPOT positron beam will serve as a tool for both fundamental and industry-oriented applied materials research.

[1] Corresponding author: smtbeck@gmail.com

Journal of Physics: Conference Series **505** (2014) 012026 doi:10.1088/1742-6596/505/1/012026

2. Design of the SPOT Positron Beam

The SPOT beam was designed under several safety guidelines: (1) Use high efficiency frozen gas moderator to considerably reduce the required positron source activity, (2) Allow usage of short-lived positron source to avoid long term contamination hazard, and (3) Use an isolated electrostatic beam-line that operates on a high-voltage platform, while keeping both the source chamber and target at zero potential. Additionally, we want the design to allow for a simple usage with either short-lived or long-lived positron sources. A cross section of the SPOT beam design is shown in Fig. 1. The design is such that positrons emitted from the source pass through a frozen gas moderator, which is kept at ground potential. The low energy positrons are then pre-accelerated to 50 eV on a high-voltage platform. A 90° bend section select the slow positrons and filter out higher energy positrons. The final positron acceleration is done by a 10-step acceleration unit, while the target (sample) is kept at ground potential. A constant low magnetic field, produced by a set of solenoids, is applied along the beam line, keeping the positrons focused along the beam direction.

Overall, the SPOT beam is designed for both safety and ease of use. All high voltage elements will be internal to the beamline and moderator structures. The beamline has been designed to allow changing the beam energy without changing any of the magnetic fields that collimate and bend the beam.

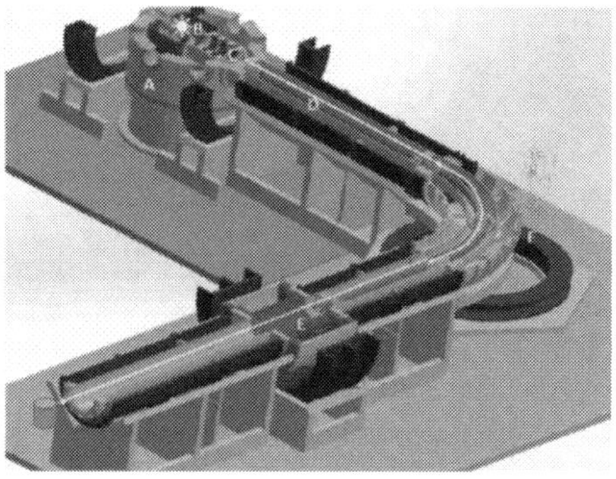

Figure 1: Top view of the cross section of SPOT design scheme. The source and beam propagation direction are marked in yellow. The moderator chamber (A) contains the moderator (B) and a pre-accelerator section (C). The beam-line consists of three parts. The first (D), from the exit of the moderator chamber to the acceleration unit (E), is made of concentric pipes. Coils, in blue, form a uniform magnetic field along the beam direction. Helmholz coils are used to maintain the same field inside the moderator chamber, in the target chamber (Not shown) and to bend the direction of the beam in 90° (F).

2.1. Positron source

The primary driver of SPOT's design is to make it a useful tool for the materials industry. As such, we examine the ability to use short lived positron sources, such as ^{18}F (half-life of ~110 minutes). This source is commonly used in PET imaging and is therefore well regulated legally and is readily available wherever PET imaging systems are in use. By using a short-lived source, a short term highly active source can be used to increase the positron flux on target. More importantly, due to its short half-live, the radiation hazard is limited only to the time of measurements. Efforts are currently made to develop the process of source production, in a dedicated source container, and the method of safely handling it in order to position it in the moderator chamber.

2.2. Positrons moderator – challenges and solutions

Comparative studies show that frozen gas moderators (especially Neon) have a much higher moderation efficiency compared to tungsten foil moderators [5]. Therefore, to reduce the required source activity, the SPOT beam will use a high efficiency frozen gas moderator. The moderator design combines both transmission and reflection geometries. It consists of a copper mesh window and a conical shaped, oxygen free copper piece, coupled to a cold head. The oxygen-free copper is an excellent thermal conductor and therefore the gas is expected to freeze on its surface and on the mesh. The conical shape is expected to further enhance the efficiency [6]. One disadvantage of frozen gas moderators is that positrons thermalize via scattering processes, and thus exit with wider energy distribution compared to that of positrons moderated in tungsten [7]. Similarly to other slow positron beams [2], a pre-acceleration unit will give the moderated positrons an initial low energy, of 50 eV, with which they can pass the bending section of the beam. The main challenges in the moderator design relate to the questions of how to keep low temperature in the vicinity of the frozen gas, near the active positron source, and how to mount the pre-accelerator close to the moderating unit and apply voltage to them. A design that addresses both requirements was developed using the COMSOL and SIMION simulation packages [8,9]. The first was used to simulate the temperature profiles in the vicinity of the moderation unit, with the influence of other neighbouring materials and the positron source, whereas the latter was used to design the geometry of the pre-accelerator unit. Fig. 2 shows the spatial and energy distributions of the accelerated positrons, as produced by the SIMION simulation. The pre-accelerator will consist of few ring shaped electrodes, separated by rings of insulating Teflon. Voltage to the rings and to the copper unit will be applied by wires with feed –through connections to the vacuum chamber. In order to keep the entire cold-head grounded, a sapphire spacer is added between the cold head and the pre-acceleration unit. We choose to use sapphire as, at low-temperatures, it is a good electrical insulator as well as a good thermal conductor.

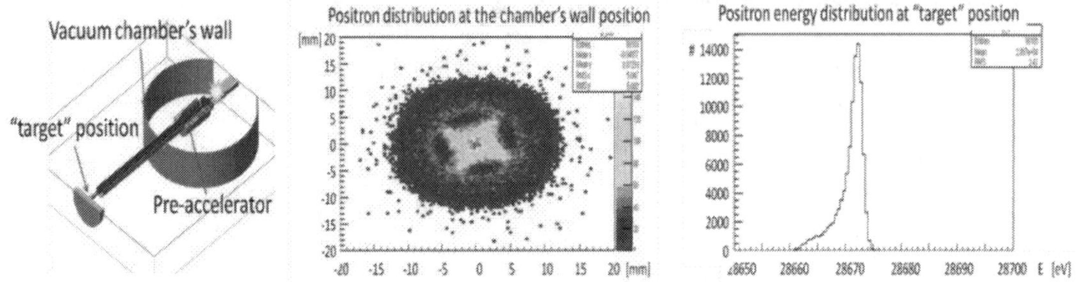

Figure 2: Left: A scheme of the pre-acceleration unit within the moderator vacuum chamber, showing positron trajectories calculated by SIMION for the shown setting. Right plots: Spatial (left) and energy (right) distributions of the same positrons. The energy shown is ~28keV, with FWHM of ~3eV. This simulation was performed before the bending was introduced.

2.3. The design process of the beam line

The electrostatic beam-line uses magnetic and electric fields to select moderated positrons and transport them from the moderator to the target, while accelerating them to the desired energy (see Fig.1). The beam-line is designed such that the target (sample) is kept at ground potential. This reduces the hazard of electrical shock at this readily accessible position. To allow the target to be kept at ground potential, all units upstream of the acceleration unit have to be kept on high-voltage platform, where the high-voltage is defined by the final positron acceleration voltage. This means that the moderator assembly and the pre-accelerator unit, as well as the bending section, should all be kept at high-voltage. Because of the uniform base voltage, the pre-acceleration of low energy positrons and their selection by the bend is indifferent to its absolute value. The demand of keeping the moderator chamber at ground potential, but the outgoing positrons at high potential presented a complicated design challenge. The solution chosen is to use two concentric tubes from the exit of the moderator to the accelerator unit (see Fig. 1). The inner tube will be held at high potential, while the outer tube will be grounded. This scheme required a special design of the inner tube, such that the vacuum within it can be maintained. A constant solenoid magnetic field of ~9 mT will be applied along the beam-line, which will collimate low-energy positrons along the beam propagation direction. In the 90° bend section, a transverse magnetic field will be used to bend the low energy positrons in the beam direction and to filter them from positrons with higher energies. A 10 step acceleration unit, downstream of the bend, will accelerate the particles from the desired voltage to zero. The planned HV will vary from 50 eV to 30-50 keV. The exact field configuration was determined following extensive beam-line simulation work, using two packages: GEANT4 and COMSOL [10,8]. The first, GEANT4, was used to define the constant magnetic field along the beam and a perpendicular field in the bending section, in accordance with the possible geometry in the laboratory room. Lead by the design of the coils at the SPONSOR beam [2], we calculated the structure and number of coils needed for the SPOT beam, as well as the required currents, to perform the uniform magnetic field of ~9 mT. COMSOL, a multiphysics solver program includes a beam-line simulation package. Its inputs are realistic physics observables, such as coils current and electric potential, and the program solves for the resulting electric and magnetic fields. A complete magnetic field map, including fringe fields, is calculated. These fields are then used to calculate more precisely the transport of the particles from the moderator to the target, as shown in Fig 3.

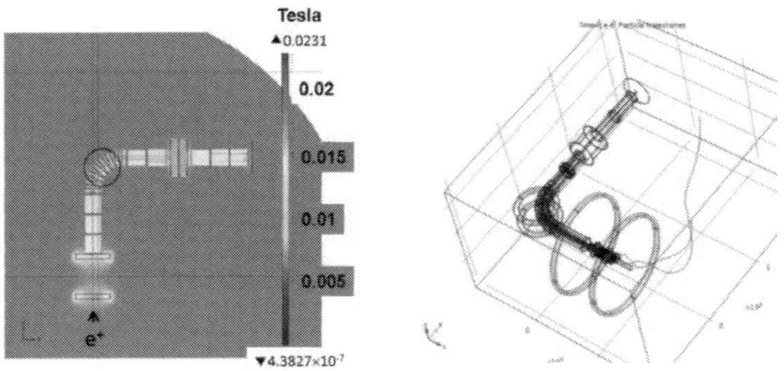

Figure 3: Left: details of magnetic field magnitude along the beam-line, calculated by COMSOL. Right: positron trajectories, calculated by COMSOL, being transport from the source, through the beam-line, to the target position. Colors represent positron energies, from low (blue) to higher (red) energies.

3. Summary

The design of the SPOT beam provided many challenges for which we developed realization concepts and designed solutions using the simulation packages COMSOL, SIMION and GEANT4. To gain high efficiency the design is based on frozen Ne as a moderator. A special pre-accelerator unit mounted within the moderator chamber will extract the moderated positrons and direct and focus them into an initial beam. Outgoing mono-energetic slow positrons, with energy of ~50eV, will be selected for further acceleration by a bending section. The acceleration to the target will be done by a 10-step acceleration unit. Coils along the beam will form a constant magnetic field of ~9mT in order to keep the positron trajectories spiral along the beam direction. Safety considerations led to a design of a grounded target and a high voltage platform internal to the moderator chamber and to the beam line, while all outer parts are kept at ground potential too. The design allows usage of either ^{22}Na source, available to purchase, or costum made ^{18}F source which, due to its short half-life, presents much less safety requirements. The implementation of the moderator assembly and the beam line are under-way. First measurement with the SPOT beam is expected in the coming year.

Acknowledgments

We are thankful to the SPONSOR group at HZDR, Dresden, Germany, the group at IFIN HH, Bucharest, Romania and the group at ANU, Australia, for instructive visits and for many fruitful discussions. The design presented in this paper was benefited from the experience of these groups.
SPOT is funded by the Kamin program of the Israeli ministry of Economics.

References

[1] Dupasquier A, Mills A P Jr. (Eds.) 1995 *Positron Spectroscopy of Solids* (IOS, Amsterdam).
[2] Anwand W, Brauer G, Butterling M et al. 2012, *Defect and Diffusion Forum* **331** 25-40.
[3] GBAR collaboration 2011, *Proposal to measure the Gravitational Behaviour of Antihydrogen at Rest GBAR*, CERN-SPSC-2011-029 / SPSC-P-342, 30/09/2011.
[4] May-Tal Beck S, Brauer G, Anwand W et al. 2004, *Materials Science Forum* **445-446**, pp 495-7.
[5] Mills A P Jr. and Gullikson E M, 1986, *Appl. Phys. Lett.* **49** 1121.
[6] Khatri R, Charlton M, Sferlazzo P et al., 1990 *Appl. Phys. Lett.* **57** 2374.
[7] Wu Y C, Chen Y Q, Wu S L et al. 2007, *Physica Status Solidi C* **4** 4032.
[8] COMSOL Multiphysics®, COMSOL 4.3, COMSOL AB, Stockholm, Sweden (2012).
[9] SIMION version 8.1.1.29, 12/29/2012, Scientific Instrument Services, Inc., Ringoes, NJ.
[10] Agostinelli S et al 2003, *GEANT4 – a simulation toolkit, Nucl. Instr. Methods A* **506** 250-303.

Enhancement and Transformation of the Phase Space Density of the NEPOMUC Positron Beam

C Piochacz[1,2], E Erdnüß[2], T Gigl[1,2], N Grill[1] and C Hugenschmidt[1,2]

[1] Technische Universität München, Physik Department, Lehrstuhl E21, James-Franck-Straße, D-85748 Garching, Germany

[2] Technische Universität München, Forschungs-Neutronenquelle Heinz Maier-Leibnitz, Lichtenbergstraße 1, D-85748 Garching, Germany

E-mail: `christian.piochacz@frm2.tum.de`

Abstract. At NEPOMUC, four different experiments are installed permanently and one beam port is open for positron experiments provided by external users. To match the different demands of these experiments on the phase space density of the positron beam, the established NEPOMUC remoderator was relocated and two novel beam switches were installed to provide the primary high-intensity beam or the remoderated high-brightness beam within short switching times. At the present state, the remoderated beam has a mean diameter of 1.85(35) mm and an intensity of $3.0(3) \cdot 10^7\,e^+/s$. For the transformation of the longitudinal phase space density a new buncher setup was developed which converts the continuous positron beam into a pulsed beam.

1. Introduction

The demands on a positron beam depend strongly on the experiment which should be performed. The highest intensity is in most cases desired to gain high statistic within a reasonable time or to investigate time dependent processes. However, some experiments have special requirements on the phase space volume (PSV). If laterally resolved measurements with high resolution are in the scope of interest at least the diameter and the radial momentum have to be sufficiently small, hence the radial part of the PSV has to be regarded. If the positron beam should be bunched to short pulses the energy spread has to be accordingly low and therefore the longitudinal part of the PSV is of special interest. Due to the theorem of Liouville the minimal attainable spatial focus is determined mainly by the radial PSV and the refractive poser of the lens; and the smallest temporal focus is determined by the dimension of the longitudinal PSV and the effective energy modulation performed by the buncher. The easiest way to reduce the PSV is given by apertures and energy filters which can be used to crop small parts of the PSV. In some cases the comparably high intensity losses caused by these tools might be justifiable. However, more preferable is the technique of remoderation by which not only the PSV is reduced but also the phase space density is enhanced and therefore the intensity loss is much lower. Nevertheless, there are situations, e.g. if highest resolution should be attained, where a trimming of the PSV is unavoidable and appropriate tools have to be provided for both, the longitudinal and the radial PSV.

At NEPOMUC, most experiments need the brightness enhanced positron beam generated by the remoderator [1]. However, there are experiments planed and have be already performed which barely depend on a small phase space volume but benefit from the higher intensity of the primary beam [2,3]. Therefore, an easy switching between these two beam options is mandatory to enable a broad range of experiments. This is now realized in a new setup which is presented in the first part of this paper. Another basic demand on a positron beam is the impact energy which enables depth resolved measurements. Usually the variation of the implantation energy is achieved by altering the potential of the source or of the sample; hence the variation is attained by a variation of the potential difference. At NEPOMUC the source potential is limited to 1 kV due to the restrictions of insulators in the high radiation field of the nuclear reactor. There are also experimental constraints, where the sample has to be potential free e.g. if re-emitted charged particles after positron impact are in the scope of interest. In these experiments the total energy of the beam has to be raised. This is only possible by time dependent fields acting on a pulsed beam.

For this purpose a novel system was developed, which converts the continuous positron beam into a pulsed beam and enhances afterwards the total energy of the beam [4]. Recently, the bunching system has been assembled and first measurements have been performed. The bunching units consist of a high efficiency pre-buncher which uses a sawtooth-like function with a repetition rate of 5 MHz and a double-gap main-buncher using a high amplitude sine wave function. While in a previous study [4] the concept of the whole setup is explained in-depth, in this paper the focus is on the results of time dependent particle tracing simulations and technical details.

2. Central setup for beam selection, enhancement and characterization

Due to larger reorganizations in the experimental hall of the FRM II also most of the positron experiments and the beam lines had to be relocated or rebuild from scratch. Especially the inevitable relocation of the remoderator opened the possibility for an entirely new setup for the characterization and enhancement of the positron beam. With this new setup two beam switches were introduced (4, 8 in figure 1) which enable a computer-controlled toggling between the primary and the remoderated positron beam. These switches use the superposition of a transversal and a longitudinal magnetic field to guide the positrons adiabatically from one end to the other. Special attention was drawn to an aberration free guidance through the new setup. Electrodes within the whole setup, which continues the potential tube of the source beam tube, enables the variation of the kinetic energy and provides therefore the basis for an adiabatic beam transport from the source through the setup. Computer simulations of the PSV development through the beam switches showed that besides appropriate low kinetic energy special attention has to be paid to a constant longitudinal guiding field [5]. This requires well dimensioned coils and proper positioning of the μ-metal shielding.

The remodertator setup itself underwent only minor changes for the improvement of the vacuum and a residual gas analyzer was installed. In addition, the magnetic field termination was electrically insulated from ground potential, advancing the imaging of the primary positron beam toward the remoderation crystal. With the two beam monitors (3, 10 in figure 1) the primary and the remoderated beam can be observed. The beam monitors consists of a carriage which can be moved by a stepper motor to one of four slots. The first slot contains a combination of a micro channel plate and a phosphor screen. By a mirror in the same slot, the intensity distribution of the beam can be observed by a camera. The measurement of the total beam intensity is done with a calibrated NaI-detector which has a collimated view onto an annihilation target mounted at the second slot of the carriage. The last two positions are used for different-sized apertures which can be used to crop the diameter of the beam.

The fully computer controlled adjustment of all beam guiding parameters and the permanently installed beam monitors simplified the recommissioning of the relocated remoderator considerably. The remoderated beam is slightly elliptic shaped were the major and minor axis have a length of about 2.2 mm (FWHM) and 1.5 mm (FWHM) respectively

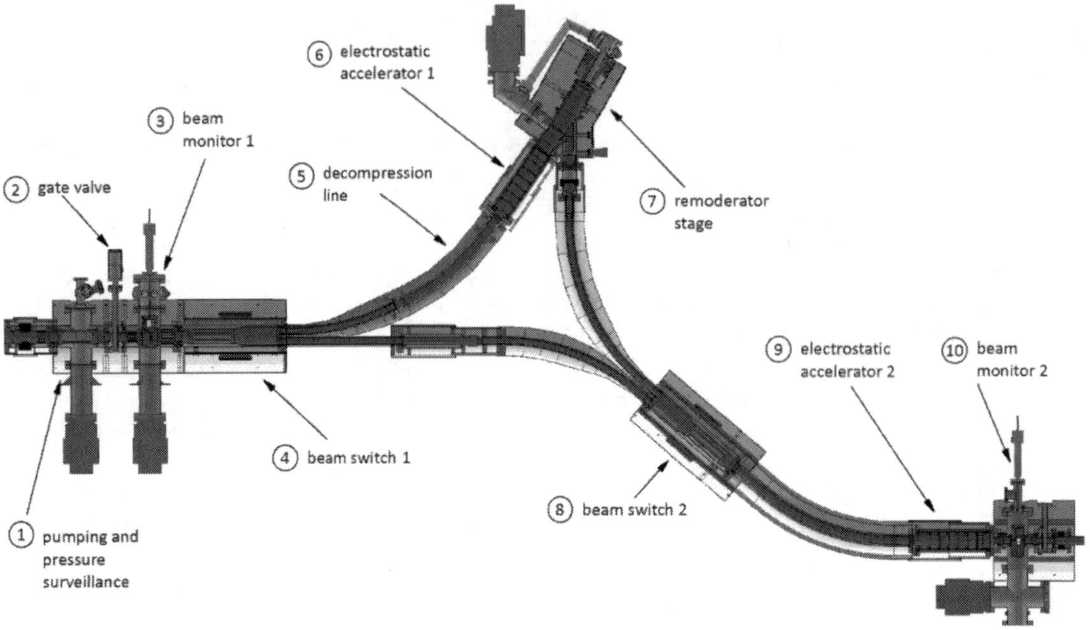

Figure 1: Overview of the new setup for the analysis (3, 10) and brightness enhancement (7) of the NEPOMUC positron beam. The two beam switches (4, 8) allow the computer controlled toggling between the high-intensity and the brightness-enhanced beam.

(see figures 2a, b). 90 % of the total intensity of the remoderated beam is within a diameter of roughly 6 mm (see figure 2c). The total intensity of the remoderated beam was determined to $3.0(3) \cdot 10^7\,e^+/s$. The total efficiency of the remoderator setup is about 3 % and is about half of the value attained with the former setup. It is expected to increase efficiency by using computer-controlled optimization routines.

3. The positron beam energy elevation project

3.1. Motivation

At the NEPOMUC positron beam facility, a comprehensively equipped SUrface SPECTrometer (SusSpect) exists which was used mainly for Auger electron spectroscopy (AES) by using positrons, electrons or x-rays for the initial excitation. It offers the possibility to clean the samples with an argon sputter gun and anneal them at a heater stage [6, 7]. Layer systems can be produced by molecular beam epitaxy (MBE) and the surface topology can be investigated with a scanning tunneling microscope. For AES and X-ray photoelectron spectroscopy (XPS) the hemispherical analyzer is used in the electron mode. This analyzer has also a positron mode allowing angle resolved reemitted positron spectroscopy (RPS). First measurements have been already performed with a sodium source in transmission geometry and with the primary positron beam in reflexion geometry [8]. During the beam measurements the positron impact energy was kept at 1 keV. However, for comprehensive studies a variable impact energy up to several keV is crucial. As the sample has to be grounded and the maximum bias voltage of NEPOMUC is limited to 1 kV the possible implantation energy is restricted up to now. For this reason the bunching and elevator setup was developed.

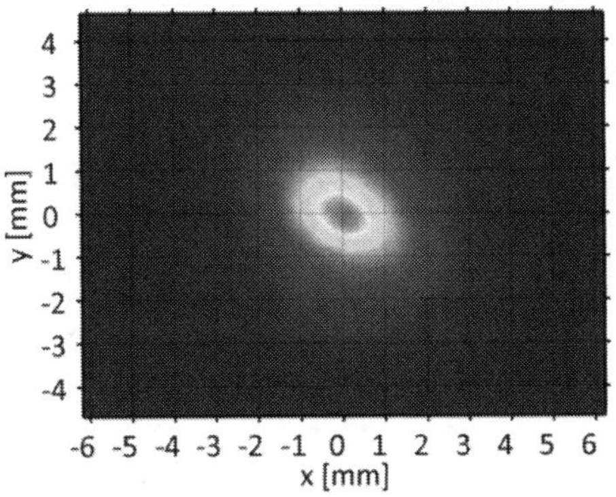

(a) Two dimensional intesity distribution.

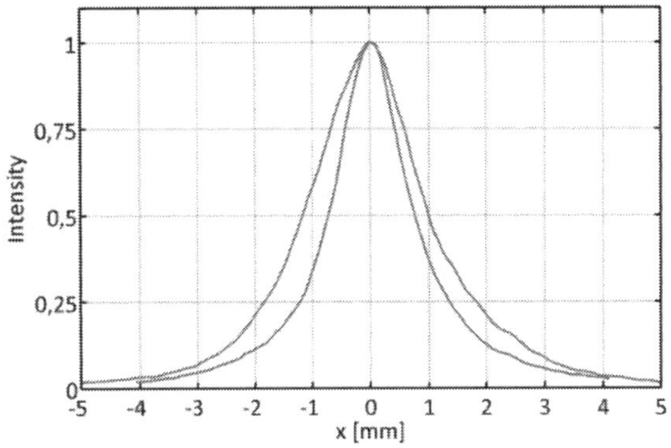

(b) Intensity profiles along the major and minor axis.

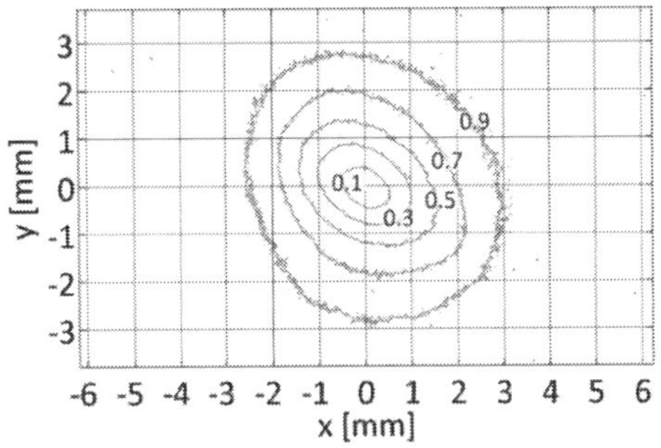

(c) The plotted lines enclose the given percentage of the total beam intensity.

Figure 2: Different representations of the intensity distribution of the remoderated beam measured at the second beam monitor of the NEPOMUC beam line facility. The total positron beam intensity was determined with a calibrated NaI-detector to about $3.0(3) \cdot 10^7\, e^+/s$.

3.2. Design and first experimental results of the bunching units

In the presented setup work two different bunchers consecutively. The pre-buncher has a repetition rate of $f_{pb} = 5\,\text{MHz}$ and utilizes an arbitrary, sawtooth-like function. To attain the highest energy transfer to the positrons, the energy modulation should occur instantaneously. Therefore, the gap on which the energy modulation occurs has a small diameter of 7 mm to keep the fringe fields and hence the transit time effects low (see figure 3). To avoid a wrong energy modulation at the opposite gap of the modulation electrode the potential difference is spread over eight additional electrodes on which the sawtooth function is applied with a linearly reduced amplitude. If the velocity of the positrons through this equalization drift is matched to its length, the transit time equals one period of the sawtooth function and the energy modulation averages to zero.

The used arbitrary function generator (Agilent 81150A) can provide a maximal peak to peak voltage of $U_{pb} = 20\,\text{V}$ in the high voltage mode. With Liouville's theorem and by supposing an initial energy spread of the continuous beam ΔE_i of about 1 eV the minimal time focus τ_{pb} can be estimated to:

$$\tau_{pb} \approx \frac{\Delta E_i}{eU_{pb}} \frac{1}{f_{pb}} \approx 10\,\text{ns} \tag{1}$$

This value is confirmed perfectly by time dependent particle tracing simulations (see figure 5a).

A buncher driven by a sine wave function can provide much higher modulation voltages by resonant amplification and hence produce much sharper time foci. The main buncher is built up equivalently to the approved sine wave function bunchers in the Munich Scanning Positron Microscope (SPM) [9] and the Pulsed Low-Energy Positron-beam System (PLEPS) [10]. In this design the center electrode acts as the main capacity in a resonant circuit and a coil outside the

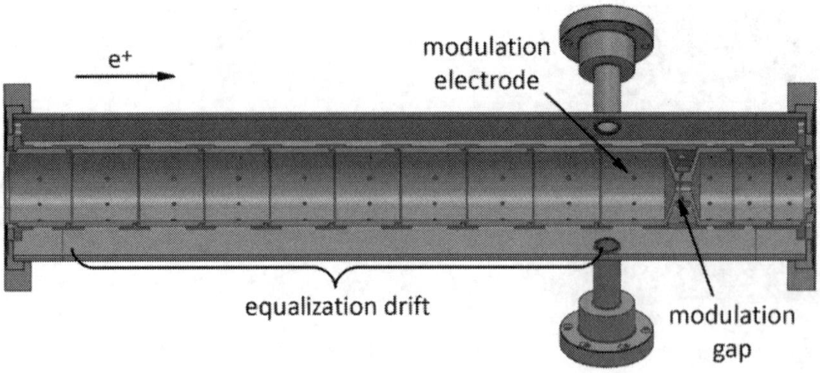

Figure 3: Sectional view of the pre-buncher. The so-called equalization drift consists of eight additionally introduced electrodes which elongate the time dependent potential difference to the modulation electrode over a longer distance. By this the energy modulation averages to zero. Hence, an effective energy modulation occurs only at the gap after the modulation electrode.

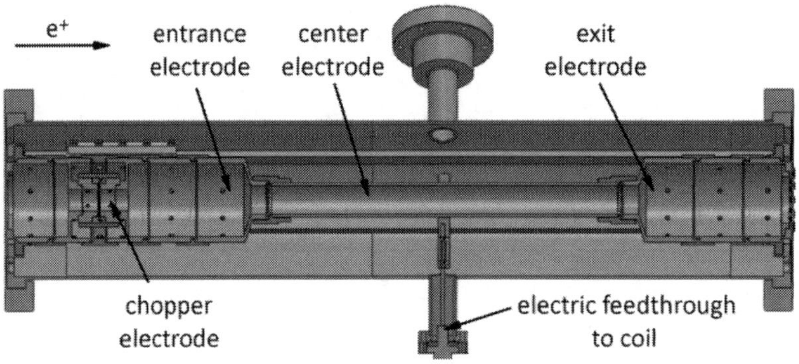

Figure 4: Sectional view of the main-buncher and chopper. The coil, which forms together with the center electrode a resonant circuit, is placed outside the vacuum.

vacuum provides the main inductance (see figure 4). The linear part of a sine wave function, which leads to a correct energy modulation with low aberrations, is about one third of the repetition time. The temporal extension of the pre-bunched beam fits therefore perfectly into the acceptance window of about 17 ns provided by a 20 MHz sine wave buncher. With the high bandwidth mode of the arbitrary function generator a rise time of about 2 ns can be attained, resulting in a very high efficiency of the bunching system. However, in this mode the output voltage is limited to 10 V leading to a time focus of the pre-buncher not lower than about 20 ns that could also be confirmed by simulations (see figure 5b). This time focus is slightly more than the optimum acceptance window of the main-buncher but well below the half sine wave. Consequently, the total time focus will increase slightly due to spherical aberrations but no additional background will be generated.

Although the background of the pre-buncher is low and smooth it is also modulated by the main-buncher which works at a four times higher repetition rate. This would lead to much lower but observable side peaks in the total time spectrum. Although this would not harm the suitability for the energy elevator but might be inappropriate for timing measurements in the future. Therefore, a simple retarding field chopper has been foreseen to clean the time spectrum (see figure 4).

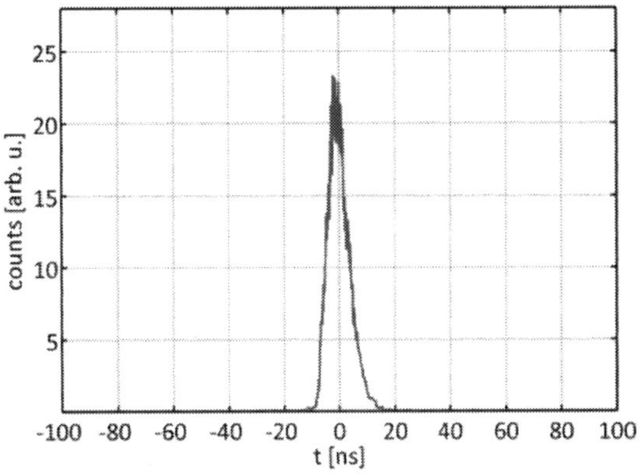

(a) Time focus of the prebuncher. Used values: Initial energy spread $\Delta E_i = 1\,\text{eV}$, modulation voltage of $U_{pb} = 20\,\text{V}$, ideal sawtooth-like function.

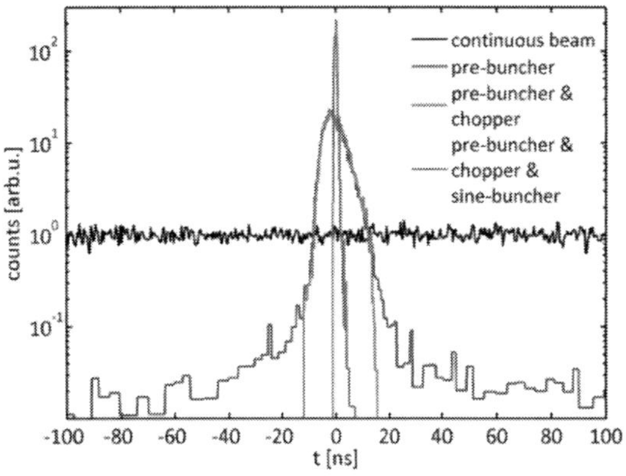

(b) Used simulation values: Initial energy spread $\Delta E_i = 1\,\text{eV}$, modulation voltage of $U_{pb} = 10\,\text{V}$, ideal functions for pre-buncher and chopper.

Figure 5: The simulated time focus attained with the proposed setup. The used ideal functions for pre-buncher and chopper result in a sharp time peak and very low background. For the visualization of the background, a variable bin size was used in sub-figure (b).

3.3. Progress of the elevator project

The time-dependent particle tracing simulations demonstrated that the attainable time focus of the pre-buncher is short enough for the acceptance window of the main-buncher and that the simple approach for the chopper is well suited to clean the timing spectrum (see figure 5b). The pre-buncher, main-buncher and the drift lines are manufactured, assembled and electrically connected. First adjustment measurements showed, that the main-buncher components act as a resonating circuit with a Q-value of about 80 and reaches therefore sufficient high modulation voltages. Preliminary timing spectra attained solely with the main-buncher showed a time focus of about 2.2 ns. The performance of the pre-buncher was below the expectations mainly due to an imperfect electrical matching of the electrodes of the equalization drift. Due to the promising bunching results, the manufacturing of the elevator itself will be started soon. A completion of the whole setup and full work capability is planed till August 2014.

References

[1] Piochacz C, Kögel G, Egger W, Hugenschmidt C, Mayer J, Schreckenbach K, Sperr P, Stadlbauer M and Dollinger G 2008 *Applied Surface Science* **255** 98–100
[2] Legl S and Hugenschmidt C 2007 *physica status solidi (c)* **4** 3981–3984
[3] Hugenschmidt C, Brunner T, Legl S, Mayer J, Piochacz C, Stadlbauer M and Schreckenbach K 2007 *phys. stat. sol. (c)* **4** 3947–52 URL `http://dx.doi.org/10.1002/pssc.200675757`
[4] Piochacz C and Hugenschmidt C 2013 *Journal of Physics: Conference Series* **443** 012093
[5] Hugenschmidt C, Piochacz C, Reiner M and Schreckenbach K 2012 *New Journal of Physics* **14** 055027
[6] Mayer J, Schreckenbach K and Hugenschmidt C 2007 *Physica Status Solidi C Current Topics* **4** 3928–31
[7] Zimnik S, Hugenschmidt C and Lippert F 2014 *Journal of Physics: Conference Series (this issue)*
[8] Mayer J, Lippert F, Piochacz C and Hugenschmidt C *(to be published)*
[9] Piochacz C, Egger W, Hugenschmidt C, Kögel G, Schreckenbach K, Sperr P and Dollinger G 2007 *phys. stat. sol. (c)* **4** 4028–31
[10] Sperr P, Egger W, Kögel G, Dollinger G, Hugenschmidt C, Repper R and Piochacz C 2008 *Appl. Surf. Sci.* **255** 35–8

A Full Electrostatic High-Brightness Slow Positron Beam Apparatus Using a Remoderator in Conjunction with a Hemispherical Energy Analyzer

Kazuaki Nagumo[a*1], **Masamitsu Hoshino**[b], **Hiroshi Tanaka**[b] **and Yasuyuki Nagashima**[a†]

[a] Department of Physics, Tokyo University of Science, 1-3 Kagurazaka, Shinjuku, Tokyo 162-8601, Japan

[b] Department of Material and Life Science, Sophia University, 7-1 Kioicho, Chiyoda, Tokyo 102-8554, Japan

E-mail: * kazuaki_nagumo@imr.tohoku.ac.jp, † ynaga@rs.kagu.tus.ac.jp

Abstract. The development of a full electrostatic slow positron beam apparatus at the Tokyo University of Science is described. The performance of different types of material as the primary moderator and the energy resolution of the beam are detailed.

1. Introduction

High quality slow positron beams are becoming increasingly necessary for various research fields using positrons[1]. Some of them, which include spin material research[2], positron microscopy[3][4], and positron scattering studies require or would benefit from the use of electrostatic beams. Recently, we have developed a full electrostatic slow positron beam apparatus at the Tokyo University of Science. This new apparatus uses a remoderator for brightness enhancement in conjunction with an electrostatic hemispherical energy analyzer. The details of the apparatus and its applications are presented in this paper.

2. Summary of the slow positron beam system

A schematic diagram of the slow positron beam apparatus is shown in figure 1. The slow positrons are produced from a ^{22}Na source with an activity of 400 MBq in combination with a tungsten mesh moderator as described in 2.1. The slow positrons emitted from the moderator are accelerated up to 5 keV by an electric field applied between the moderator and a tungsten mesh grid situated 5 mm in front of the moderator and are transported using several electrostatic lens elements made of aluminum alloy. The beam is deflected using a mirror analyzer in order to eliminate high-energy positrons and γ-rays from the positron source. The beam, of 5 keV, is then transported with further lens elements and focused onto a remoderator through an aperture of 5 mm in the outer shell of the hemispherical energy analyzer made of stainless steel. The radius of the outer shell is 44 mm and that of the inner shell is 33 mm.

[1] Present Address: The Oarai Center, Institute for Materials Research, Tohoku University, 2145-2 Narita, Oarai, Ibaraki 311-1313, Japan

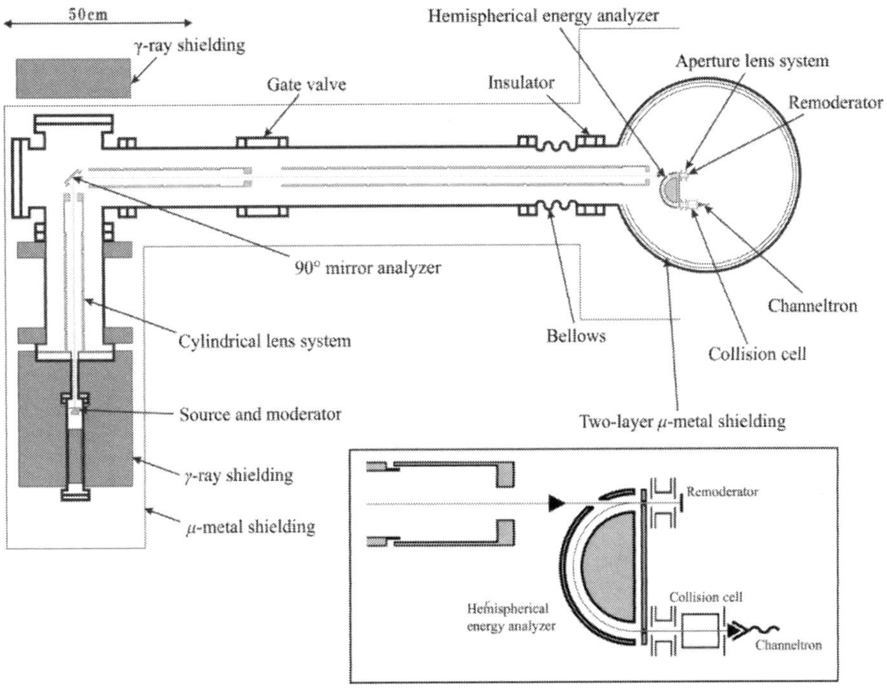

Figure 1. Schematic diagram of the full electrostatic high-brightness slow positron beam system[5].

The positrons reemitted from the remoderator are deflected and energy selected by the hemispherical analyzer so that back-scattered high-energy positrons from the remoderator are eliminated. In order to guide the high-brightness slow positron beam from the exit of the hemispherical analyzer, electrostatic lens elements of molybdenum discs with a central aperture are used. The slow positrons are transported to a channeltron (BURLE 5901) for detection through a gas cell used in the measurement studies of the total scattering cross section for positrons[5][6].

In order to avoid geomagnetic effects, the beam transport chambers are enclosed in a single layer μ-metal shielding box and two-layer μ-metal shielding surrounded the experimental chamber.

2.1. Performance of different types of primary moderators

In the present study we have investigated the moderation efficiencies of three different types of material for the primary moderator: a single-crystal (100) tungsten foil, a polycrystalline tungsten foil and a tungsten mesh. The single-crystal and the polycrystalline foils of 2 μm thickness were purchased from Aarhus University. The mesh moderator is composed of 10 overlapping tungsten meshes which were thinned to about 10 μm in diameter by an electro-polishing treatment[7]. They were annealed at 2400 K for a few minutes in a separate chamber by passing an electric current through a 25 μm tungsten foil wrapping the meshes. Then, they were placed 2 mm in front of the source and were used in transmission geometry.

The intensity and the diameter of the slow positron beam obtained from the three different moderator types were measured at the remoderator position, without the hemispherical analyzer in place, using a micro-channel plate with a phosphor screen. The results are given in table 1.

Table 1. Positron rates and beam diameters of three types of moderators measured at the remoderator position by a micro-channel plate with a phosphor screen.

Primary Moderator	Positron Rate(count/s)	Beam Diameter(mm)
W(100)	1800	2
W polycrystalline	3200	2
W mesh	7000	3–4

Though the beam diameter was the same for the W(100) and the polycrystalline W, the count rate for the W(100) was lower than that for the polycrystalline W. The highest positron rate was obtained using the W mesh though the diameter was larger than the others as the angle of divergence was bigger. The W mesh was selected as the primary moderator as a higher intensity is a greater requirement at this stage.

2.2. Brightness enhancement
The diameter and divergence of most positron beams are generally limited by the geometry of the primary sources and moderators employed. High quality beams require not only small diameter but also the minimum possible spread in both angle and energy. This is quantified by the brightness, R, of the beam by

$$R = \frac{S}{\sin^2\theta d^2 E}, \tag{1}$$

where d is the beam diameter, E is the positron energy, S is the beam intensity, and θ is the angular divergence. The brightness of a particle beam is usually limited by the beam emittance and Liouville's theorem, which states that the volume occupied in phase space of the beam is constant under the influence of conservative forces. Thus the diameter of the beam can be reduced but at the expense of the angular divergence. During the positron remoderation process non-conservative forces are present and therefore the limitations of Liouville's theorem do not apply. Positrons are focused to a small spot and remoderated, resulting in a small sized beam with a low angular spread and a narrow energy distribution. Although the number of positrons is reduced by the remoderation process, the intensity is far higher than that of a beam produced using a collimator or an energy analyzer. In the present system, a tungsten (100) single-crystal foil of 2 μm thickness is employed as the remoderator in the reflection geometry. It was previously annealed at 2400 K for a few minutes in a separate chamber by passing an electric current through a 25 μm tungsten foil wrapping. The slow positron beam, of 5 keV, is focused onto the (100) surface of the crystal and the remoderated positrons are transported and energy-analyzed by the electrostatic hemispherical analyzer.

The use of this arrangement was motivated by the work of Fischer *et al*[8], who measured the energy distribution of reemitted positrons from various metals using a hemispherical energy analyzer. The advantage of using the combination of a remoderator and a hemispherical analyzer is that the beam diameter and the divergence are small enough to be for an adequately intense beam to be transmitted through the gas cell for the purposes required.

2.3. Energy resolution
The kinetic energy distribution of the positrons reemitted from the remoderator surface was measured using the hemispherical energy analyzer. In order to attain a high energy resolution with this small analyzer, an aperture of 1 mm diameter was placed at the exit of the analyzer and the mean pass energy was set at 3 eV, resulting in an analyzer resolution of 40 meV. The energy spectrum of the reemitted positrons from the remoderator can be obtained by sweeping

the potential of the remoderator, is shown in figure 3 of [5]. If we neglect the contact potential between the remoderator and the spherical analyzer, the positron work function for the W single-crystal remoderator is determined to be $-(3.0 \pm 0.2)$ eV. The full-width at half-maximum (FWHM) of the peak is 250 meV[5].

3. Application of the full electrostatic high-brightness slow positron beam

Recently, the first magnetic field-free measurements of the total cross sections (TCS) for positron scattering have been performed using this positron beam system[5][6]. The TCS values were determined by the measurement of the transmitted beam intensities with and without the target gas in the collision cell. Corrections for foward scattered positrons, which are needed for measurements made using a magnetic beam, are not necessary in this case. This beam can also be used for the investigation of solid surfaces, taking advantage of the magnetic field-free characteristic and the low divergence.

References

[1] Schultz P and Lynn K 1988 *Rev. Mod. Phys.* **60** 3 701.
[2] Gidley D *et al* 1982 *Phys. Rev. Lett.* **49** 24 1779.
[3] David A *et al* 2001 *Phys. Lev. Lett.* **87** 6 067402.
[4] Fujinami M *et al* 2008 *Analytic. Sci. J.* **24** 73.
[5] Nagumo K *et al* 2011 *J. Phys. Soc. Jpn.* **80** 06431.
[6] Nagumo K *et al* 2012 *Eur. Phys. J. D* **81** 66.
[7] Saito F *et al* 2002 *Appl. Surf. Sci* **194** 13.
[8] Fischer D and Lynn K 1986 *Phys. Rev.* **33** 7 4479.

Positron Beam Characteristics at NEPOMUC Upgrade

C Hugenschmidt[1,2], H Ceeh[2], T Gigl[2], F Lippert[2], C Piochacz[1,2], M Reiner[2], K Schreckenbach[1,2], S Vohburger[1], J Weber[2] and S Zimnik[1,2]

[1] Technische Universität München, Forschungs-Neutronenquelle Heinz Maier-Leibnitz, Lichtenbergstraße 1, D-85748 Garching, Germany

[2] Technische Universität München, Physik Department, Lehrstuhl E21, James-Franck-Straße, D-85748 Garching, Germany

E-mail: hugen@frm2.tum.de

Abstract. In 2012, the new neutron induced positron source NEPOMUC *upgrade* was put into operation at FRM II. Major changes have been made to the source which consists of a neutron-γ-converter out of Cd and a Pt foil structure for electron positron pair production and positron moderation. The new design leads to an improvement of both intensity and brightness of the mono-energetic positron beam. In addition, the application of highly enriched ^{113}Cd as neutron-γ-converter extends the lifetime of the positron source to 25 years. A new switching and remoderation device has been installed in order to allow toggling from the high-intensity primary beam to a brightness enhanced remoderated positron beam. At present, an intensity of more than 10^9 moderated positrons per second is achieved at NEPOMUC *upgrade*. The main characteristics are presented which comprise positron yield and beam profile of both the primary and the remoderated positron beam.

1. Introduction

The principle of the reactor based positron source NEPOMUC at FRM II is based on the emission of high-energy prompt γ-rays after thermal neutron capture in ^{113}Cd. The released high-energy γ-radiation can convert into positron-electron pairs in a structure of Pt foils that also leads to the emission of mono-energetic positrons (see e.g. [1]). After 1250 days of reactor operation at the nominal power of 20 MW the lifetime of the beam tube was reached due to the burn-up of ^{113}Cd. Therefore, the inclined beam tube SR11 was replaced by an upgraded version of NEPOMUC using Cd enriched with 80% ^{113}Cd which leads to an extension of the operation time to 25 years. In addition, a new design of the positron beam extraction was developed in order to increase both intensity and brightness of the primary positron beam. The new geometry at NEPOMUC *upgrade* with electric lenses and magentic field coils including calculations of positron trajectories are presented in detail elsewhere [2, 3].

Due to major construction work in the experimental hall in 2010 and 2011 the positron beam line and several spectrometers had to be reinstalled. Subsequently, the beam line was reconstructed and a new beam switching and remoderation unit (see e.g. [4]) was installed in order to allow quick toggling between the primary high-intensity and the high-brightness remoderated positron beam [2]. In 2012, NEPOMUC *upgrade* started operation successfully.

Technical details like the simulation of temperature field in the tip of the beamtube and the measured temperature increase as function of reactor power are presented in [3]. Until now, the spectrometers PLEPS, CDBS, PAES, and the open beam port have been connected to the positron beam line.

In this work, the new beam monitor at NEPOMUC *upgrade* is elucidated that allows the determination of both intensity and shape of the primary positron beam. The measured beam characteristics of both the primary and the remoderated positron beam generated by a W(100) crystal in back reflection geometry are presented.

2. Experimental: Positron beam monitor and annihilation target

The positron beam parameters like shape and intensity can be determined at the first accessible place outside the reactor shielding. For this purpose, a beam monitor and an Al target, which are mounted on an UHV-manipulator, can be moved onto the axis of the beam line (see Figure 1). The beam monitor consists of a micro-channel plate (MCP) detector with a phosphorus screen and 45° mirror. The image on the mirror is recorded by a CCD-camera through an UHV-window.

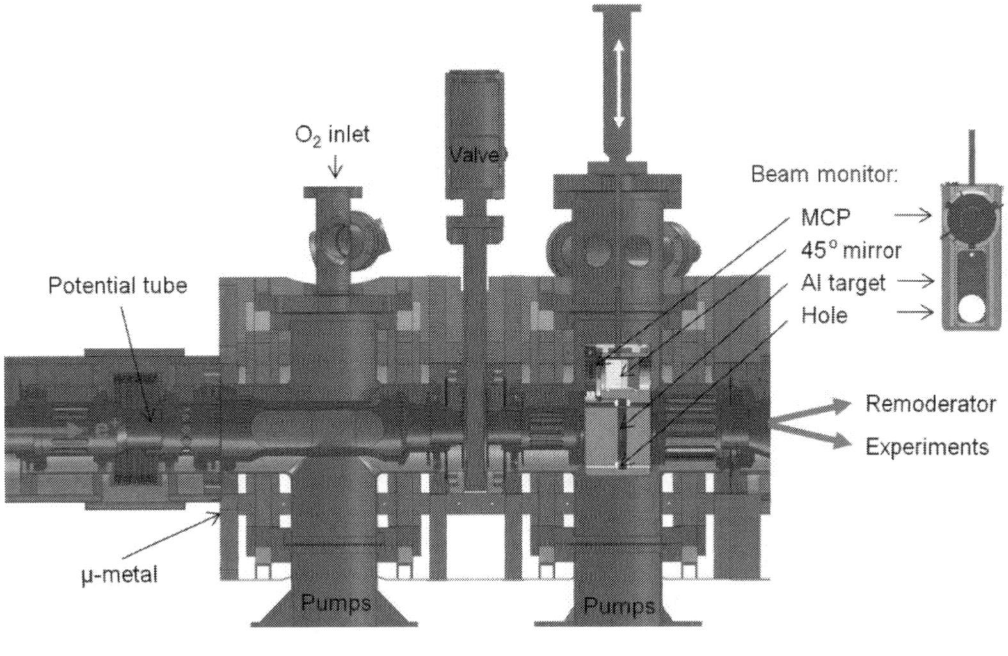

Figure 1. Characterization of the primary positron beam: Movable beam monitor (MCP detector with phosphorus screen and 45° mirror) and Al target for intensity measurements.

By moving the Al target into the beam the non-reflected positrons annihilate in the target leading to the emission of 511 keV annihilation radiation. Hence, the detection of annihilation quanta enables the determination of the positron beam intensity. For this, the γ-spectrum is recorded with a NaI(Tl)-scintillator (3" ×3") placed perpendicular to the beam axis at a distance of 1.05 m. The Al target is sligthly tilted with respect to the field of view of the detector in order to minimize selfabsorption of annihilation quanta in the target. The detector is collimated by 10 cm thick Pb bricks with a holesize of 25 mm. A Pb plate of 10 mm and an additional Cu plate of 10 mm (for absorption of characteristic x-rays from Pb) were put between the target and the detector in order to reduce the dead-time of the detector.

Without beam, a ^{22}Na-calibration source encapsulated in Al can be mounted at the target position in order to record an annihilation spectrum with known intensity with the same experimental constraints. Hence, systematic errors such as γ-attenuation in the stainless steel wall of the beamline or Compton scattering in the lead shielding could be minimized.

3. Measurements and results

The parameters of the primary positron beam at NEPOMUC *upgrade* were first measured at a kinetic energy of 540 eV in a longitudinal magnetic guide field of 5 mT. The shape of the beam was recorded with the MCP-detector and a CCD-camera. The resulting diameter of the primary beam amounts to $d^+_{\text{FWHM}} = 9.3\,mm$ (see figure 2).

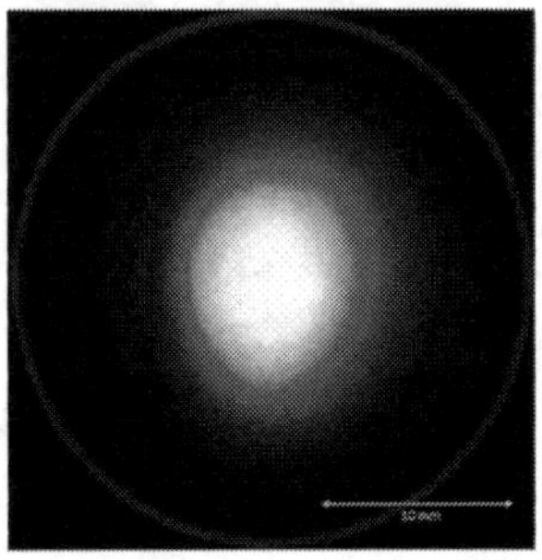

Figure 2. Beam shape at E_{kin}=540 eV in a longitudinal magnetic field of $B_{\|}$=5 mT with a diameter of d^+_{FWHM} =9.3(1.0) mm.

For the determination of the beam intensity, i.e. positron yield Y^+_{mod} the annihilation radiation at the Al target was detected and compared with a ^{22}Na calibration source.

$$\text{Y}^+_{\text{mod}} = \left(\frac{\text{I}^{\text{beam}}_{pp}}{\text{I}^{\text{cal}}_{pp}} \cdot \text{A} \cdot 0.9\right) \cdot (1-\text{r})^{-1} \cdot (1+\text{F}_{3\gamma}) \cdot (1+\text{F}_{\text{Pileup}}) \tag{1}$$

The number of 2γ annihilation events is calculated from the background-subtracted areas of the photopeaks for the beam $\text{I}^{\text{beam}}_{pp}$ and for the calibration source $\text{I}^{\text{cal}}_{pp}$. The positron yield of the ^{22}Na source is given by the branching ratio for positron emission of 0.9 times its activity $\text{A} = 2.22 \cdot 10^5$ Bq. The second term in equation 1 accounts for the fraction of back-scattered positrons which amounts to $\text{r} \approx 12\,\%$ for 540 eV positrons impinging on Al [5].

Ortho-positronium (o-Ps) emitted from the surface leads to a considerable contribution of 3γ annihilation events that is observed by a strongly asymmetric annihilation photopeak and higher event rate at γ-energies below 511 keV. Detailed analysis of the spectrum leads to the result that about $F_{3\gamma}$=0.466(7) of the intensity of the 511 keV photopeak corresponds to γ's of the 3γ-decay of o-Ps. Despite the application of Pb and Cu absorbers in front of the detector pile-up events are still observed. The last term in equation 1 takes into account that annihilation events are recorded as pile-ups with a fraction of F_{Pileup}=0.076(3) of the 511 keV photopeak.

Later the primary beam energy was set to E_{kin}=1 keV in order to increase the remoderation efficiency at the W(100) remoderator. Shape and yield of the primary beam are not expected to change signficantly by increasing its kinetic energy. The measured values of the positron yield Y^+_{mod} and the beam diameters d^+ for both the primary and the remoderated positron beam are summarized in Table 1.

4. Summary and outlook

After the successful replacement of the beamtube, NEPOMUC *upgrade* enables the continuation of a large variety of studies using the high-intensity positron beam as well as the development of new experiments. These experiments range from high-sensitive and elemental selective

	Positron beam	
	primary	**remoderated**
E_{kin} $[eV]$	540 (1000)	20
$B_{\parallel}$ $[mT]$	5.0	4.0
Y^{+}_{mod} $[s^{-1}]$	$1.14(7) \cdot 10^{9}$	$3.0(3) \cdot 10^{7}$
d^{+}_{FWHM} $[mm]$	9.3(1.0)	1.85(35)
d^{+}_{FWTM} $[mm]$	15.4(1.3)	4.8(7)

Table 1. Measured beam parameters at NEPOMUC *upgrade* for the primary moderated and the remoderated positron beam at the kinetic energy E_{kin} in the longitudinal magnetic guide field $B_{\parallel}$. Y^{+}_{mod} is the yield of (re)moderated positrons and d^{+} the beam diameter at full width of half (tenth) maximum (index FWHM and FWTM respectively; errors in parentheses). Recently, the primary beam energy was set to E_{kin}=1 keV.

investigations in surface physics over defect spectroscopy in thin films and solids to fundamental studies in atomic and plasma physics. At NEPOMUC *upgrade* the intensity of the mono-energetic positron beam was determined to $1.14(7) \cdot 10^{9}$ moderated positrons per second. Presently, the remoderated positron beam is used for most positron beam applications. In the upcoming beamtime it is planned to further improve the brightness and the intensity of the beam by implementing a automatic down-hill simplex procedure for adjusting the high number of magnetic field coils and electrostatic lenses.

Acknowledgments

The authors thank the group of G Dollinger at the UniBW München for their continuous support to operate PLEPS at NEPOMUC. Financial support provided for the improvement of the instruments PAES and CDBS (project nos. 05KI0WOB and 05K13WO1) within the integrated research initiative PosiMethod (05K2010) by the BMBF and for the experimental determination of the electronic structure by using 2D-ACAR (Transregional Collaborative Research Center TRR80) by the DFG is gratefully acknowledged.

References

[1] Hugenschmidt C, Schreckenbach K, Stadlbauer M and Straßer B 2005 *Nucl. Instr. Meth. A* **554**(1-3) 384–91
[2] Hugenschmidt C, Piochacz C, Reiner M and Schreckenbach K 2012 *New Journal of Physics* **14** 055027
[3] Hugenschmidt C, Ceeh H, Gigl T, Lippert F, Piochacz C, Pikart P, Reiner M, Weber J and Zimnik S 2013 *Journal of Physics: Conference Series* **443** 012079
[4] Piochacz C and Hugenschmidt C 2013 *Journal of Physics: Conference Series* **443** 012093
[5] Puska M and Nieminen R 1994 *Rev. Mod. Phys.* **66** 841–97

Positron beam facility at Kyoto University Research Reactor

Q Xu[1], K Sato[1], T Yoshiie[1], T Sano[1], H Kawabe[1], Y Nagai[2], K Nagumo[2], K Inoue[2], T Toyama[2], N Oshima[3], A Kinomura[3] and Y Shirai[4]

[1] Research Reactor Institute, Kyoto University, Kumatori-cho, Osaka 590-0494, Japan
[2] The Oarai Branch, Institute for Materials Research, Tohoku University, Ibaraki 311-1313, Japan
[3] National Institute of Advanced Industrial Science and Technology, Ibaraki 305-8568, Japan
[4] Department of Materials Science and Engineering, Graduate School of Engineering, Kyoto University, Kyoto 606-8501, Japan

E-mail: xu@rri.kyoto-u.ac.jp

Abstract. A positron beam facility is presently under construction at the Kyoto University Research Reactor (KUR), which is a light-water moderated tank-type reactor operated at a rated thermal power of 5 MW. A cadmium (Cd) – tungsten (W) source similar to that used in NEPOMUC was chosen in the KUR because Cd is very efficient at producing γ–rays when exposed to thermal neutron flux, and W is a widely used in converter and moderator materials. High-energy positrons are moderated by a W moderator with a mesh structure. Electrical lenses and a solenoid magnetic field are used to extract the moderated positrons and guide them to a platform outside of the reactor, respectively. Since Japan is an earthquake-prone country, a special attention is paid for the design of the in-pile positron source so as not to damage the reactor in the severe earthquake.

1. Introduction

Positron annihilation spectroscopy is a powerful tool for detecting vacancy-type defects in condensed matter. Monoenergetic positrons are widely used to study thin layers, surfaces and near surface effects in solids. There are three methods for the generation of positron beam [1]. β^+ active isotopes such as ^{22}Na are mainly employed as the positron source in slow positron beam facilities. The beam intensity depends on the radio activity of isotopes and the high beam intensity is not easy to be gotten because of the difficulty for obtaining high activity isotopes. The linac-based positron source can increase the positron beam intensity [2, 3]. Positrons are generated in a high-Z target, which is bombarded by electrons with energies of several tens of MeV. This system, however, is cost a lot for the construction of high energy and high intensity electron accelerator.

The in-pile positron source can provide a monoenergetic positron beam of high intensity, where positrons are generated by pair production of high energy of γ–rays [4, 5]. Though the beam is continuous one, it is possible to make high intensity pulse beam using beam buncher. The application of positron beam can be expanded with increasing the beam intensity. For example, the microstructural changes in deformed and irradiated materials can be dynamically measured. These factors motivated the development of a high-intensity positron beam facility at Kyoto University Research Reactor (KUR). In this paper, the design of a reactor-based slow positron beam facility is described, and the safety assessment of the in-pile positron source is also discussed because Japan is earthquake country.

2. Positron source and positron beam facility at KUR

In order to promote materials science research using positron beams, an in-pile positron source is now in production at the B-1 hole (20 cm in diameter) of KUR, which is a light-water moderated tank-type reactor operated at maximum thermal power of 5 MW. The cross sectional view of the reactor is shown in Figure 1 [6]. The thermal neutron flux and the γ–ray flux at the positron source position are about 1.5×10^{12} n/cm^{2}s and 10^5 Gy/h, respectively, at 5 MW. Figure 2 shows the draft of positron beam line and material irradiation chamber.

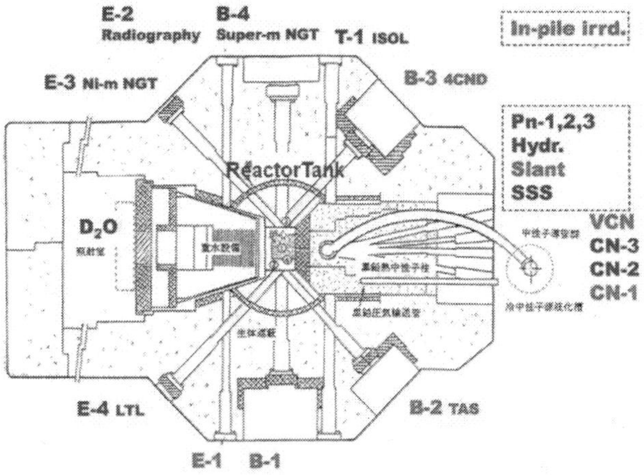

Figure 1. Cross sectional view of KUR.

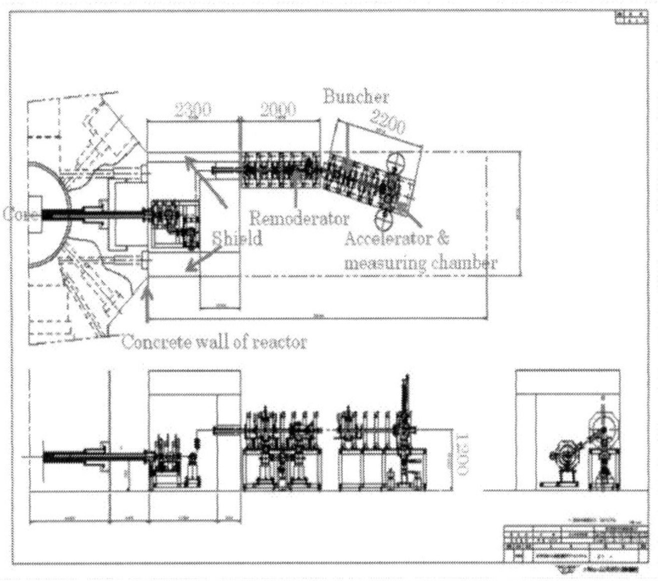

Figure 2. Draft of positron beam line and material irradiation chamber.

Figure 3 shows a schematic view of the in-pile positron source. A W foil capped with a Cd shroud is used to generate a positron beam. In addition to the γ–rays from the reactor core, high energy γ–rays are generated by the ^{113}Cd (n, γ) ^{114}Cd reaction, since the cross-section of thermal neutron capture in ^{113}Cd (the abundance of ^{113}Cd in natural Cd is 12.22%) is up to 26000 barns. On average, 2.3 photons with energies above 1.5 MeV are emitted per captured neutron [7]. The present design, a 1-mm-thick Cd cap covered by an Al plate is used. Though ^{113}Cd in natural Cd burns up by neutron capture, enough ^{113}Cd remains even after five years of KUR operation of 1500 h with an average power of 2 MW per year. According to reference 7, the optimal thickness for positron production in the W converter is about 0.2 – 0.8 g/cm^2, which corresponds to a W thickness of 0.1 – 0.4 mm. The peak energy of positrons produced by pair production is 800 keV. Well-annealed W foils with a mesh structure are used as a moderator to moderate the positrons with high energy. Figure 4 shows a depth profile of stopping probability for positrons with an energy of 800 keV. The depth profile was calculated by the following equation [8, 9].

$$P(x,E) = -\frac{d}{dx}\exp(-x^m / x_0^m) \qquad (1)$$

Here, $x_0=(\alpha/\rho)E^n$, ρ is the material density in g/cm^3, E is the incident positron energy in keV, and x is the distance from the surface in cm. We used the Makhovian parameters of α=4.0 μg/cm^2/keV$^{-1.6}$, m=2, and n=1.62 determined by Vehanen et al [9]. The calculation indicated that the optimum thickness of W foils is about 250 μm for positron moderation. But 0.2 mm in thickness of W foil is selected as converter shown in Figure 3 in consideration of heat generation of W induced by γ radiation. As described above, the thermal neutron flux at positron source is 1.5×10^{12} n/cm^{2}s at 5 MW. γ ray flux with energy higher than 1.022 MeV needed for pair production is estimated to be 8.4×10^{12} photos/cm^{2}s using the model calculation code (MCNP: Monte Carlo N-Particle-transport-calculation [10]). Though we have not estimated the slow positron flux yet, the γ ray flux is almost the same as that at the Hoger Onderwijs Reactor in the Reactor Institute Delft [8.8×10^{12} photos/cm^{2}s]. A positive voltage of several tens of eV is applied at the electrical lens to extract the moderated positrons. Then, the positron beam is magnetically guided in a solenoid field of 7 mT and compensation fields. In order to reduce the background of fast neutrons and γ–radiation from the reactor core, the positron beam passes two bends in shields consisting of polyethylene, concrete and lead block. The former two materials are to moderate the fast neutrons, and the latter is to shield the γ–radiation. After passing these bends, the beam is guided to samples in the irradiation chamber and accelerated up to 30 keV.

3. Safety assessments of positron beam facility

Earthquake resistant beam lines, especially in-pile positron sources, are required in KUR. Therefore, we design an in-pile positron source with a tough structure and ensure that the positron source would not damage the reactor in a severe earthquake. The beam line is fixed using more than twenty anchor bolts of 16 mm in diameter. Furthermore, in order to maintain the allowable stress of the Al alloy tube of the positron source, the temperature of the Al tube was designed to be below 473 K at in-pile position. Finally, to decrease the production of ^{41}Ar by absorption of neutrons (^{40}Ar (n, γ) ^{41}Ar) in air and reduce the background γ–radiation in the reactor room, the composition and configuration of the shield are designed based on calculations using the MVP2 code (Monte Carlo Codes for Neutron and Photon Transport [11]). Figure 5 shows positron beam facility at Kyoto University Research Reactor.

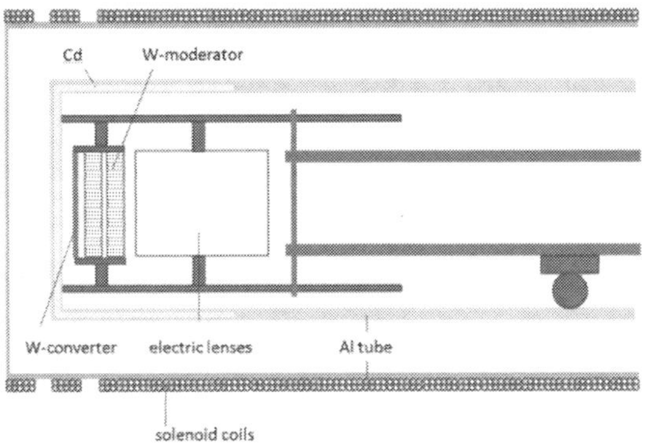

Figure 3. Schematic diagram of the in-pile positron source at KUR.

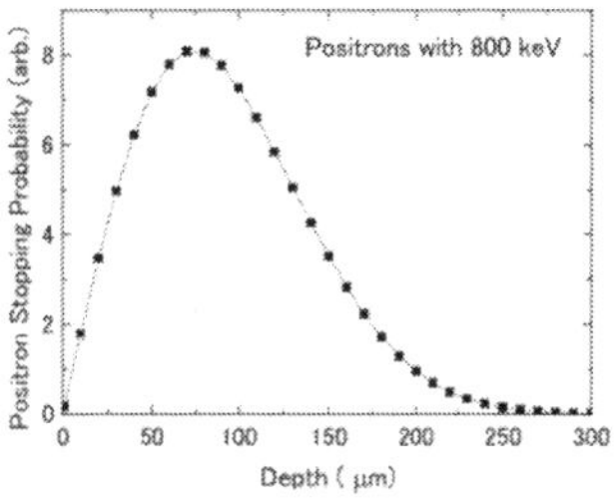

Figure 4. Depth profile of positron stopping probability for energy of 800 keV.

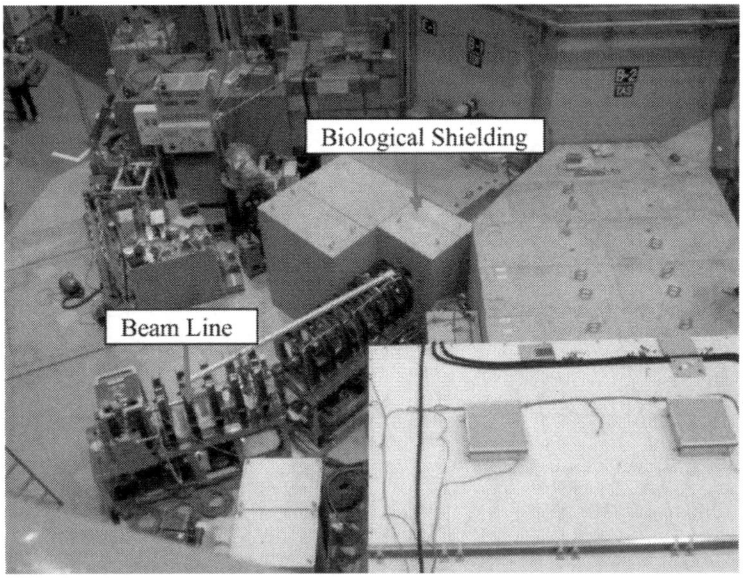

Figure 5. Positron beam line at Kyoto University Research Reactor

4. Summary and outlook

The main parts of the positron beam facility at KUR were made in the last year. After adjustment of the magnetic coils, the in-pile components will be installed in the B-1 hole at KUR by the end of 2013. Positron generation tests and coincidence Doppler broadening measurements will be carried out in early 2014. Then beam buncher is introduced for the production of pulse positron beam, and the positron annihilation lifetime and age-momentum correlation can be measured in the facility within two years.

Acknowledgments
This work includes the results of "research and development on degradation prediction of structural materials in nuclear reactors based on microstructural damage mechanisms" entrusted to Tohoku University by the Ministry of Education, Culture, Sports, Science, and Technology of Japan (MEXT). The authors are also deeply grateful to Japan Positron Science Society (Prof. T. Hyodo, Prof. Y. Nagashima, Dr. K. Wada and so on) for cooperation, and Prof. C. Hugenschmidt, Prof. H. Schut and Dr. C. Piochacz for helpful discussions.

References
[1] Lynn K G and Jacobsen F M 1994 *Hyperfine Interactions* **89** 19
[2] Kurihara T 2001 *Mat. Sci. For.* **363-365** 433
[3] Suzuki R, Ohdaira T, Mikado T, Uedono A, Ohgaki H, Yamazaki T and Tanigawa S 1997 *Mat. Sci. For.* **255-257** 114
[4] Veen A van, Labohm F, Schut H, Roode J de, Heijenga T and Mijnarends P E 1997 *Appl. Surf. Sci.* **116** 39
[5] Hugenschmidt C, Kögel G, Repper R, Schreckenbach K, Sperr P, Straβer B and Triftshäuser W 2002 *Nucl. Instr. Meth. Phys. Res. B* **192** 97
[6] http://www.rri.kyoto-u.ac.jp/facilities/kur.
[7] Krusche B and Schreckenbach K 1990 *Nucl. Instr. Meth. Phys. Res. A* **295** 155
[8] Makhov A.F. 1960 *Soviet Phys. Solid State* 2 1942
[9] Vehanen A., Saarinen K., Hautojarvi P., Huomo H. 1987 *Phys. Rev. B* **35** 4606
[10] Briesmeister J F et. al 2000 *LA-13709-M* Los Alamos National Laboratory
[11] Nagaya Y et. al 2005 *JAERI 1348* Japan Atomic Energy Research Institute

Positron bunching system for producing positronium clouds into vacuum

L Penasa[1], L Di Noto[1], M Bettonte[1], S Mariazzi[1], G Nebbia[2] and R S Brusa[1]

[1]*Dipartimento di Fisica, Università di Trento and TIFPA-INFN, Via Sommarive 14, I-38123 Povo, Trento, Italy*
[2]*INFN Padova, Via Marzolo 8, 35131 Padova, Italy*

E-mail: luca.penasa@unitn.it

Abstract. A magnetic transport line, a magnetic field terminator and a positron buncher were designed and built to focus low-energy positron pulses from a Surko-type accumulator on a porous target. The 25 electrodes buncher, which produces a parabolic potential, was designed to implant 5 ns positron bunches with a spot of 3 mm into a target held at cryogenic temperature. These pulses will be used to obtain cooled Ps clouds into vacuum for laser excitation in spectroscopy experiments. By using high-voltage fast switches and a proper mu-metal shield the requirement to form Ps in a free (magnetic and electric) field region was satisfied compatibly with the request of injecting positrons at energies of 5-9 keV. The optical design, the electrical circuitry of the buncher and the construction solutions of the whole apparatus will be presented and explained.

1. Introduction

A new apparatus was designed and built in order to process pulses of $10^7 - 10^8$ positrons previously stored in a Surko-type accumulator every 100 seconds. Re-bunching and focusing these pulses on a nanochannelled Si sample [1] for producing cooled Ps into vacuum is the task of the system [2, 3]. Ps in vacuum will be used for spectroscopy experiments such as the excitation of Ps in the n=3 state and from n=3 to Rydberg states [4] in a free field region.

The apparatus will be connected, through a cross and a valve, to the first section of the AEgIS (Antimatter Experiment: gravity, Interferometry, Spectroscopy) positron transfer line that links the positron accumulator to the AEgIS main magnets. The apparatus will run for Ps studies during the down time of the AEgIS experiment [5], ie. when the antiproton are not supplied for the antihydrogen production.

The accumulator integrates at its end a parabolic potential buncher stage that can deliver positron pulses with a time width lower than 8 ns at an energy from 50 to 400 eV. Having to cross the first magnetic transport section of the AEgIS transfer line (~ 60 cm), the positron pulses reach the entrance of the new apparatus with about 15 ns time duration. The designed optics will allow to re-bunch the pulses to less than 5 ns in a spot of ~3 mm at the target position.

In section 2 an overview of the optics is given; in section 3 the electronic and mechanical design of the buncher is shown and explained, while in the final section 4 results of the electronic tests are reported.

2. Transport and focussing optics

The optical system of the apparatus can be divided into three different stages (see figure 1).

The first stage will extract the positron pulses transported through the AEgIS transfer line by a field of 0.1 T and inject them into the buncher (second stage). It is made by two magnetic coils and two electrodes. The coils, positioned at the two sides of the entrance valve, extend the magnetic field of the transfer line to a μ-metal electrode. This last, due to its proper shape, works as a magnetic field terminator reducing it from 0.1 T to 0.02 T in 5 mm. The μ-metal electrode (-500 V) with the second one (-1500 V) and the buncher-electrodes (held at -600 V) produce an electric field that focalizes the positrons in the middle of the buncher itself.

The buncher, second stage, is composed by 25 electrodes so that the buncher length becomes longer than the positron cloud length (~ 20 cm). When all positrons enter the buncher, a parabolic potential (2 kV amplitude) is applied between the first and the last electrode, in order to compress the positron bunches in space and time. In order to keep the target at ground potential and to give the positrons the desired energy, the parabolic function is superimposed to a bias voltage of 5-9 kV . The parabolic potential and the bias voltage are obtained by a HV pulse (10 kV) of 20-30 ns FWHM and a rise time of 4 ns.

The third stage is a lens, formed by a last electrode and the target kept at ground, which focalizes the positrons on the target at the desired energy (5- 9 keV). In order to obtain a free electric field zone near the sample where Ps atoms are emitted, the last electrode is switched to ground potential in about 10 ns.

The apparatus (figure 2) will be shielded by a μ-metal box with double walls (1.5 mm thick) with a 1 cm gap in between. The magnetic field near the target region is expected to be < 2 gauss when the 5 T magnet of the AEgIS experiment is on.

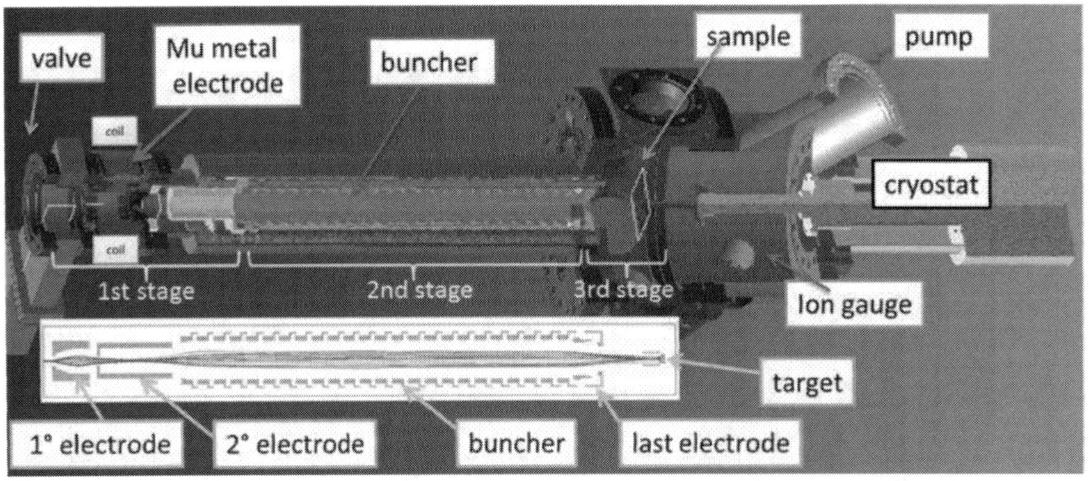

Figure 1. The apparatus layout and the simulation of the positrons trajectory. The simulation take into account the magnetic field distribution along the transfer line before the apparatus and both the static and the time varying potential of the optical system. Start positron pulse conditions at the output of the accumulator are: energy = 80eV ± 10%, duration = 5 ns. Final conditions on the target are: energy = 8 keV ± 10%, pulse duration = 5ns, spot diameter = 3 mm.

3. The buncher design

The design must take into account mechanical and electrical constraints. The buncher length is 40 cm, the electrodes have an outer diameter of 56 mm and are inserted in a chamber of 100 mm inner diameter. With a 20-30 ns FWHM pulse and 4 ns rise time the buncher behaves as a transmission line. The chamber and the buncher are like a coaxial cable with a peculiar inner conductor. The mechanical layout (resistors, chamber, electrodes, materials) establishes the characteristic parameters of the transmission line. In particular the design of the buncher takes into account the following parameters:

- the capacitance between electrodes must be as small as possible in order to have a high impedance for the HV pulse;
- the capacitance between the electrodes and the chamber must be small in order to have a high propagation velocity of the pulse;
- the ratio between the electrodes and the chamber diameters (ie. the outer conductor of the "coaxial line") must have the right value to match the buncher with the characteristic impedance of the system (Z_0=75 Ω);
- the resistors between electrodes must have a small inductance (L_r < 2-3 nH) to avoid electric oscillations;
- the total resistance of the buncher (R_b=16 Ohm) must sustain the 120 A current to reach a 2 kV voltage drop.

In order to produce the parabolic potential, the electrodes of the buncher are connected to each other by proper resistors in such a way that the series of voltage drops generates the desired function. All resistors were realized with a gold electro-deposition on a high resistivity silicon support (figure 3) and then soldered on a kapton-gold strip with indium (figure 4). Three of these strips are wired to the electrodes along the buncher (figure 5). They are placed at 120° to reduce the magnetic field due to the high 120-130 A current flowing through them.

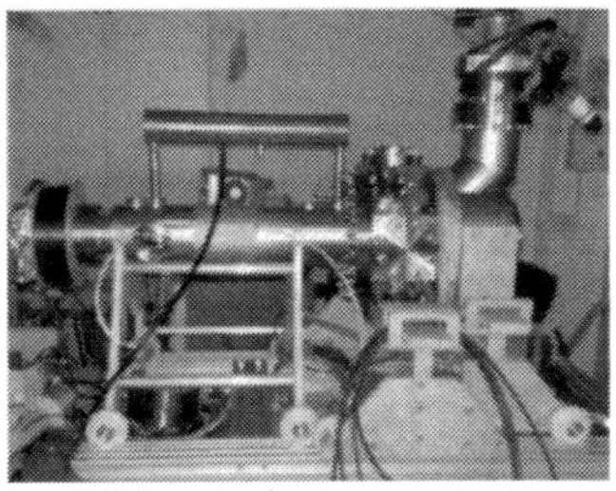

Figure 2. Apparatus

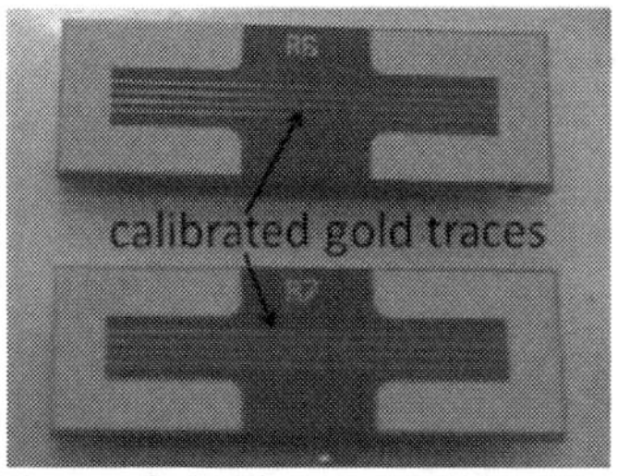

Figure 3. Resistors on Si support

Figure 4. Resistors soldered on kapton-gold strip

Figure 5. The buncher

It is important to notice that the superposition of the parabola and the bias is realized by using only one HV fast switch in a "voltage divider configuration", where an input resistor Ri and an output resistor Ro are used to change the bias voltage value and to match the load with the impedance of the system Z_0=75 Ω. Different values of Ro correspond to different values of the bias potential of the buncher and different values of the positron energy on target. To keep the buncher network matched, the sum of Ri, Ro and Rb (total resistance of the buncher), must be equal to the characteristic impedance of the circuit Z_0 (figure 6).

In order to supply the buncher with a pulse of 10-11 kV with 120-130 A current and a fast rise time the solution we decided to use a fast switch produced by Behlke (HTS-12-150-UF). Due to its maximum operating voltage of 12 kV it is not possible to place at the output of the switch a resistance equal to the characteristic impedance of our network. In this configuration the buncher-system must be electrically matched with particular care since the switch cannot absorb a wave eventually reflected and moreover multi-reflected waves could distort the bunching of positrons. In order to avoid the

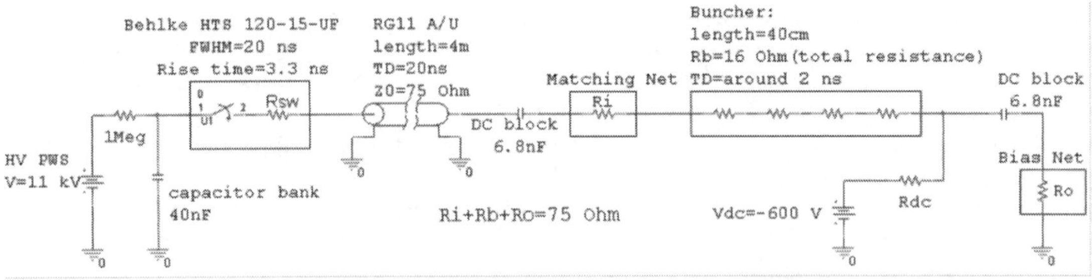

Figure 6. Electrical layout of the buncher system

problem of multiple reflections along the buncher a coaxial cable with Time Delay = 30 ns has been connected between the switch and the buncher.

To surge the high current, a HV capacitor bank of 40nF is placed before the switch. In order to verify the matching of the system, the shape of the pulses and the voltage drops between the resistors of the buncher, the electrical behavior of the system has been simulated with PSpice.

4. Results

Tests of the described system have been performed at different voltages setting Ri=0 Ω and Ro=59 Ω. To check the voltage drop from the first to the last electrode of the buncher a low voltage pulse has been used (figure 7). We found, as desired, a voltage drop of 22% from the input to the output voltage which corresponds, at the nominal working condition, to about 2 kV. A test has been performed at 10 kV (figure 8) in order to verify that the resistors of the buncher can sustain the nominal current of the system (around 120 A). The difference between the supply voltage and the voltage amplitude of the pulses is due to the internal resistance of the switch (Rs=10-15 Ω).

The next series of tests are now planned at CERN using the real positron bunches delivered by the Surko-trap of the AEgIS experiment.

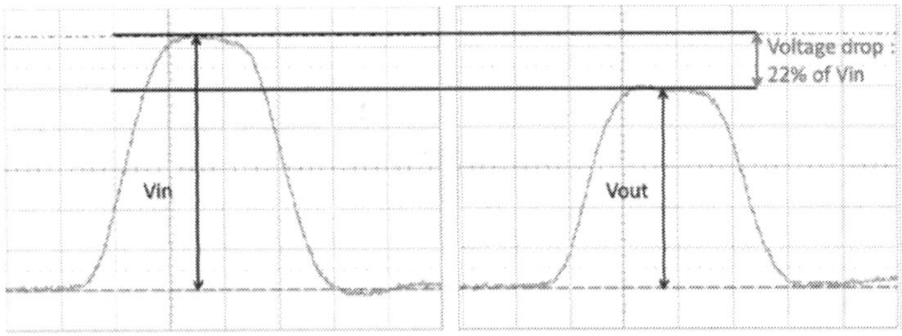

Figure 7. Test at low voltage. Vin is the voltage of the pulse measured on the first electrode of the buncher, Vout is the voltage measured on the last electrode.

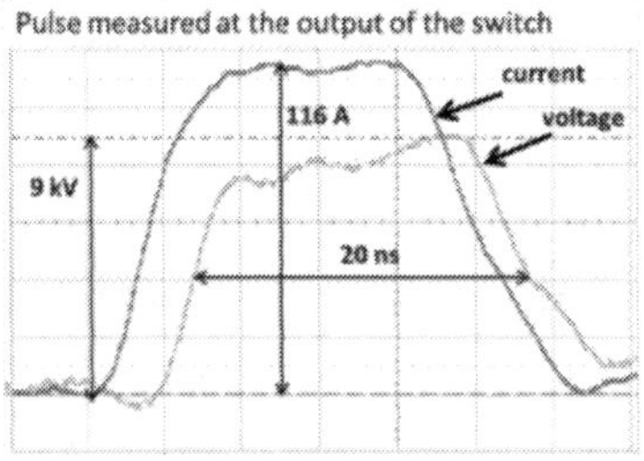

Figure 8. Test at 10 kV supply voltage

References

[1] Mariazzi S, Bettotti P, Larcheri S, Toniutti L and Brusa R S 2010 *Phys. Rev. B* **81** 235418

[2] Mariazzi S, Bettotti P and Brusa R S 2010 *Phys. Rev. Lett.* **104** 243401

[3] Brusa R S and Dupasquier A 2010 *Positronium emission and cooling*, in *Physics with many Positrons*, Dupasquier A, Mills A P jr and Brusa R S eds, (North Holland, Amsterdam) p 245

[4] Castelli F, Boscolo I, Cialdi S, Gianmarchi M, Comparat D 2008 *Phys. Rev. A* **78** 052512

[5] A. Kellerbauer et al. *Nucl. Instr. Meth. Phys. Res.* 2008 B 266, 351-356

Journal of Physics: Conference Series **505** (2014) 012032 doi:10.1088/1742-6596/505/1/012032

Positronbeam for μm Resolved Coincident Doppler Broadening Spectroscopy at NEPOMUC

T Gigl[1,2], C Piochacz[1,2], M Reiner[1,2] and C Hugenschmidt[1,2]
[1] Technische Universität München, Physik Department, Lehrstuhl E21, James-Franck-Straße, D-85748 Garching, Germany
[2] Technische Universität München, Forschungs-Neutronenquelle Heinz Maier-Leibnitz, Lichtenbergstraße 1, D-85748 Garching, Germany

E-mail: thomas.gigl@frm2.tum.de

Abstract.
A new high resolution coincidence Doppler spectrometer is under development at the positron beam facility NEPOMUC at the research neutron source FRM II. A Ni(100) re-moderation foil will be installed at the new spectrometer in transmission geometry for producing a positron beam with high brightness. Therefore, a magnetic lens for focusing the positrons onto the re-moderator and an electrostatic lens system for focusing the beam onto the sample have to be designed and simulated. The aim is to reach a beam diameter below 10 μm at the sample position for high resolution experiments. The current status of the project will be described and presented.

1. Introduction

Doppler broadening spectroscopy with positrons is a well-established non-destructive tool for the investigation of lattice defects in solids. Therefore, a (C)DB spectrometer is operated at the NEutron induced POsitron source MUniCh (NEPOMUC) which provides a high intensity positron beam of about $10^9\,e^+/s$. With the spectrometer it is possible to detect lattice defects which are caused for example after irradiation or by plastical deformation. Lateral resolved measurements are possible by focusing the positron beam and moving the sample with two stepper motors. For depth resolved measurements the sample can be biased up to -30 kV. The sample can be heated up to about 800 K in-situ for studying thermal annealing processes in thin layers [1], heavy deformed metal samples [2] or it can be cooled in order to study phase transitions i.e. in ionic liquids [3] and shallow trapping sides [4]. With CDBS the chemical surrounding of vacancies or small metallic segregations can be analysed [5]. In the near future investigations on technical alloys will be performed. The formation and growth of precipitations in aluminium based alloys in early state and the analysis of the elemental distribution in grains and grain boundaries of polycrystalline metallic alloys are of great interest. Therefore the beam diameter of the current instrument has to be reduced considerably.

2. Experimental set-up of the new CDB

At the current spectrometer a focused beam spot on the sample of about 300 μm is reached with apertures and an electrostatic lens system. For reducing the spot size of the beam at the sample

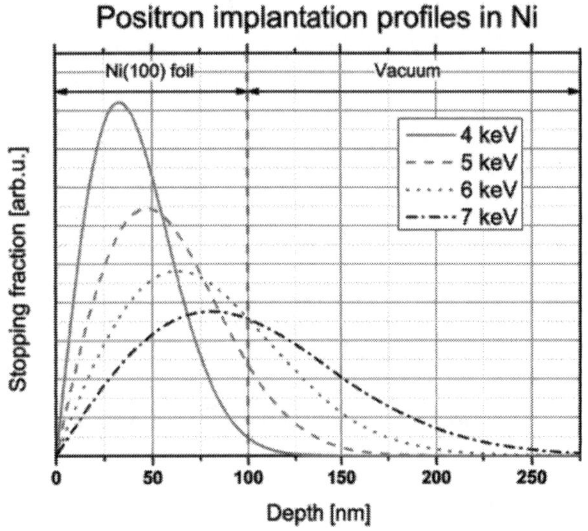

Figure 1: Makhovian profiles for positrons in nickel for different implantation energies. An energy of about 5 keV is expected to result in a maximum yield of re-emitted positrons.

position there are in principle two possibilities. The first one is to use an optimized electrostatic lens system in order to get a better focused beam. However, this method is technically limited due to geometric constraints mainly the minimal distance from the lens to the sample. The second possibility is to apply the technique of re-moderation to enhance the brightness of the beam. The incident positron beam, which occupies a certain phase space volume (PSV), is focused onto a moderator. Due to Liouville's theorem the transversal momentum must grow in order to keep the PSV constant and to compensate the reduction in real space. By entering the material the momentum of the positrons is reduced due to inelastic scattering and hence the PSV gets smaller. When materials with negative positron work function, i.e. tungsten or nickel, positrons which diffuse to the surface are emitted and can be accelerated with an electrostatic lens system. The result is a smaller positron beam with low energy spread and small angular divergence.

For the transmission re-moderator a Ni(100) foil with a thickness of 100 nm will be used. According to the Makhovian profile (see figure 1), a positron implantation energy of about 5 keV must be applied to maximize the amount of stopped positrons in the Ni(100) foil and to re-emit efficiently re-moderated positrons on the back side of the foil. At two research facilities in Japan this principle of re-moderating a positron beam was successfully realized [6]. The new design of the beam guidance to the CDB spectrometer at NEPOMUC is shown in figure 2. It consists of the magnetic beam guidance (beamline of NEPOMUC), an accelerator stage, the brightness enhancement system with the magnetic lens, the re-moderation foil and the electrostatic lens system. Two piezo sample positioners will be installed in the vacuum chamber enabling a high accuracy positioning of the sample. In order to achieve a minimal spot size of below 10 μm it is planned to mount an aperture for the incident beam and an additional magnetic single pole lens below the sample. For depth profiling the sample potential will be adjustable between about -5 kV and -30 kV. A heater enables the in-situ annealing and surface cleaning of the re-moderation foil. For monitoring the beam diameter at the site of the re-moderation foil a micro channel plate (MCP) can be inserted instead of the foil without breaking the vacuum.

3. Simulation of the new CDB design

At the current project status simulations for the new CDB spectrometer are performed with COMSOL multiphysics. Most important are the simulations for the magnetic lens (see figure 3) and the electrostatic focusing system below the re-moderation foil. The accelerated positrons

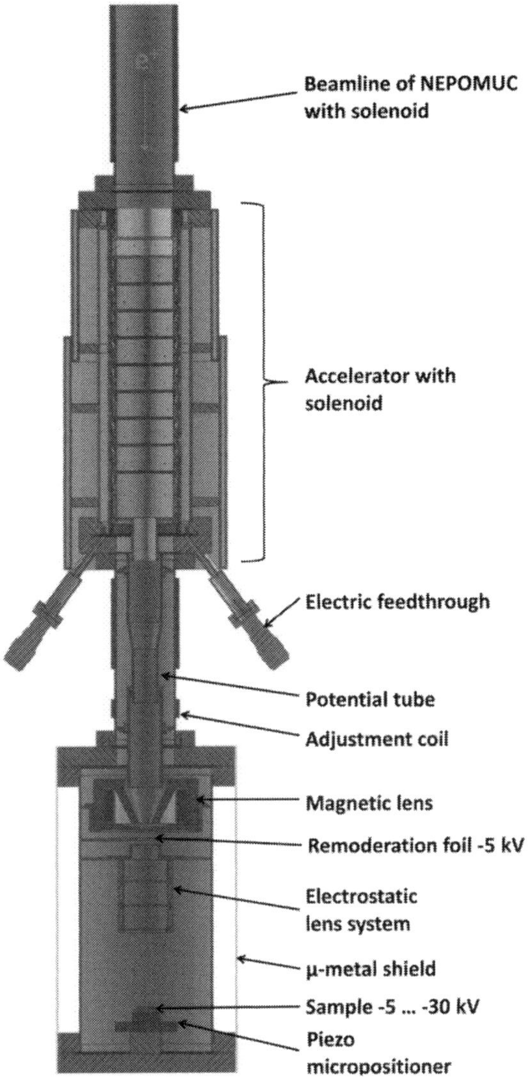

Figure 2: Design of the new CDB spectrometer at NEPOMUC. The incident positron beam is guided magnetically from the source to the instrument and is then accelerated to an implantation energy of 5 keV for a maximum re-moderation efficiency. The beam then passes an adjustment coil for illuminating the magnetic lens which focusses the beam onto the re-moderation foil. By three electrostatic lenses the re-moderated positrons are accelerated and formed to a beam which is focused onto the sample. For scanning the sample in 2D, two piezo micropositioners are mounted in the sample chamber. For depth dependent measurements the potential of the sample can be adjusted between -5 kV and -30 kV.

are focused onto the re-moderation foil by a magnetic field of about 500 Gs which is produced by two specially designed pole pieces of the magnetic lens. According to the simulations (see figure 3) the beam diameter at the position of the re-moderation foil is reduced by a factor of 10, i.e. starting with a beam diameter of 1.5 mm a beam spot of 150 µm is expected at the Ni-foil. The magnetic field at the Ni(100) foil is reduced by a factor of 100 due to the geometry of the pole pieces. They will be further improved so that the magnetic field will be close to zero at the position of the re-moderation foil and hence cannot affect the electrostatic focusing onto the sample. Simulations of the electrostatic lens system are in progress. It will consist of three or four electrostatic electrodes (see figure 2) to extract the positrons from the re-moderation foil and generate a beam which is focused onto the sample. After additional focusing with a magnetic single pole lens below the sample and additional apertures a reduction in diameter of a factor of 20 is estimated to be achievable leading to a diameter $< 10\,\mu m$.

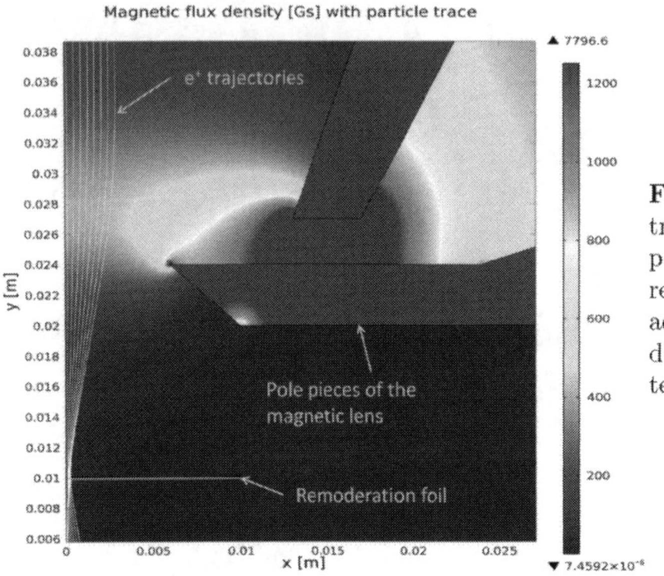

Figure 3: Simulation of the positron trajectories within the magnetic lens performed with COMSOL. The current and the pole pieces of the lens are adjusted for a focused beam spot at a distance of 1 cm and the beam diameter is reduced by a factor of 10.

4. Outlook

For the new design of the CDB spectrometer at NEPOMUC first simulations for the design of the magnetic lens have been made. Further simulations for the electrostatic lens system are in progress. The simulations deliver promising results of the beam spot diameter at the sample position but they have to be further adapted to the beam parameters of the re-moderated beam of NEPOMUC. At the end of 2013 and the beginning of 2014 the construction of the UHV components and the transmission re-moderator is planned. First experiments for testing the components will be performed at the open beam port of NEPOMUC in 2014. The final setting up of the new CDB spectrometer is planned to be in 2015.

Acknowledgments

Financial support provided for the improvement of the instruments PAES and CDBS (project nos. 05KI0WOB and 05K13WO1) by the BMBF is gratefully acknowledged.

References

[1] Reiner M, Pikart P and Hugenschmidt C 2012 *Physics Procedia* in press
[2] Oberdorfer B, Steyskal E M, Sprengel W, Puff W, Pikart P, Hugenschmidt C, Zehetbauer M, Pippan R and Würschum R 2010 *Phys. Rev. Lett.* **105**(14) 146101
[3] Pikart P, Hugenschmidt C and Schreckenbach K 2009 *phys. stat. sol. (c)* **6** 2487–9
[4] Stadlbauer M, Hugenschmidt C, Schreckenbach K and Böni P 2007 *Phys. Rev. B* **76** 174104
[5] Gilles R, Hofmann M, Johnson F, Gao Y, Mukherji D, Hugenschmidt C and Pikart P 2011 *Journal of Alloys and Compounds* **509** 195 – 9
[6] Oshima N, Suzuki R, Ohdaira T, Kinomura A, Narumi T, Uedono A and Fujinami M 2008 *J. Appl. Phys.* **103** 094916

Journal of Physics: Conference Series **505** (2014) 012033 doi:10.1088/1742-6596/505/1/012033

Spin polarizations of positron beams generated using electrostatic and magnetic transportation systems with ^{68}Ge and ^{22}Na sources

M Maekawa, Y Fukaya, H Zhang, H Li and A Kawasuso
Advanced Science Research Center, Japan Atomic Energy Agency
Watanuki 1233, Takasaki, Gunma Japan

E-mail: maekawa.masaki@jaea.go.jp

Abstract. Spin polarizations of electrostatic positron beams generated using ^{68}Ge and ^{22}Na sources with tungsten moderators were 47 % and 30 %, respectively. A comparable spin polarization (27 %) was obtained with much reduced beam diameter (0.5 mm), when electromagnetic lenses, a ^{22}Na source and a tungsten moderator were used. Replacing the tungsten moderator with a solid neon moderator in this system, the beam flux was significantly enhanced with maintaining the spin polarization. The Doppler broadening of annihilation radiation spectra of polycrystalline Fe measured using the above beams showed clear asymmetry upon field reversal.

1. Introduction

The pair-annihilation probability of positron and electron charges depend on their relative spin directions. Therefore, if both positrons and electrons are spin-polarized, positron-electron momentum distribution shows asymmetry upon spin reversal. We proposed that the Doppler broadening of annihilation radiation (DBAR) technique with spin-polarized positrons can be used for studying ferromagnetic band structures [1, 2]. Such spin-polarized positron annihilation spectroscopy (SP-PAS) will be a potential tool in current spintronics study. Considering the fact that novel spin-phenomena occur at surfaces and interfaces and in thin films, a spin-polarized positron beam needs to be developed [3-5]. To obtain a highly spin-polarized positron beam, it is important to use β^+-emitters with higher energy endpoints (E_{max}). We have demonstrated the spin polarization of an electrostatic positron beam using a ^{68}Ge isotope (E_{max} = 1.9 MeV) was 47 % [6, 7]. In this study, we first compare the spin polarizations of positron beams using ^{68}Ge and ^{22}Na sources in electrostatic transportation systems. We also examined the spin-polarization of the positron beam generated by ^{22}Na source, magnetic lenses and tungsten/solid neon moderators.

2. Positron beam apparatus

Four types of positron beam apparatus were constructed: (i) ^{68}Ge source + tungsten moderator + electrostatic transportation, (ii) ^{22}Na source + tungsten moderator + electrostatic transportation, (iii) ^{22}Na source + tungsten moderator + electromagnetic transportation and (iv) ^{22}Na source + solid neon moderator[8, 9] + electromagnetic transportation. These are schematically summarized in figure 1 and 2. The source strengths and moderator details are also listed in table 1.

The longitudinal spin polarizations of the above positron beams were measured from the magnetic quenching of the positronium in fused silica, which is simpler method than measuring the positronium decay rate[7, 10, 11]. The strength and direction of the magnetic field were controlled by changing the permanent magnets. Furthermore, the Doppler broadening of the annihilation radiation spectra of polycrystalline Fe were also measured.

3. Results and discussion

The magnetic field dependences of S parameters from the fused silica and estimated spin polarizations

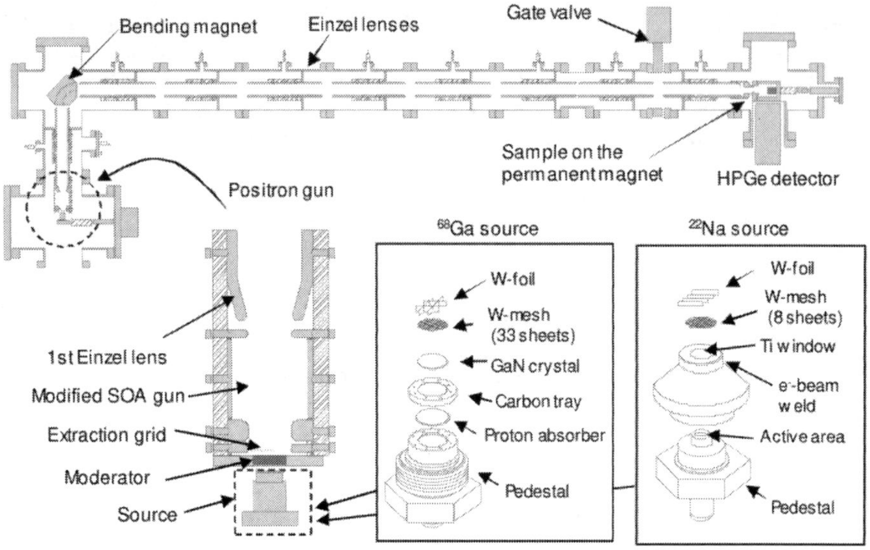

Figure 1. Schematics of the apparatus using ^{68}Ge and ^{22}Na source with the tungsten moderator and electrostatic lenses : type (i) and (ii).

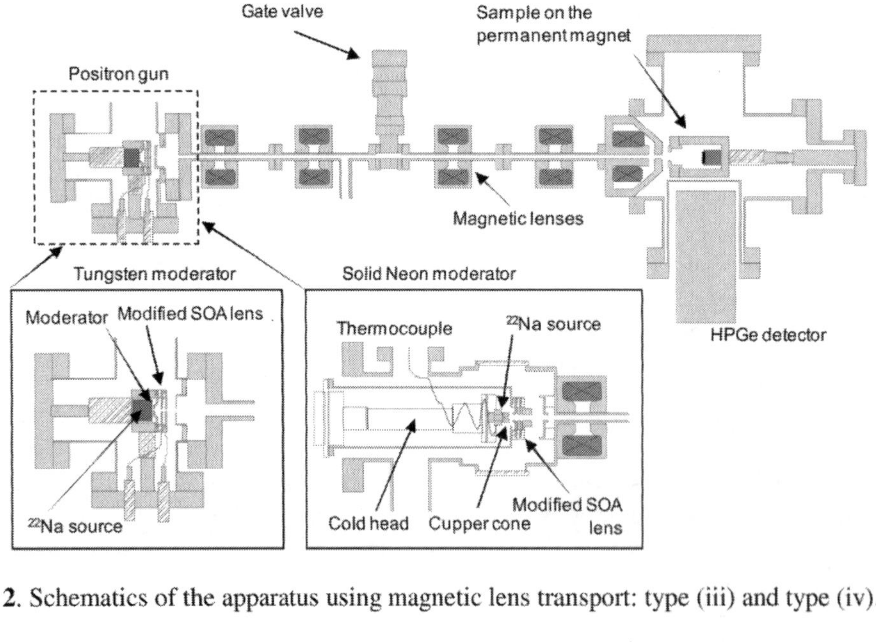

Figure 2. Schematics of the apparatus using magnetic lens transport: type (iii) and type (iv).

for each beam are shown in figure. 3. The spin polarization of the type (i) positron beam is estimated to be 47 ± 8 %, which is in good agreement with the theoretical calculation. Details of this analysis is described in the previous paper [7]. This value is ~20 % higher than that of the type (ii) ^{22}Na-based positron beam (30 ± 3 %). From this result, the ^{68}Ge source is superior to the ^{22}Na source.

In the case of the electromagnetic beam transport, spin polarization is expected to be lost due to the fringing magnetic field, which is across to the beam trajectory. However, spin polarizations of type (iii) and (iv) beams are 28 ± 4 % and 27 ± 4 %, respectively, that are comparable to the case of electrostatic beam (type (ii)). Probably, since the directions of the positron spin and magnetic fields of magnetic lenses are parallel, the fringing fields are relatively weak near the beam axis. In the electromagnetic beam systems, the beam diameter is better focused (0.5 mm in diameter) as compared to the electrostatic systems (2~5 mm). Furthermore, when solid neon moderator was used (type (iv)), the positron beam intensity is much higher than in the case of tungsten moderator. Thus, in the electromagnetic system, without losing longitudinal spin polarization, much better positron beam is obtained.

Figure 4 shows the differential DBAR spectra of a well-annealed Fe sample in a magnetic field. This differential spectra $[N_\uparrow(p)-N_\downarrow(p)]$ was obtained by altering the field polarity. The subscript ↑ or ↓ indicates whether the positron polarization and the magnetic field direction was parallel or anti-parallel. The finite differential intensity means that there is a field-reversal asymmetry, which arises from the enhanced annihilation between the spin-up positrons and spin-down 3d unpaired electrons [12,13]. The differential amplitude of (b) is almost 2.3 times higher than that of (a). This is because the differential amplitude of the differential DBAR spectra is proportional to the sample magnetization and the positron spin polarization[1] (0.91T/0.55T × 47%/30% ≅ 2.5). This result also shows that our positron beam was sufficiently spin-polarized.

4. Summary

From the comparison between two electrostatic positron beams using ^{68}Ge and ^{22}Na sources, it is concluded that sources with higher energy endpoints are preferred in generating highly spin-polarized positron beams. Electromagnetic transportation is suitable to generate well focused and longitudinally

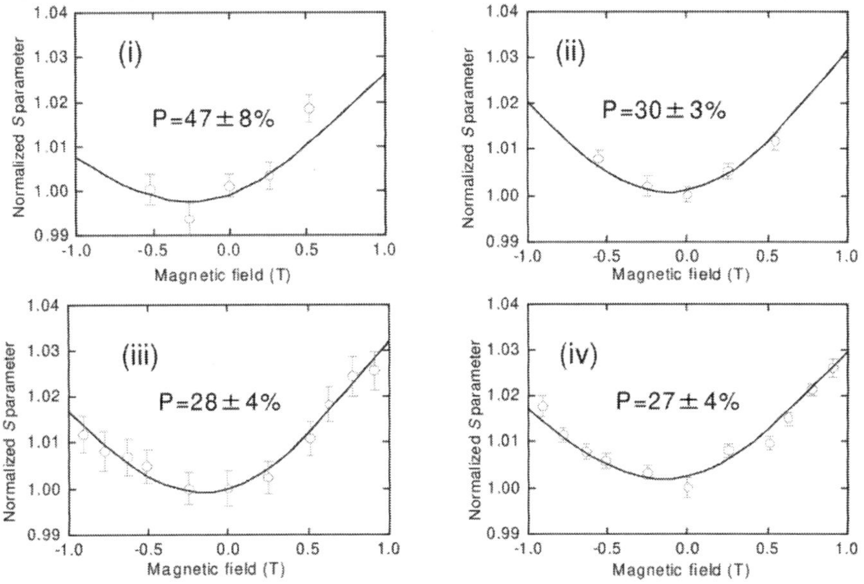

Figure 3. The magnetic field dependences of S parameters from a fused silica obtained using the slow positron beams of the apparatus type (i)~(iv). Theoretical curves of best fitting and estimated spin polarizations are also shown.

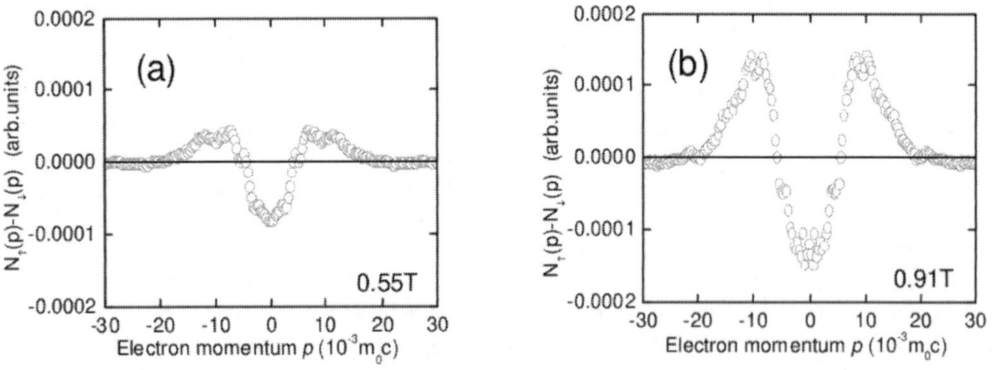

Figure 4. Differential DBAR spectrum of the well-annealed Fe sample in a magnetic field. (a) and (b) were obtained using type (i) and (iv) apparatus, respectively.

Table 1. Positron beams and their spin polarization.

Type	Source	Moderator	Transpotation	Flux (e^+/s)	Spin polarization (%)
(i)	^{68}Ge (250 MBq)	Tungsten mesh (33 sheets)	Electrostatic	~2×10^3	47 ± 8
(ii)	^{22}Na (20 MBq)	Tungsten mesh (8 sheets)	Electrostatic	~2×10^3	30 ± 3
(iii)	^{22}Na (220 MBq)	Tungsten mesh (16 sheets)	Magnetic	~1×10^3	28 ± 4
(iv)	^{22}Na (88 MBq)	Solid neon	Magnetic	~8×10^4	27 ± 4

spin-polarized positron beam. Positron beam generated using solid neon moderator has a comparable spin polarization to that using tungsten moderator.

References

[1] Kawasuso A, Maekawa M, Fukaya Y, Yabuuchi A and Mochizuki I 2011 *Phys. Rev.* **B 83** 100406.
[2] Kawasuso A, Maekawa M, Fukaya Y, Yabuuchi A, Mochizuki I 2011 *Phys. Rev.* **B 85** 024417.
[3] Gidley D W, Köymen A R, Capehart T W 1982 *Phys. Rev. Lett.* **49** 1779-1783.
[4] Coleman P G and Kallis A 2011 *J. Phys. Conf. Ser.* **262** 012016.
[5] Kawasuso A and Maekawa M 2008 *Appl. Surf. Sci.* **255** 108-110.
[6] M Maekawa, Y Fukaya, Y Yabuuchi and A Kawasuso 2011 *J. Phys. Conf. Ser.* **262** 012035.
[7] Maekawa M, Y Fukaya, Y Yabuuchi A, Mochizuki I and Kawasuso A, 2013 *Nucl. Instr. Methods* **B 308** 9-14.
[8] Mills Jr A P and Gullikson E M 1986 *Appl. Phys. Lett.* **49** 1121.
[9] Maekawa M and Kawasuso A 2008 *Appl. Surf. Sci.* **255** 39–41.
[10] Nagai Y, Nagashima Y, Kim J, Itoh Y and Hyodo T 2000 *Nucl. Instr. Methods* **B171** 199-203.
[11] NagashimaY and Hyodo T 1990 *Phys. Rev.* **B 41** 3937–3941.
[12] Mihalisin T W and Parks R D 1966 *Phys. Lett.* **21** 610–611.
[13] Genoud P, Singh A K, Manuel A A, Jarlborg T, Walker E, Peter M and Weller M 1988 *J. Phys.* **F 18** 1933–1947.

Tomographic Positron Annihilation Lifetime Spectroscopy

A Wagner[1], W Anwand[1], M Butterling[1], F Fiedler[1], F Fritz[1,2], M Kempe[1,2] and T E Cowan[1,2]
[1]Institute of Radiation Physics, Helmholtz-Zentrum Dresden-Rossendorf, Bautzner Landstr. 400, 01314 Dresden, Germany
[2]Technische Universität Dresden, 01062 Dresden, Germany

E-mail: a.wagner@hzdr.de

Abstract. Positron annihilation lifetime spectroscopy serves as a perfect tool for studies of open-volume defects in solid materials such as vacancies, vacancy agglomerates, and dislocations. Moreover, structures in porous media can be investigated ranging from 0.3 nm to 30 nm employing the variation of the Positronium lifetime with the pore size. While lifetime measurements close to the material's surface can be performed at positron-beam installations bulk materials, fluids, bio-materials or composite structures cannot or only destructively accessed by positron beams. Targeting those problems, a new method of non-destructive positron annihilation lifetime spectroscopy has been developed which features even a 3-dimensional tomographic reconstruction of the spatial lifetime distribution. A beam of intense bremsstrahlung is provided by the superconducting electron linear accelerator ELBE (Electron Linear Accelerator with high Brilliance and low Emittance) at Helmholtz-Zentrum Dresden-Rossendorf. Since the generation of bremsstrahlung and the transport to the sample preserves the sharp timing of the electron beam, positrons generated inside the entire sample volume by pair production feature a sharp start time stamp for lifetime studies. In addition to the existing technique of in-situ production of positrons inside large (cm^3) bulk samples using high-energy photons up to 16 MeV from bremsstrahlung production, granular position-sensitive photon detectors have been employed. The detector system will be described and results for experiments using samples with increasing complexity will be presented. The Lu_2SiO_5:Ce scintillation crystals allow resolving the total energy to 5.1 % (root-mean-square, RMS) and the annihilation lifetime to 225 ps (RMS). 3-dimensional annihilation lifetime maps have been created in an offline-analysis employing well-known techniques from PET.

1. Introduction

Positron Annihilation Lifetime Spectroscopy (PALS) allows for studies of crystal lattice defects on the nanometer scale and at low defect concentrations, open volumes in polymers, porosity and others. Several techniques have been developed over the last decades which make use of either kinematical observables of the annihilation radiation of positrons with electrons from the sample materials and of the annihilation lifetime of positrons after injection into the sample material. Especially, long positron annihilation lifetimes beyond a few ns are caused by voids and porous structures inside the sample which may be of high relevance for material failure and/or structural failure. In the following we will discuss a new method which allows correlating positron annihilation lifetime studies (PALS) with a three-dimensional tomographic analysis of a bulk sample. The well-established PALS technique makes use of radioactive positron sources like ^{22}Na where the positron emission is accompanied by electromagnetic transitions from excited states in the daughter nuclei. Because photon emission is prompt with respect to the positron emission the time resolution of the additional photon detector adds to the overall achieved accuracy in lifetime measurements. This additional contribution can in turn be avoided using an accelerator-based source of positrons which furthermore offers the advantage of possible adjustments of the source strength and the time structure. We will describe in the following how a versatile source of positrons using a superconducting electron accelerator has been realized and used in position-resolved positron annihilation lifetime spectroscopy.

2. Positron production using a superconducting electron linear accelerator

The Helmholtz-Zentrum Dresden-Rossendorf operates the superconducting electron accelerator ELBE [1] which delivers energies up to 40 MeV and average beam currents up to 1.6 mA. It acts as a source of secondary radiation like coherent infrared light from free-electron lasers, THz radiation from undulators and dipole magnets, photo-neutrons produced inside a liquid-lead target, bremsstrahlung, and positrons. The layout of the facility and the various end stations are shown in Fig. 1. The superconducting technology allows adjustment of the electron beam time structure allowing versatile time-of-flight experiments (neutrons) or lifetime experiments as described below. The micro-pulse repetition rate can be selected as 2^n divisors of 26 MHz with $n = 0...8$. Typically, the micro-pulse repetition rate for positron annihilation lifetime experiments is chosen to be 26 MHz or 13 MHz with micro-pulse intervals of 38 ns or 77 ns, respectively. The micro-pulse width has been measured to be less than 5 ps using electro-optical sampling [2].

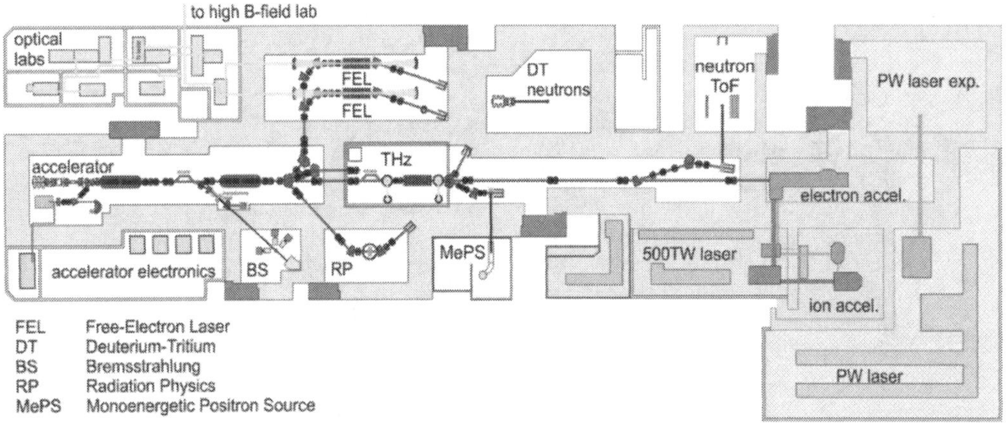

Figure 1. Layout of the ELBE centre for high power radiation sources. The overall width is about 100 m. Experiments discussed here took place at the bremsstrahlung facility labelled BS. Colour-shaded areas indicate high-power laser setups which are currently being developed for electron beam-laser interactions [3].

3. Tomographic Positron Annihilation Lifetime Spectroscopy

The gamma-induced positron annihilation lifetime facility [4, 5] has been extended by a set of position-sensitive photon detectors which will allow reconstructing a three-dimensional image of the distribution of positron lifetimes inside bulk samples [6]. Two pixelated photon detectors each made from 13 × 13 crystals of Lu_2SiO_5 of 4 × 4 × 20 mm^3 volume [7] have been set. Each detector is equipped with 4 photomultiplier tubes and dedicated preamplifier and discriminator electronics have been developed in-house in order to achieve a timing resolution of 530 ps (FWHM) for the combined mean-time of all channels. The photon energy deposition is calculated using the sum of all four charge-integrated signals and individually calibrated crystal responses. The obtained energy resolution

Journal of Physics: Conference Series **505** (2014) 012034 doi:10.1088/1742-6596/505/1/012034

is 11.4% (FWHM) at 511 keV photon energy for both detectors, respectively. Signal partitioning between the four photomultiplier tubes of one detector allows identification of the crystal in which the photon has interacted. Figure 2 shows a sketch of the setup with both detectors fixed perpendicular to the direction of the incoming photon beam.

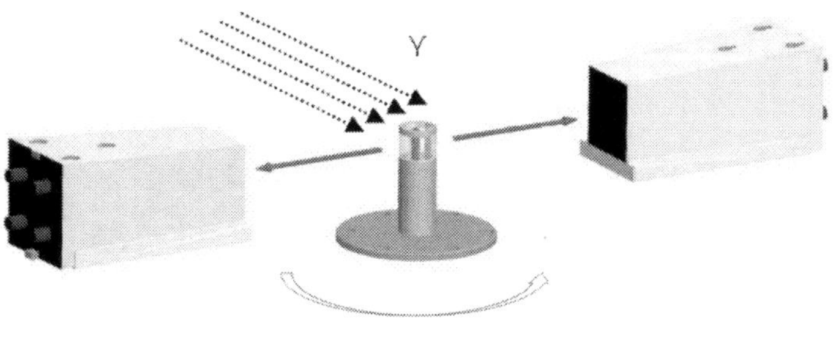

Figure 2. Sketch of the positron annihilation lifetime tomography setup. The bremsstrahlung beam hits the sample which is mounted on a rotational stage allowing for 3D image reconstruction. Two pixelated Lu_2SiO_5 detectors detect both annihilation quanta in coincidence.

In the case of two single-pixel events between the two detectors the line-of-response is calculated as the 3-dimensional connection between the two pixel positions. The simplest image reconstruction is performed for a two-dimensional distribution which is reconstructed by calculation the point of intersection between the line-of-response and a plane centred in between both detectors.

The system has been tested with a two-dimensional structure made from Si/SiO_2 which had not been rotated during the experiment. Equal-sized pieces of monocrystalline Silicon and microscope slides of dimensions 12.5 × 25 × 0.8 mm^3, see Fig. 3, have been fixed in between two thin-walled Kapton sheets and mounted parallel to the beam direction. Fig. 4 shows the projected images derived from correlated events between both detectors. The annihilation lifetime has been derived from the time difference between the mean time of all 8 photomultiplier signals and the radio-frequency of the accelerator.

Figure 3. 2D target consisting of Si wafer material (dark) and microscopy slides (light) wrapped into Kapton foils.

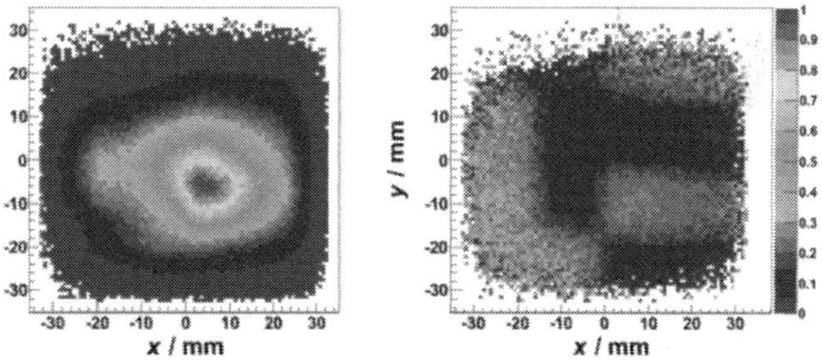

Figure 4. Two-dimensional distribution of two-photon annihilation integrated over all positron annihilation lifetimes (left) and the ratio of intensities gated for annihilation lifetimes in excess of 225 ps (1 σ) by all lifetimes (right).

The lifetime-integrated distribution shows no distinct features except an enhancement in the centre of the image due to the increased solid angle for correlated detection. Gating on positron lifetimes in excess of 225 ps (1 σ of the timing resolution) clearly discriminates for regions inside the sample with enhanced formation of o-Ps, namely SiO_2. Both materials have been selected in order not to emphasize areas with different pair production yields. The timing resolution has not been corrected for the inaccuracy of time-zero in the direction of the beam (x-axis in Fig. 4) which amounts to about 48 ps (RMS). Taking the mean timing of both detectors as the annihilation lifetime the inaccuracy due to different positron production sites perpendicular to the beam cancels. From Fig. 4 a lateral position resolution of 2.8 mm has been obtained by fitting one of the boundaries with the Gauss error function. Two three-dimensionally structured examples have been employed for further studies. One sample consists of a 25 mm diameter PTFE (Teflon) cylinder with embedded slabs made from Copper, Iron, and Aluminium having the same volume ($12\ mm^2 \times 25$ mm) but different geometrical shapes, see Fig. 5 a).

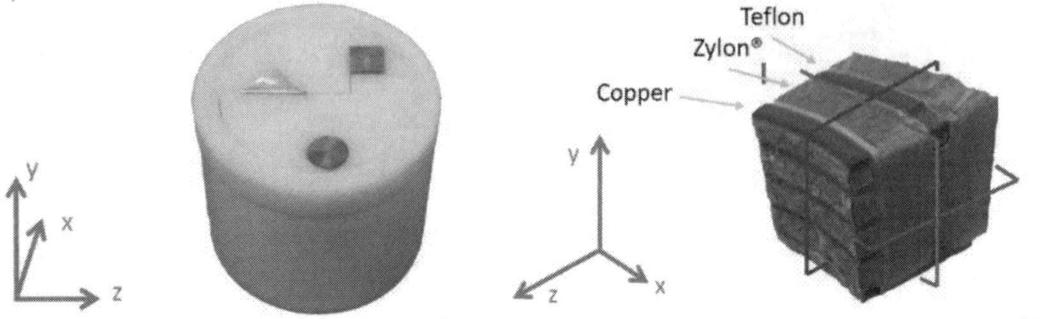

Figure 5. Two of the samples studied using tomographic positron annihilation lifetime spectroscopy. Picture a) shows a PTFE cylinder of 2.5 cm diameter with inserts made of Al, Cu, and Fe with different shapes. Picture b) shows a cut-out part of a high-field magnet coil which is composed of copper wires and reinforcing materials consisting of PTFE (Teflon) and Poly(p-phenylen-2,6-benzobisoxazol) (Zylon®). Lines indicate the planes of intersection used in the analysis.

A second sample has been cut from a high-field magnet coil resembling a 3-D structure composed of copper wires, reinforcing Zylon®, and again Teflon, see Fig. 5 b). The structure is manufactured with the aim to withstand phases of high mechanical pressure during ms-long periods where magnetic field strengths of up to 91.4 T are generated inside the coils with peak currents reaching 100 kA from a capacitor bank with 50 MJ stored energy [8].

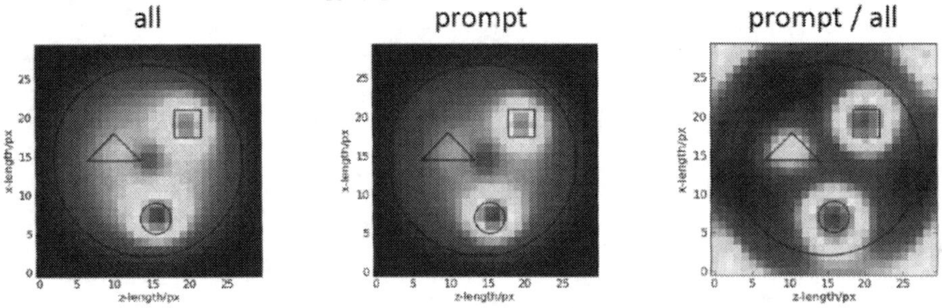

Figure 6. Reconstructed distribution of annihilation events for different regions of annihilation lifetimes after thirty iterations of the MLEM algorithm. Left picture shows all measured annihilations while the central picture is gated on a time interval between −1 ns to 2 ns (prompt). The right picture shows the ratio between prompt and all events discriminating for short annihilation lifetimes. The boundaries of the sample materials are indicated as thin lines.

In order to obtain a reconstructed distribution of annihilation events, the Maximum Likelihood Expectation Maximization [9] (MLEM) has been employed. The algorithm allows reconstructing iteratively a 3-D distribution of events from the set of all measured annihilation and the system matrix which comprises all possible combinations of detector pixels (169 × 169), rotation angles (180), image voxels (30 × 30 × 30) resulting in an overall size of 138 billion matrix cells. Two reconstructed distributions which are obtained from full (Fig. 6, left) and gated on only prompt annihilation events (Fig. 6, centre) are again used to generate a contrast enhanced annihilation-lifetime sensitive source distribution shown in Fig. 6, right side.

As the last case shown here, we study with the same procedure as described above the "real-world" sample shown in Fig. 5 b). While having a much more complicated internal structure and a variety of different materials with significant differences in positron annihilation lifetimes the procedure as described above still delivers contrast-enhanced images of the annihilation lifetime distributions after convergence of the MLEM algorithm. Fig. 7 shows cuts in the z, y, and x-planes. Here, the ratio between the prompt and the full lifetime spectrum reveals structures with rather long positron annihilation lifetimes indicating open (porous) regions of the structure.

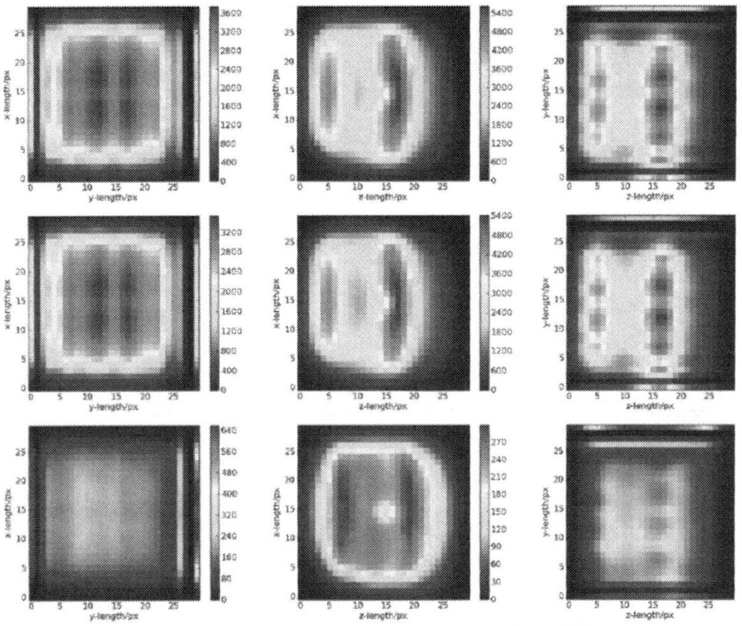

Figure 7. Reconstructed distribution of annihilation events for different regions of annihilation lifetimes after convergence of the MLEM algorithm for the sample shown in Fig. 5 b). The three columns represent different intersections of the three-dimensional reconstruction. The three rows are gates on the integral lifetime (top row), gated on prompt emission (middle row), and the ratio between prompt and all events (lower row).

As the last setup in the analysis a complementary approach for a more quantitative analysis can be done which aims at determining the positron annihilation lifetime distribution for a selected voxel. Lifetime distributions for all voxels (here 30^3) have been created by summing-up all list-mode events weighted by the intersection probability of each line-of-response for all rotational angles measured. Two selected lifetime distributions obtained in this way are shown in Fig. 7. It should be noted that also contributions from all voxels along the line-of-response contribute to the lifetime distribution; it is only that for a given voxel the statistical weight is enhanced.

Both distributions show a long lifetime component with a component of about 1.5 ns lifetime constant which is common for materials with significant positronium production such as polymers or porous structures. Clearly, both distributions show a significantly different intensity of the long lifetime component.

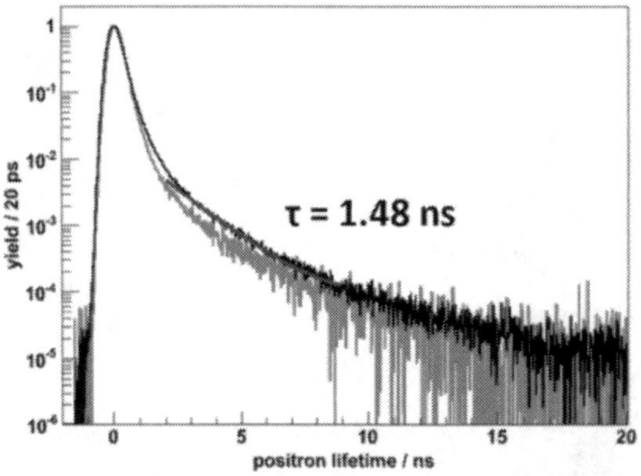

Figure 8. Positron annihilation lifetime distribution for two selected voxels from the reconstruction shown in Fig. 7. One of the distributions is parameterized by an exponential decay with 1.48 ns lifetime.

The newly developed system complements earlier developments of a positron annihilation microprobe which enabled high-resolution defect analysis at surfaces [10] or a high-energy positron beam for PALS studies in bulk material [11]. Further developments of the presented system aim at improving the position resolution by using smaller scintillator crystals and improved timing resolution by employing digital silicon photomultiplier readout.

Acknowledgements
The authors like to thank R. Sobiella, J. Wosnitza, and S. Zherlitsyn (Dresden High Magnetic Field Laboratory at HZDR) for providing and preparing the coil sample.

References
[1] Gabriel F, Gippner P, Grosse E, Janssen D, Michel P, Prade H, Schamlott A, Seidel W, Wolf A, Wünsch R 2000 *Nucl. Inst. Meth. Phys. Res.* **B 161** 1143.
[2] Kaya C, Schneider C, Al Shemmary A, Seidel W, Kuntzsch M, Bhattacharyya J, Mittendorff M, Evtushenko P, Winnerl S, Staat G, Helm M, Stojanovic N, Michel P, Gensch M 2012 *Appl. Phys. Lett.* **100** 141103.
[3] Jochmann A, Irman A, Bussmann M, Couperus JP, Cowan TE, Debus AD, Kuntzsch M, Ledingham KWD, Lehnert U, Sauerbrey R, Schlenvoigt HP, Seipt D, Stöhlker T, Thorn DB, Trotsenko S, Wagner A, Schramm U 2013 *Phys. Rev. Lett.* **111** 114803.
[4] Butterling M, Anwand W, Brauer G, Cowan TE, Hartmann A, Jungmann M, Kosev K, Krause-Rehberg R, Krille A, Schwengner R, Wagner A 2010 *Phys. Status Solidi* **A 207** 334.
[5] Butterling M, Anwand W, Cowan TE, Hartmann A, Jungmann M, Krause-Rehberg R, Krille A, Wagner A 2011 *Nucl. Inst. Meth. Phys. Res.* **B 269** 2623.
[6] Wagner A, Anwand W, Butterling M, Cowan TE, Fiedler F, Kempe M, Krause-Rehberg R 2012 *Defect and Diffusion Forum* **331** 41.
[7] Siemens: Biograph® PET scanner: www.medical.siemens.com.
[8] Zherlitsyn S, Wustmann B, Herrmannsdörfer T, Wosnitza J 2012 *IEEE Transactions on Applied Superconductivity* **22** 3.
[9] Shepp LA, Vardi Y 1982 *IEEE Transactions on Medical Imaging* **1** 113.
[10] Haaks M, Bennewitz K, Bihr H, Männig U, Zamponi C, Maier K 1999 *Appl. Surf. Sci.* **149** 207.
[11] Oshima N, Kuroda R, He C, Kinomura A, Ohdaira T, Toyokawa H, Suzuki R, Hayashizaki N, Hattori T 2007 *Phys. Status solidi* **C4** 4023.

Journal of Physics: Conference Series **505** (2014) 012035 doi:10.1088/1742-6596/505/1/012035

First results of a new positron-accumulation scheme using an electron linac and a Penning-Malmberg trap

P Grandemange[1], P Comini[2], P Debu[2], P Dupré[1], L Liszkay[2], D Lunney[1], P Pérez[2], J-M Rey[2], J-M Reymond[2], N Ruiz[2], Y Sacquin[2] and B Vallage[2]

[1] CSNSM/IN2P3/CNRS, Université de Paris Sud, 91405 Orsay, France
[2] SPP/IRFU/CEA, Saclay, 91191 Gif sur Yvette, France

E-mail: pierre.grandemange@cea.fr

Abstract. The *G*ravitational *B*ehaviour of *A*ntihydrogen at *R*est experiment - GBAR - is designed to perform a direct measurement of the weak equivalence principle on antimatter by measuring the acceleration ($\bar{g}$) of antihydrogen atoms in free fall. Its originality is to produce $\overline{H}^+$ ions and use sympathetic cooling to minimize the initial velocity. These ions are produced using charge exchange reactions with a dense positronium cloud, created by an intense pulse of electron-linac-produced positrons that are accumulated in a Penning-Malmberg trap.

The GBAR experiment at CERN [1] intends to measure the acceleration of neutral antihydrogen atoms in the Earth's gravitational field. Such a measurement would provide the first direct test of the weak equivalence principle with antimatter. To achieve sufficient accuracy on $\bar{g}$, $\overline{H}$ atoms must be cooled to $\sim$ µK temperatures. Based on a proposal by J. Walz and T. W. Hänsh [2], this can be achieved by Coulomb interaction with a co-confined, laser-coolable species – on the condition that the antihydrogen is also charged. The challenge for GBAR is to produce $\overline{H}^+$ ions by means of a two-step reaction $\bar{p} + Ps \rightarrow \overline{H} + e^-$, and $\overline{H} + Ps \rightarrow \overline{H}^+ + e^-$, where $\bar{p}$ is an antiproton and Ps is positronium. The excess positron is then laser-detached to perform the free-fall measurement with $\overline{H}$.

The antiprotons are delivered by ELENA - the *E*xtremely *L*ow *En*ergy *A*ntiproton ring - which will provide $6 \times 10^6\,\bar{p}$ at a kinetic energy of 100 keV every 100 seconds. In order to produce at least one $\overline{H}^+$ ion per $\bar{p}$ pulse, an extremely dense positronium cloud of $10^{12}\,Ps/cm^3$ is required meaning $2 \times 10^{10}\,e^+$ must strike a porous-silica converter within the lifetime of ortho-positronium ($\sim$ 140 ns) every 100 s. This is presently impossible, since the most intense ^{22}Na radioactive sources achieve $\sim 10^7\,e^+/s$ slow DC positron flux, and the best accumulation rate achieved is a quasi-linear load of $1.2 \times 10^9\,e^+$ in 2.5 hours [3]. Therefore a new accumulation technique is under development, using a pulsed positron beam produced by a linear electron accelerator feeding a Penning-Malmberg trap, provided by RIKEN collaborators. The trap is pre-loaded with a large cloud of electrons that cool by synchroton radiation, which in turn cool the injected positron pulses.

The slow positron source located at CEA Saclay consists of five elements, shown in Fig. 1.

The accumulation process is an adaptation of the technique developed by Oshima et al. [4]. The four-step positron-accumulation scheme is illustrated in Fig. 2. A dense electron plasma is loaded in large and deep harmonic potential well with a electron gun placed upstream the trap.

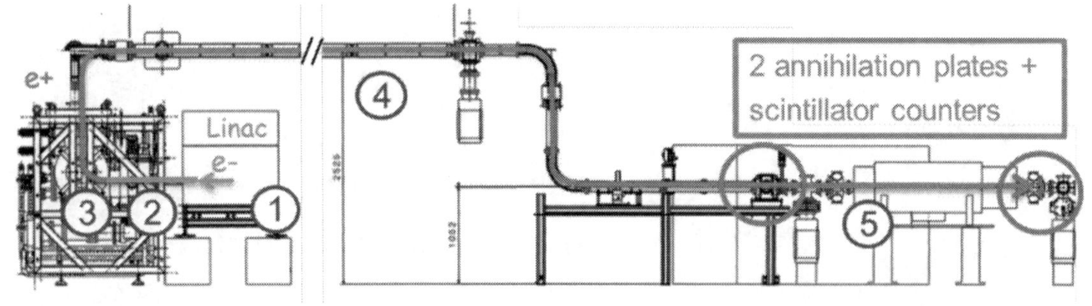

Figure 1. The pulsed-positron-beam accumulator at CEA Saclay. (1) A 4.3 MeV electron linac with a 20 to 200 Hz repetition rate, electron peak current of 140 mA, and effective pulse length of 2.5 µs; (2) a water cooled $e^- \rightarrow e^+$ 1 mm-thick tungsten converter plate; (3) a moderator that slows the e^+ beam to the eV range; (4) a 16 m-long solenoidal beamline creating an 8 mT adiabatic transport magnetic field; (5) the Penning-Malmberg trap consisting of 27 electrodes in a 5 T superconducting magnetic field.

The confined electron plasma cools by synchroton radiation. The positrons are accelerated to 1 keV through the magnetic field gradient into the trapping region where the entrance electrode (-11) is switched up to prevent reflected positrons from the downstream potential barrier from exiting the trap (a round trip takes about 80 ns). The positrons are then cooled down by Coulomb interaction within the electron plasma. When they have lost sufficient longitudinal energy, they are trapped between the electron plasma potential and the downstream potential barrier.

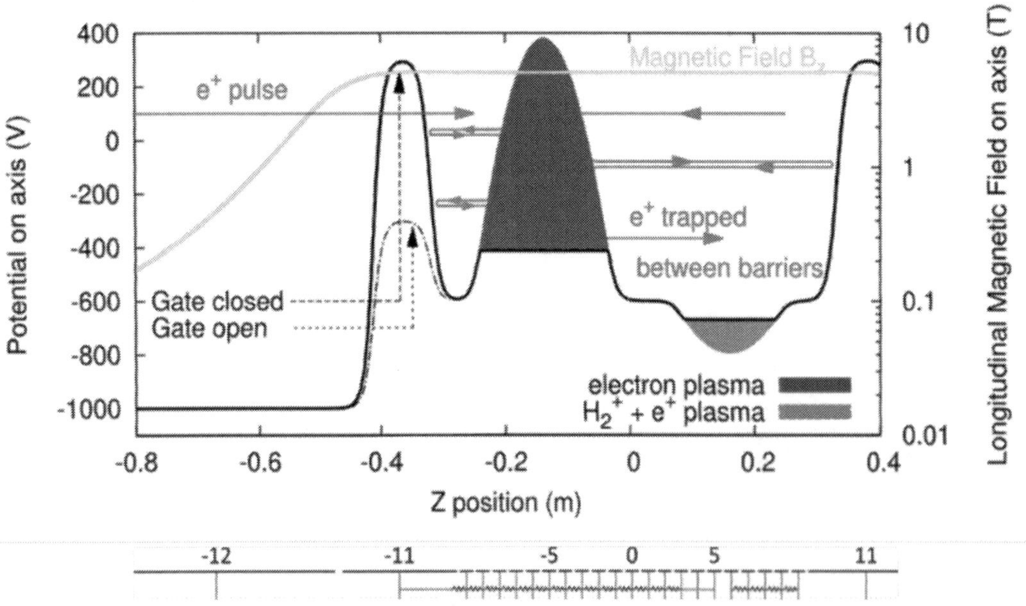

Figure 2. Positron accumulation scheme showing the potential distribution within the trap [5] and corresponding electrode geometry at the bottom.

Positron accumulation consists of repeating steps 2-4 for each positron pulses delivered to the trap. The overall cooling time should be less than 3.3 ms, which can be achieved with an electron plasma density of $10^{11}\,e^-/cm^3$ [5]. To squeeze all the positrons into the trap before closing the potential gate, the pulse must be compressed from the original 2.5 µs length to $\sim$ 80 ns. The compression of the pulse is realized by applying a dynamic potential on the moderator during the positron pulse generation and forming a sharp time focus in the middle of the trap.

The electron plasma density achieved at RIKEN has been reproduced and adapted for the present positron accumulation technique. Over $2 \times 10^{10}\,e^-$ have been confined for more than 500 s in a 950 V harmonic potential well of 300 mm length, corresponding to the $10^{11}\,e^-/cm^3$ density required to cool the positrons in the millisecond time range.

Shown in Fig. 3 is the annihilation signal from positrons extracted from to the trap and sent to the downstream scintillator after a 20 s positron accumulation and 500 ms storage time (left). No signal is detected after a 500 ms confinement time when the e^+ well is suppressed (right) or if the electron plasma is not preloaded before the accumulation sequence. These observations lead us to conclude that the observed growing number of trapped e^+ with the increasing number of injected positron stacks is the result of their being cooled.

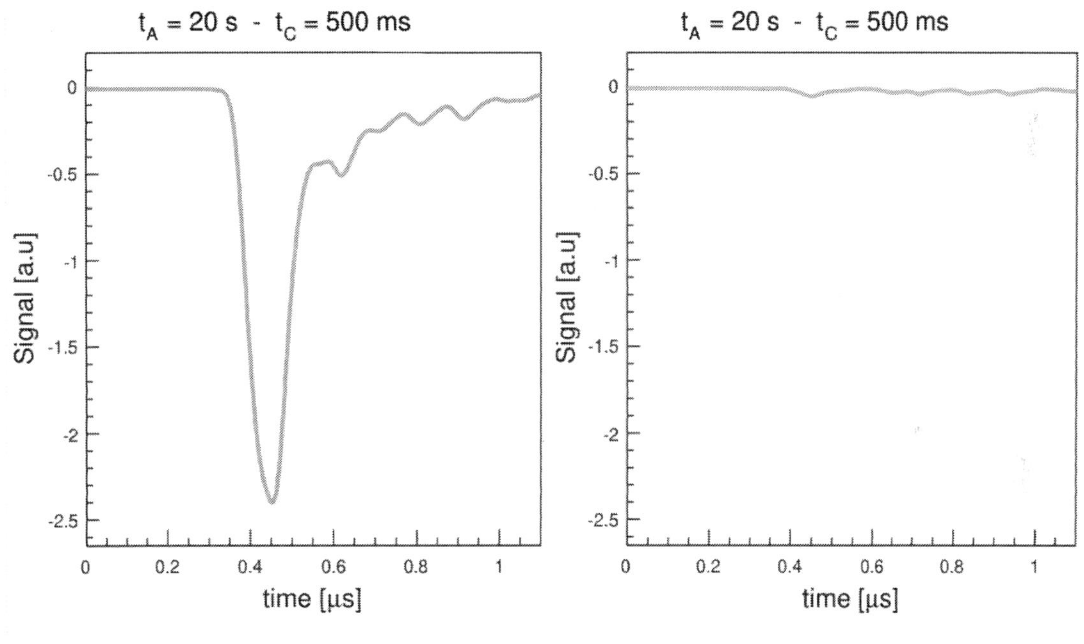

Figure 3. Positron extraction signal on the downstream scintillator counter. (left) accumulation time of 20 s and confinement time of 500 ms (right) same conditions but without the positron well.

Pulse by pulse positron accumulation has also been measured, as shown in Fig. 4. About 4400 e^+ have been trapped after 40 s, corresponding to an accumulation efficiency of 1 %. The positron injection efficiency, determined from the number of trapped positrons over the number of slow positrons per pulse, is 4 %. The main loss comes from the initial energy spread due to the isotropic positron emission from the tungsten meshes. Also, the beamline introduces an energy spread, which is worsened by the necessity of introducing a time focus. This efficiency can be improved by reducing the magnetic disruptions in the beamline and implementing a dedicated buncher before the trap.

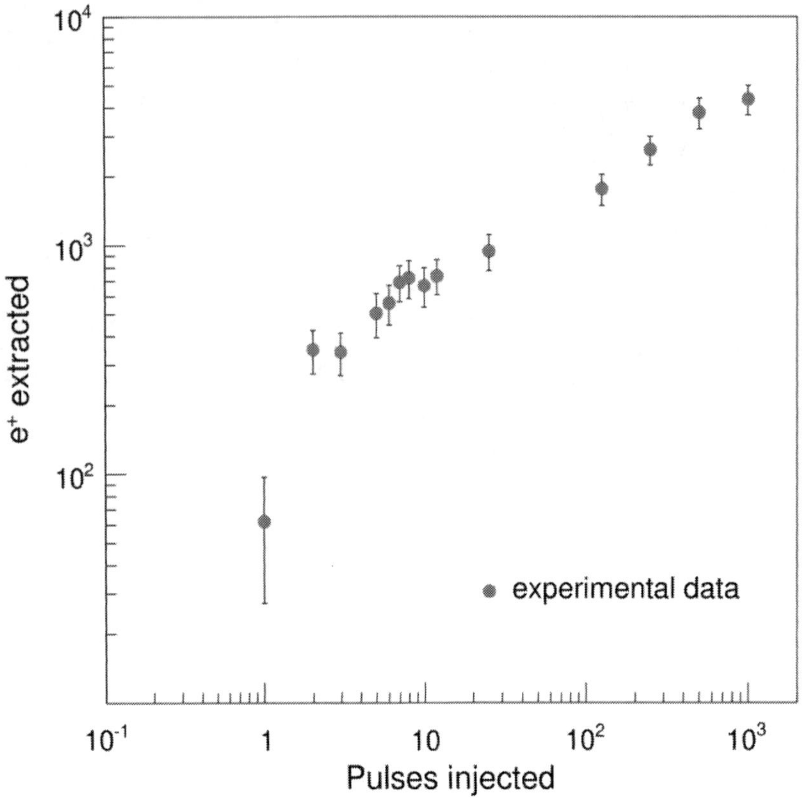

Figure 4. Number of positrons accumulated versus number of injected pulses (25 Hz).

These first results were achieved with a magnetic field of only 3 T and degraded vacuum conditions, due to problems with the cryocooler system. This system has now been serviced and will allow improved results with better vacuum and higher magnetic field.

References

[1] Chardin G, Grandemange P, Lunney D, Manea V, Badertscher A, Crivelli P, Curioni A, Marchionni A, Rossi B, Rubbia A, Nesvizhevsky V, Hervieux P A, Manfredi G, Comini P, Debu P, Dupré P, Liszkay L, Mansoulié B, Pérez P, Rey J M, Ruiz N, Sacquin Y, Voronin A, Biraben F, Cladé P, Douillet A, Gérardin A, Guellati S, Hilico L, Indelicato P, Lambrecht A, Guérout R, Karr J P, Nez F, Reynaud S, Tran V Q, Mohri A, Yamazaki Y, Charlton M, Eriksson S, Madsen N, van der Werf D P, Kuroda N, Torii H and Nagashima Y 2011 Proposal to measure the gravitational behaviour of antihydrogen at rest Tech. Rep. CERN-SPSC-2011-029. SPSC-P-342 CERN Geneva

[2] Walz J and Hänsch T W 2004 *General Relativity and Gravitation* **36** 561–570

[3] Jørgensen L, Amoretti M, Bonomi G, Bowe P, Canali C, Carraro C, Cesar C, Charlton M, Doser M, Fontana A, Fujiwara M, Funakoshi R, Genova P, Hangst J, Hayano R, Kellerbauer A, Lagomarsino V, Landua R, Lodi Rizzini E, Macrì M, Madsen N, Mitchard D, Montagna P, Rotondi A, Testera G, Variola A, Venturelli L, van der Werf D P and Yamazaki Y 2005 *Phys. Rev. Lett.* **95** 025002

[4] Oshima N, Kojima T, Niigaki M, Mohri A, Komaki K and Yamazaki Y 2004 *Phys. Rev. Lett.* **93** 1–4

[5] Dupré P 2013 *AIP Conference Proceedings* **1521** 113–122

Present status of the low energy linac-based slow positron beam and positronium spectrometer in Saclay

L Liszkay[a], P Comini[a], C Corbel[b], P Debu[a], P Grandemange[c], P Pérez[a], J-M Rey[a], J-M Reymond[a], N Ruiz[a], Y Sacquin[a] and B Vallage[a]

[a] CEA/IRFU, CEA Saclay, 91191 Gif sur Yvette, France.

[b] Laboratoire des Solides Irradiés, Ecole Polytechnique, F-91128 Palaiseau, France.

[c] CSNSM-IN2P3/CNRS, Université de Paris Sud, F-91406 Orsay, France

E-mail: laszlo.liszkay@cea.fr

Abstract. A new slow positron beamline featuring a large acceptance positronium lifetime spectrometer has been constructed and tested at the linac-based slow positron source at IRFU CEA Saclay, France. The new instrument will be used in the development of a dense positronium target cloud for the GBAR experiment. The GBAR project aims at precise measurement of the gravitational acceleration of antihydrogen in the gravitational field of the Earth. Beyond application in fundamental science, the positron spectrometer will be used in materials research, for testing thin porous films and layers by means of positronium annihilation. The slow positron beamline is being used as a test bench to develop further instrumentation for positron annihilation spectroscopy (Ps time-of-flight, pulsed positron beam). The positron source is built on a low energy linear electron accelerator (linac). The 4.3 MeV electron energy used is well below the photoneutron threshold, making the source a genuine on-off device, without remaining radioactivity. The spectrometer features large BGO (Bismuth Germanate) scintillator detectors, with sufficiently large acceptance to detect all ortho-positronium annihilation lifetime components (annihilation in vacuum and in nanopores).

1. Introduction

One of the fundamental questions of today's physics is the action of gravity upon antimatter. A difference between the free fall of matter and antimatter in the gravitational field of Earth would signal a violation of the Weak Equivalence Principle. The GBAR experiment [1], recently accepted at CERN, proposes a scheme for a direct measurement of the gravitational acceleration of the simplest antimatter atom, the antihydrogen, with 1% precision. The distinctive feature of the experimental scheme is that the extreme low energy (10 μK) that is needed for the precise determination of the gravitational constant is reached through cooling of positively charged antihydrogen atoms, which are then neutralized by photodetachment prior to the free fall measurement. The production of the ions require a positronium (Ps) cloud with unprecedented density (10^{12} Ps/cm^3) as target for the antiproton beam provided by the Antiproton Decelerator facility at CERN. A high field Penning-Malmberg trap accumulates the particles from an intense slow positron source and ejects them in an intense pulse

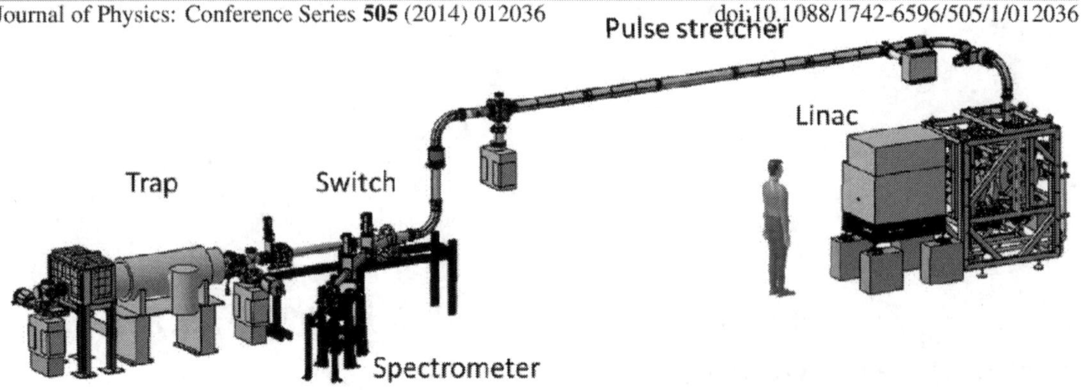

Figure 1 Schema of the positron installation in Saclay with the main constituents.

after typically 110 s. Positronium will be generated by the e^+ pulse, hitting a positron-positronium converter target.

An experimental facility (Figure 1) has been built at CEA-IRFU (Saclay, France) to test and develop the key constituents of the experimental setup, needed for the production of the dense positronium cloud. The facility consists of a linac-based slow positron generator, a high field (5 T) positron trap (a contribution from RIKEN, Japan) and a large acceptance positronium lifetime spectrometer. The spectrometer is being used in the development of the positron-positronium converter. In addition to the application in the GBAR experiment, the spectrometer will be used in positron annihilation studies of nanoporous oxide films and layers.

2. The slow positron generator

A linac based slow positron generator does not produce any radioactive contamination in the structural elements below 10 MeV electron energy. The fact that the positron source does not contain any radioactive parts in its "off" state makes it an attractive alternative to positron sources based on radioactive isotopes. Such sources can be deployed in laboratories where there is no infrastructure for the use of radioactive isotopes. The slow positron generator at IRFU [2] is based on a linac of 4.3 MeV electron energy, which is well below the photoneutron threshold of all structural materials. The accelerator produces electron bunches with an effective length of 2.5 μs at 200 Hz repetition rate and 140 mA peak current. The electron beam is incident on a water cooled tungsten target and generate positrons by pair production. The setup at CEA-IRFU contains a magnetic selector, which separates the positrons emitted from the target from the electron flow. This feature will allow the use of solid neon moderator in a later phase of development. In the present working setup, a positron moderator is placed directly behind the electron target. The positron moderator is formed by a stack of thin tungsten meshes, annealed by Joule heating in a separate vacuum chamber. Possibility of in situ heating will be added in a later phase of the project. We found optimum positron production at 12 mesh layers. Work is underway to improve the positron yield with an optimized moderator structure. At present, a sustained positron rate of $2x10^6$ e^+/s is available.

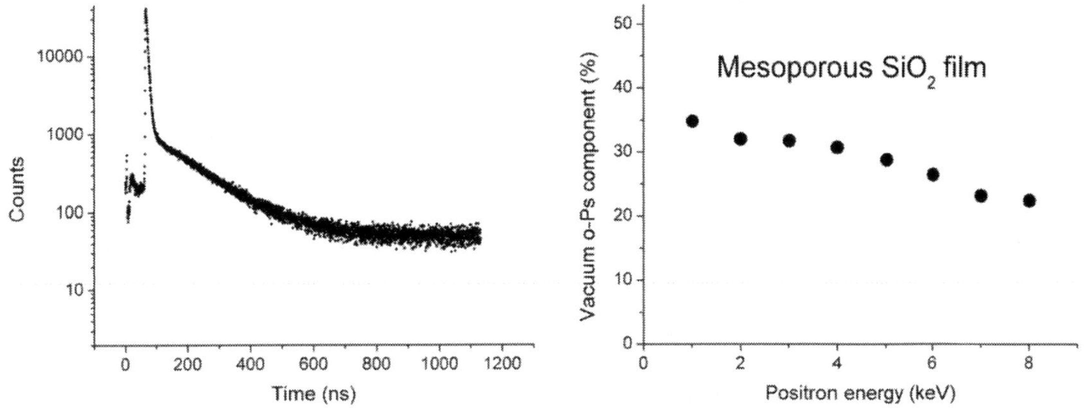

Figure 2 (a) Measured positron spectrum with intense vacuum ortho-positronium component (b) Vacuum ortho-positronium component as a function of the positron energy in mesoporous silica film. Spectra measured with the IRFU spectrometer.

3. The slow positron transport beamline

Positrons are adiabatically transported in 8 mT magnetic field, generated by solenoids wound around the vacuum tube. A positron pulse stretcher is inserted into the beamline at the exit point of the concrete bunker of the biological shield. The function of the pulse stretcher is to form a quasi-continuous beam from the 2.5 μs pulses generated by the linac. It is a low-field Penning-Malmberg trap, which consists of three electrodes. The second electrode is 4 m long, in order to accommodate the entire slow positron pulse. The other two electrodes open and close the trap, synchronized with the linac pulses. A static beam switch, based on a rotatable coil, feeds the beam into two beamlines. One line is reserved for the high field (5 T) Penning-Malmberg positron trap, used to develop the technology to trap up to $2x10^{10}$ positrons for the GBAR experiment. The second beamline feeds a large acceptance positronium lifetime spectrometer, used in the development of the positron-positronium converter for the GBAR project and for materials science studies.

4. The large acceptance ortho-positronium lifetime spectrometer

Lifetime spectrometers used routinely in positron annihilation spectroscopy can detect intensity and length of lifetime components up to a few nanoseconds. The presence of long-living ortho-positronium (o-Ps) in a system, however, makes the reliable analysis of the spectra difficult. One problem is the difference between the energy distribution of photons created in two gamma annihilation processes dominant at short lifetime components and those of three-gamma annihilation events of ortho-positronium. Correct evaluation of the spectra is possible only by using a correction factor, determined experimentally or by Monte-Carlo simulation. Measurement of positron annihilation lifetime spectra in the presence of long ortho-positronium components in a slow positron beam poses another problem. In highly porous samples, where a significant fraction of ortho-positronium escapes from the sample, the spectrum will be distorted by the difference between the detection efficiency of annihilation events in the sample and those of the escaping positronium. This is the case in all samples suitable for use as positron-positronium converter in the GBAR experiment. To tackle this problem, we have developed a special device, a large acceptance ortho-positronium lifetime spectrometer, in a collaboration between IRFU and ETH Zurich (Switzerland) [3]. A similar spectrometer has been implemented and tested (Figure 2) on the CEA-IRFU slow positron beamline described in the present paper. The spectrometer uses secondary electrons, ejected by the impinging positrons, as start signal (detected by a micro-channel plate detector) and one of the annihilation gamma photons as stop signal. Two large (55 mm diameter, 200 mm long) BGO (bismuth germanate) scintillation detectors are used for gamma detection with large acceptance. The lifetime spectrometer uses the quasi-continuous slow positron beam provided by the positron pulse stretcher, with a collimator of variable diameter to limit the beam diameter on the sample. The positron energy is set by the potential of the sample holder.

5. Development of positron-positronium converter for GBAR and study of nanoporous silica

Positron lifetime studies using the large acceptance ortho-positronium spectrometer have shown that nanoporous silica with high porosity is a promising candidate as the basis of the positron-positronium converter for GBAR [4]. Our time-of-flight studies [5] as well as laser-based measurements [6] confirmed that the energy of the reemitted o-Ps remains under 50 meV, a value that is satisfactory for the GBAR target construction. We analyzed nanoporous silica films with larger pore size (32-75 nm) to understand better the underlying physical processes and find ways to optimize pore size and density for the GBAR positron-positronium converter [7]. Our studies have also confirmed that the positron method can contribute to the better understanding of the pore properties of nanoporous silica and can serve as a basis of quality control devices for low-κ dielectrics used in microelectronics.

6. Summary and outlook

We constructed a low energy linear linac-based slow positron generator and an associated beamline at CEA-IRFU. The installation serves as a test bench for the various technologies needed to create a high density ortho-positronium cloud, essential for the formation of antihydrogen ions in the GBAR experiment. One beamline of the two feeds positrons into a high field trap, the second one provides a quasi-continuous flux of slow positrons for a large acceptance positronium lifetime spectrometer. Extension of the second beamline with other detection methods and construction of a new sample chamber is planned to extend the capability of the spectrometer to study of a wider range of materials.

Acknowledgments

This work is supported in part by Conseil Général de l'Essonne under ASTRE 2006 funding and by Research National Agency under contract number ANR-05-BLAN-0380-01. The spectrometer has been realized using a grant from the Centre Francilien de l'Innovation.

References

[1] Pérez P and the GBAR collaboration 2011 Proposal to measure the Gravitational Behaviour of Antihydrogen at Rest GBAR, CERN-SPSC-2011-029, Geneva, Switzerland

[2] Rey J-M, Coulloux G, Debu P, Dzitko H, Hardy P, Liszkay L, Lotrus P, Muranaka T, Noel C, Pérez P, Pierret O, Ruiz N, Sacquin Y 2013 J. Phys: Conf. Ser. **443** 012077

[3] Crivelli P, Gendotti U, Rubbia A, Liszkay L, Perez P 2009 Materials Science Forum **607** 251

[4] Liszkay L, Corbel C, Pérez P, Desgardin P, Barthe M-F, Ohdaira T, Suzuki R, Crivelli P, Gendotti U, Rubbia A, Etienne M and Walcarius A 2008 *Appl. Phys. Lett.* **92** 063114

[5] Crivelli P, Gendotti U, Rubbia A, Liszkay L, Pérez P, Corbel C 2010 *Phys. Rev. A* **81** 052703

[6] Cassidy D B, Crivelli P, Hisakado T H, Liszkay L, Meligne V E, Pérez P, Tom H W K and Mills A P Jr. 2010 *Phys. Rev. A* **81** 012715

[7] Liszkay L, Guillemot F, Corbel C, Boilot J-P, Gacoin T, Barthel E, Pérez P, Barthe M-F, Desgardin P, Crivelli P, Gendotti U and Rubbia A 2012 *New Journal of Physics* **14** 065009

Positronium and positronium negative ion emission from alkali-metal coated tungsten surfaces

Y Nagashima[1], K Michishio[1], H Terabe[1], R H Suzuki[1], S Iida[1], T Yamashita[1], R Kimura[1], T Tachibana[2], I Mochizuki[3], K Wada[3], A Yagishita[3] and T Hyodo[3]

[1] Department of Physics, Tokyo University of Science, 1-3 Kagurazaka, Shinjuku, Tokyo 162-8601, Japan
[2] Department of Physics, Rikkyo University, 3-34-1 Nishi-Ikebukuro, Toshima, Tokyo 171-8501, Japan
[3] Institute of Materials Structure Science, High Energy Accelerator Research Organization (KEK), 1-1 Oho, Tsukuba, Ibaraki 305-0801, Japan

E-mail: ynaga@rs.kagu.tus.ac.jp

Abstract. The emission efficiencies of positronium atoms and positronium negative ions from tungsten surfaces increase by alkali-metal coating. We have studied the positronium emission from alkali-metal coated surfaces using the positronium time-of-flight method. We have also performed the observation of the photodetachment of Ps^- emitted from a Na-coated surface and the production of an energy-tunable Ps beam employing the photodetachment technique. This paper describes our recent studies using alkali-metal coated surfaces.

1. Introduction

Recently, we found that the efficiency for the emission of the positronium negative ion (Ps^-), a bound state of one positron and two electrons [1], from tungsten surfaces bombarded with slow positrons increases by coating the surfaces with alkali-metals [2, 3]. This method has been applied for the observation of Ps^- photodetachment [4] and the production of an energy-tunable positronium (Ps) beam employing the photodetachment technique [5]. The emission efficiency of neutral Ps atoms also increases by coating the surfaces with alkali-metals. In order to investigate Ps emission from alkali-metal coated surfaces, we have started the measurements of the time-of-light (TOF) of Ps emitted from alkali-metal coated surfaces [6].

In this paper, we will review our recent experiments on Ps and Ps^- emitted from alkali-metal coated tungsten surfaces.

2. Ps emission from alkali-metal coated surfaces

Figure 1 shows the annihilation γ-ray energy spectra that indicate the emission of Ps^- from clean and Na-coated tungsten surfaces. The Ps^- ions emitted from the surfaces were accelerated by an electric field applied in front of the surface and the blue-shifted γ-rays emitted from the self-annihilation of Ps^- were detected using a Ge detector [3]. The electric field forced re-emitted positrons to accelerate back towards the surface. While the peak in the spectrum at 529keV, which indicates Ps^- emission, becomes higher by the coating, the peak at 511 keV becomes lower.

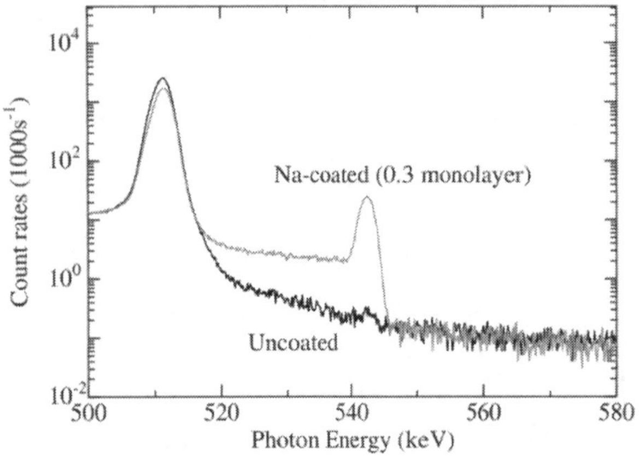

Figure 1. Annihilation γ-ray energy spectra for the Ps^- emission from uncoated and Na-coated polycrystalline tungsten surfaces [3].

This shows that the fraction of ortho-Ps, which self-annihilates into three γ-rays, increases by the coating.

The change in Ps yield by alkali-metal coating was investigated by Mills [7], Gidley *et al* [8] and Fazleev *et al* [9]. The increase in the Ps yield was attributed to a change in activation energy for the emission of Ps atoms formed from positrons trapped by the image potential at the surface. The activation energy for this process can be formulated as

$$E_a = E_b + \phi_- - 6.8\text{eV}, \quad (1)$$

where E_b is the binding energy of the positrons at the surface. The value of ϕ_- is reduced by the coating and hence E_a becomes lower and causes the positrons trapped at the surface to desorb as Ps. The emission energy of the Ps formed by this process should be lower than the emission energy of Ps formed from delocalized positrons, which can be formulated as

$$E_{\text{Ps}} = -\phi_+ - \phi_- + 6.8\text{eV}, \quad (2)$$

where ϕ_+ is the positron work function.

In order to investigate the Ps formation process, we measured the TOF of Ps emitted from a Na-coated tungsten surface [6]. A pulsed slow positron beam generated at the beam dump of an electron linac at the KEK-IMSS Slow Positron Facility [10] was incident onto the surface and the time interval between the linac pulse signal and the detection through a lead slit of the γ-rays from the o-Ps was measured [6]. The resulting TOF data are shown in figure 2. It shows that the component of the Ps emitted with E_{Ps} increases by the coating. This increase may be due to a reduction in the conduction electron density near the surface to a point suitable for the Ps formation [11, 12].

Further systematic measurements of Ps emission energy from tungsten surfaces coated with other alkali-metals are being conducted. A preliminary measurements using a Cs-coated tungsten surface shows that the yield of the lower energy component also increases.

3. Ps^- emission from alkali-metal coated surfaces

The enhancement of the Ps^- emission efficiencies by alkali-metal coating is attributed to a reduction in the surface dipole barrier, which results in an increase in the number of conduction electrons contributing to the Ps^- formation. This enhancement may also be due to the reduction in the conduction electron density near the surface to a point suitable for the Ps^- formation [11, 12].

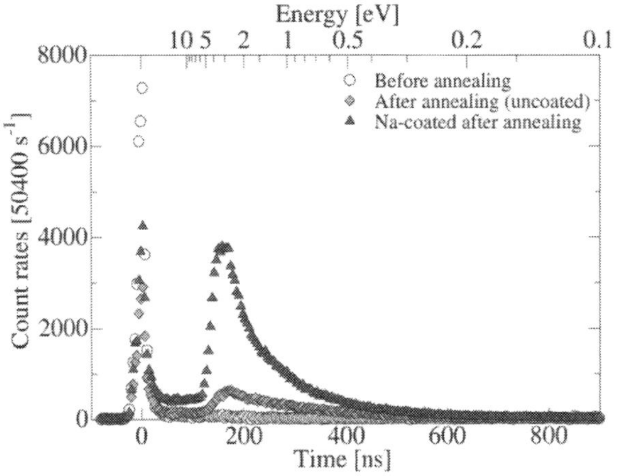

Figure 2. Time-of-flight spectra for clean and Na-coated polycrystalline surfaces [6].

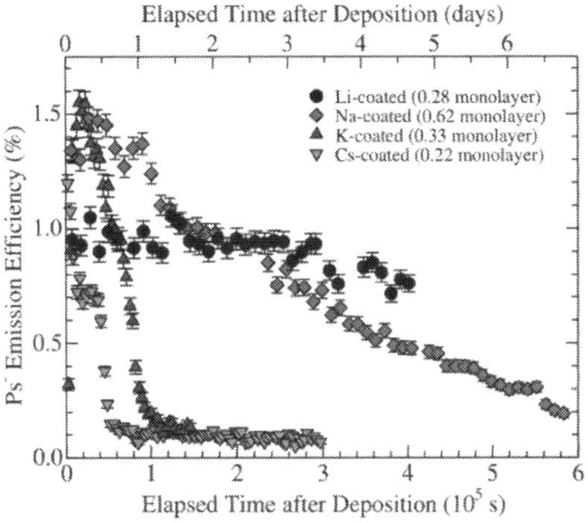

Figure 3. Ps^- emission efficiencies from Cs, K, Na and Li-coated tungsten surfaces plotted against time elapsed after the deposition.

Since alkali-metals are chemically reactive, the Ps^- emission efficiency decreases with the time elapsed after the coating, even in ultra-high vacuum conditions [2, 3]. In the case of Cs-coating, the efficiencies yield decreases to 1/10 of the maximum value in half a day. The efficiency for Na-coated surfaces is more durable than for Cs-coated surfaces because Na is less reactive than Cs. The efficiency reached a maximum value of 1.5 % after 4×10^4 s and decreased to 0.5 % after 4 days.

We have also measured the efficiency from a Li-coated polycrystalline surface. The thickness was 0.3 monolayer, which is slightly thinner than the thickness for which the electron work function is minimum. The results are plotted in figure 3. Although the maximum efficiency was lower than those for other alkali-metals, the efficiency remained almost constant for 3×10^5 s after the coating.

4. Production of an energy-tunable Ps beam using the Ps^- photodetachment technique

Recently, an experimental study for the observation of Ps^- photodetachment was performed using Ps^- emitted from a Na-coated tungsten surface [4]. We have also performed the production of an energy-tunable Ps beam employing the photodetachment technique [5].

A high-power pulsed laser light is required for Ps^- photodetachment because the Ps^- lifetime

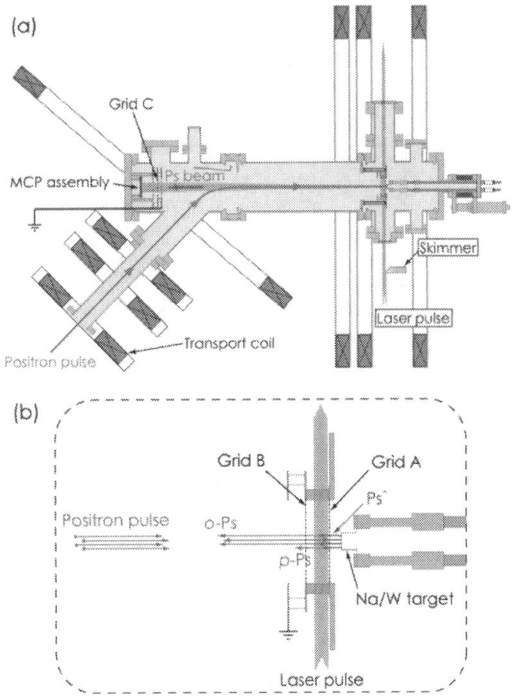

Figure 4. Schematic diagram of the energy-tunable Ps beam developed using the Ps^- photodetachment technique [5].

is as short as 479 ps. Thus a pulsed slow positron beam synchronized to the laser is also required. In this study, a pulsed slow positron beam generated at the beam dump of an electron linac at the KEK-IMSS Slow Positron Facility [10] was transported to a Na-coated polycrystalline tungsten target. The emitted Ps^- ions were accelerated by an electric field between the target and a grid located in front of the target and were then photodetached using the fundamental wave of Q-switched Nd:YAG laser (figure 4). The formed Ps atoms were detected by a micro-channel plate (MCP) located 80 cm downstream of the photodetachment region. The linac trigger pulses and the signals from the MCP were recorded using a digital oscilloscope.

The TOF spectra of Ps formed by the photodetachment are shown in figure 5. The peaks at $t = 0$ are due to the detection of the γ-rays emitted from positron annihilation at the target by the MCP and the peaks at $t \sim 25$ ns are attributed to the γ-rays emitted from positron annihilation at the chamber wall of back scattered positrons from the target. When the laser is on, another peak appears, which is due to the detection of energetic Ps produced by the photodetachment. The TOF varies by changing the Ps^- acceleration potential. This means that an energy-tunable Ps beam is produced and the Ps kinetic energy is controlled by the Ps^- acceleration. We have also observed the Ps beam spot using a position sensitive detector with delay-line anodes. The details of this work will be published elsewhere [13].

5. Future plans

Resonances of the photodetachment of H^-, a three-body system similar to Ps^-, have been observed [14]. Theoretical studies on resonances in the Ps^- photodetachment have also been performed and the energies and the cross sections have been predicted [15, 16, 17, 18]. We are planning experiments to observe the resonances.

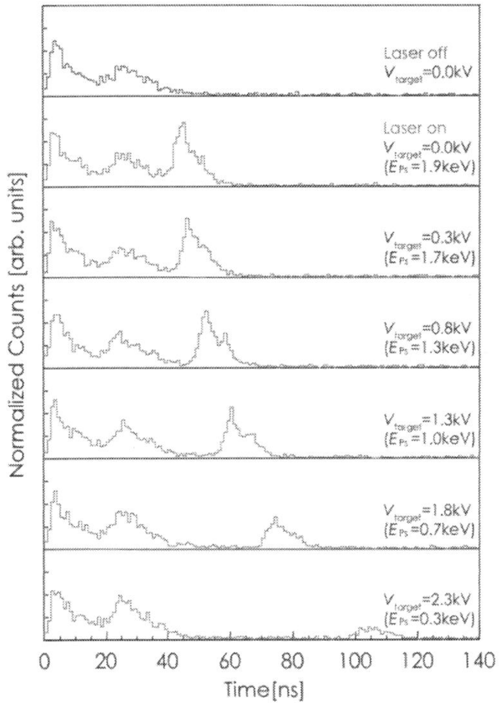

Figure 5. TOF spectra of the Ps produced by the photodetachment of Ps^- accelerated by an electric field [5].

The Ps beam may be applied for surface sciences. Recently, experimental studies on the diffraction of fast atomic projectiles during grazing scattering from surfaces have been performed [20, 21, 22]. Ps has been predicted to be an ideal probe for surfaces because the affinities for most materials are negative and hence information regarding the topmost atomic layers can be obtained [19]. We will attempt to observe reflected high-energy Ps diffraction (RHEPsD).

We also plan to produce an energy-tunable Ps beam using a pulsed positron beam extracted from a buffer-gas positron trap (Surko trap). Figure 6 shows the schematic diagram of the new Ps beam apparatus. Ps^- ions emitted from an alkali-metal coated surface opposite to the source of a tungsten thin film will be accelerated and photodetached to obtain Ps beams with energies greater than 5 keV. Another advantage of this geometry is that the length of the Ps flight path can reduced and hence the beam spot size at the sample can be smaller. The apparatus is compact and can be used in a typical university laboratory.

5.1. Acknowledgments

We thank A. Weiss, N. Fazleev, R. Greaves, H. Winter, T. Kanai and T. Azuma for valuable discussions. We also thank T. Shidara, S. Osawa, M. Ikeda and other KEK staff for their generous support for the positron beam line. This work has been performed under the approval of the Photon Factory Advisory Committee (Proposal No. 2010S2-003). It was partly supported by a Grand-in-Aid for Scientific Research (No. 244221006) from the Ministry of Education, Science and Culture of Japan.

References

[1] Mills A P Jr. 1981 *Phys. Rev. Lett.* **46** 717
[2] Nagashima Y, Hakodate T, Miyamoto A and Michishio K 2008 *New J. Phys.* **10** 123029
[3] Terabe H, Michisho K, Tachibana T and Nagashima Y 2012 *New J. Phys.* **14** 015003

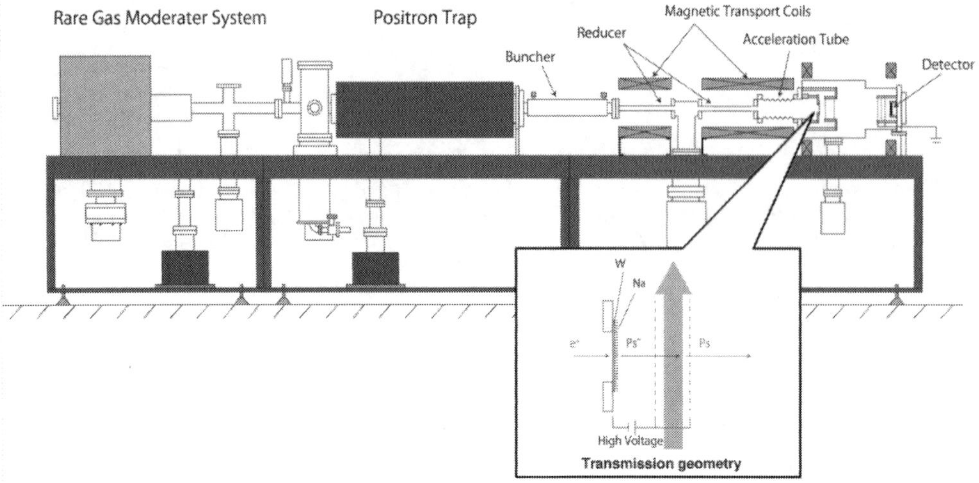

Figure 6. Conceptual design of the energy-tunable Ps beam using a buffer gas positron trap (Surko trap).

[4] Michishio K *et al* 2011 *Phys. Rev. Lett.* **106** 153401
[5] Michishio K *et al* 2012 *Appl. Phys. Lett.* **100** 254102
[6] Terabe H, Iida S, Wada K, Hyodo T, Yagishita A and Nagashima Y 2013 *J. Phys. Conf. Ser.* **443** 012075
[7] Mills A P Jr. 1983 *Positron Solid-State Physics: Proc. Int. School Phys. 'Enrico Fermi', Course 83* ed W Brandt and A Dupasquier (Amsterdam: North-Holland) p 432
[8] Gidley D W, Köymen A R and Capehart T W 1988 *Phys. Rev. B* **37** 2465
[9] Fazleev N G, Fry J L, Kaiser J H, Koymen A R, Lee K H, Niedzwiecki T D and Weiss A H 1994 *Phys. Rev. B* **49** 10577
[10] Wada K *et al* 2012 *Eur. Phys. J. D* **66** 37
[11] Lang N D 1971 *Phys. Rev. B* **4** 4234
[12] Wimmer E, Freeman A J and Weinert M 1982 *Phys. Rev. lett.* **48** 1128
[13] Michishio K *et al,* in preparation.
[14] Buckman S J and Clark C W 1994 *Rev. Mod. Phys.* **66** 539
[15] Bhatia A K and Drachman R J 1985 *Phys. Rev. A* **32** 3745
[16] Ward S J, Humberston J W and McDowell M R C 1987 *J. Phys. B: At. Mol. Phys.* **20** 127
[17] Igarashi A, Shimamura I and Toshima N 2000 *New J. Phys.* **2** 17
[18] Mills A P Jr. 2013 *Can. J. Phys.* **91** 751
[19] Canter K F 1983 *Positron Scattering in Gases* ed J Humberston and M R C McDowell (New York: Plenum) p 219
[20] Schüller A, Wethekam S and Winter H 2007 *Phys. Rev. Lett.* **98** 016103
[21] Rousseau P, Khemliche H, Borisov A G and Roncin P 2007 *Phys. Rev. Lett.* **98** 016104
[22] Winter H and Schüller A 2011 *Prog. Surf. Sci.* **86** 169

Positronium cooling at cryogenic temperature for advanced experiments

R S Brusa[1], L Di Noto[1], S Mariazzi[1] and G Nebbia[2]

[1] Dipartimento di Fisica, Università di Trento and TIFPA-INFN, Via Sommarive 14, I-38123 Povo, Trento, Italy
[2] INFN Padova, Via Marzolo , 35131 Padova, Italy

E-mail: brusa@science.unitn.it

Abstract. New Ps spectroscopy measurements, formation of antihydrogen for antimatter-matter comparison experiments, production of Ps beams require the efficient production of cooled positronium in vacuum. At present the most efficient positron-positronium converters are silica based ordered or disordered porous materials, in which formed Ps decreases its kinetic energy by collisional cooling. Recently new positron-positronium converters based on oxidized nanochannels in silicon were found to be very promising because of the tunability of the nanochannel size, which allows to overcome the limits imposed to the Ps cooling by the quantum confinement. With these converters, Ps with temperatures as low as 150 K was detected in vacuum by a TOF apparatus. The Ps formation, quantum confinement, collisional cooling and emission into vacuum from nanochanneled silicon will be discussed in light of recent results.

1. Introduction

Positronium (Ps), the metastable electron-positron bound state, can be formed by injecting a slow positron beam with energy ranging from few eV to some keV in different targets. In metals or semiconductors, Ps is produced at the surface and not in the bulk, where it gives a local enhancement of electron density that prevents a stable binding with a single electron. In insulators Ps can be formed in the bulk and/or at the surface. The basic principles of Ps formation and emission are well described and discussed in the reviews of Ref. 1-3.

In the last years, the availability of intense positron bunches with the Surko trap [4] has allowed Mill's research group to study Ps-Ps interactions [5] finding evidence of positronium molecule (Ps_2) formation [4, 6], to perform Ps_2 optical spectroscopy [7] and Ps excitation in Rydberg states [8].

At present several groups have increased their research efforts at looking both for materials giving high Ps yield and for efficient Ps cooling. In particular, new measurements and experiments demand cooled Ps in vacuum or in closed cavities.

Cold Ps, reducing the Doppler broadening of the Ps atomic transitions, will allow more precise Ps spectroscopy measurements. Indeed, it is also necessary to start with a pre-cooled Ps, at least to 90 K, to make for the first time a Ps laser cooling at the recoil temperature $T_r = 295$ mK [6].

In the AEgIS (Antimatter Experiment: gravity, interferometry, spectroscopy) experiment running at the AD (Antiproton Decelerator) hall at CERN, an antihydrogen ($\bar{H}$) cold beam will be produced by charge exchange between antiprotons and cooled Ps excited in Rydberg states [9, 10]: cold Ps [11] is essential to increase the cross section of the reaction $\bar{p} + Ps^{*} \rightarrow \bar{H} + e^{-}$. The first milestone of AEgIS is the test of the weak equivalence principle of General Relativity by measuring the gravitational interaction between antimatter as $\bar{H}$ and matter.

High yield $e^{+} \rightarrow Ps$ converters will make Ps beams feasible. These can be used to fill magnetic traps for positron-electron plasma studies [12]; furthermore, the proposed Ps free fall experiment [13] needs a cold Ps beam. Ps accumulation (at least 10^{18} Ps/cm^{3}) and cooling in cavities is a step towards the Ps Bose Einstein condensation [4, 6].

The Ps yield in metals and semiconductors can be increased up to 100% by heating the sample above room temperature (RT) and activating the thermal emission of Ps. In this case, positrons trapped at the surface by image charge forces, forms Ps [1-3]. Unfortunately, with these targets there is no way to combine efficient conversion with low Ps kinetic energy (few tens of meV). Recently, another Ps emission channel from clean surface of p and n type Si (111), Si (100) crystals was discovered [14]. Energetic Ps emission ($\langle E_{\perp Ps}\rangle$=160 meV, temperature $T = \frac{\langle E_{\perp Ps}\rangle}{k_B}$ =1860 K, where $E_{\perp}$ is the component of the kinetic energy perpendicular to the surface of the sample), occurs by an exciton-like surface state, and this emission can be laser induced.

Actually, the most efficient way to obtain cool Ps is to use silica nanostructured materials with disordered pores or ordered cavities with sizes that allow to overcome the temperature limit imposed by the quantum confinement [15]. In silica, up to 80% of implanted e^{+} forms Ps [16]. Ps can be emitted into nano-pores or nano-cavities from bulk or surface states. By time of flight (TOF) measurements [17], kinetic energies peaked at 1 eV and 3 eV were attributed to Ps emitted from these two states respectively. Successively, the emitted Ps starts to lose energy by collision with the walls of the cavities, and if these cavities are connected with the surface, Ps can escape into the vacuum with a reduced kinetic energy. In the best cases Ps is emitted after complete thermalization at the sample temperature.

In this paper we will discuss Ps cooling with particular emphasis to Ps formation, diffusion and cooling in ordered oxidized nanochannels in Si [18-20]. In this type of samples, overcoming the limit of Ps confinement, we observed Ps escape into vacuum with a thermal distribution at 150 K ($\langle E_{\perp Ps}\rangle$= k_bT = 13 meV) [20]. Characterization of the samples was carried out with different positron annihilation spectroscopy (PAS) techniques and measurements of Ps velocity with a TOF apparatus.

2. Experimental

2.1. Samples

An electrochemical etching procedure, commonly used to produce porous silicon, was optimized to synthesize single-crystals nanochannels oriented perpendicular to the surfaces in p-type Si (100). Important parameters are the density of the doping atoms (Si resistivity), the etching current and the time of etching in a H_2O-HF-ethanol solution. At last an induced moderate temperature oxidation in air is necessary to form a silica layer on the internal surface of the channels. The silica layer is the region where Ps is formed and emitted into the nanochannels. Typical sample synthesis parameters are reported and discussed in Ref. 18-19. This method allows tuning the size of the nanochannels, extending about 2 μm in depth, from 4-5 nm to 80-100 nm. The size tuning is obtained by iterating the etching-oxidation steps starting in smaller channels; in this way the nanochannels spatial density remains constant. In this paper we will discuss results obtained from samples with 5-8 nm sized nanochannels.

The morphology of the cavity is expected to influence the Ps cooling process. In comparison with ordered straight nanochannels, in porous SiO_2 sol-gel films produced by spin coating of tetraethoxysilane (TEOS) with sacrificial porogens [21- 23], a more or less connected distribution of pores is obtained after curing the samples at 300-400 °C. In some cases holes of smaller diameter interconnect the disorderly distributed quasi-spherical pores [23].

2.2. PAS techniques

Information about Ps formation, pick-off annihilation and diffusion in nano-channelled silicon samples was obtained by 2γ-3γ annihilation ratio spectroscopy (3γ-PAS) at the Trento slow positron beam [24] and by annihilation Lifetime Spectroscopy (LS) at the Pulsed Low Energy Positron System (PLEPS) [25] at the high intense NEutron induced POsitron source MUniCh (NEPOMUC) [26].

2.3. TOF

TOF measurements were done at the Trento slow positron beam [20]. We derived Ps kinetic energy ($E_{\perp Ps}$) measuring the time t between e^+ implantation in the target, determined by detecting the secondary electrons with channeltrons, and *o*-Ps annihilation in flight at a z distance from the target. *o*-Ps annihilation γ-rays were detected with NaI(Tl) detectors behind a 100 mm long lead shield, with a 5 mm slit. Details in Ref. 20. An improved apparatus [27] is now set up at the NEPOMUC facility.

3. Results and discussion

3.1 Positronium yield

Ps formation was obtained by measuring the oPs lifetime reduced by pick-off with PLEPS and three gammas oPs annihilation by 3γ-PAS, in a sample in which the nanochannels were capped with ~30 nm of TiO_2. The intensity oPs lifetime reduced by pick-off and the oPs 3γ annihilation fraction ($F_{3\gamma}(E_+)$) are, above 4 keV e^+ implantation energy E_+ (the nanochannel region), ~ 15 % and ~ 25 %, respectively (see figure 1a).

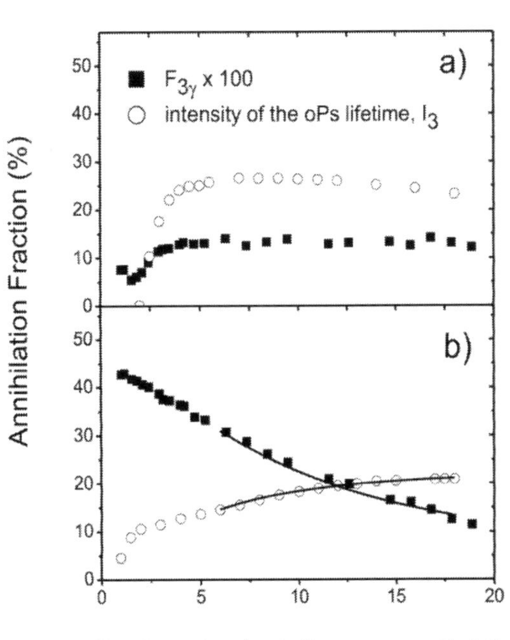

Fig 1: a) intensity I_3 of the oPs reduced lifetime by pick-off (open circles), oPs annihilation into 3 gammas $F_{3\gamma}$ (E_+)x100 (full squares), as measured in 30 nm TiO_2 capped nanochannelled Si sample, as a function of the positron implantation energy; above 4 keV annihilations are into nanochannels. b) Same quantities as in a) but measured in nanochannelled Si sample without capping.

From the lifetime spectra we found that ~ 20 % of implanted e^+ annihilates in Si or as pPs. The remaining ~ 40 % annihilates in the silica layers covering the nanochannels walls and e^+ in surface states with a lifetime of ~400 ps [28]. The low fraction of e^+ annihilating in Si is due to the long e^+ diffusion length (L_+) in Si, about 200 nm, more than ten times the interspacing length among nanochannels. With this L_+ the majority of positrons, after thermalization, reaches the silica layer. Indeed, the efficient production of Ps is caused by the high probability of e^+ reaching the silica layer and the favorable surface/volume ratio in the converter.

3.2 Positronium out-diffusion

Information about the fraction of Ps escaping into vacuum and evaluation of Ps diffusion length L_{Ps} were extracted by analyzing the 3γ-PAS measurements done in an un-capped nanochannelled Si sample. In figure 1b we report the oPs 3γ fraction for the un-capped sample. The shape of the $F_{3\gamma}(E_+)$ curve is characteristic of Ps escaping into the vacuum [18]. At low e^+ implantation energy (< 1 keV), about all Ps (~ 42 %) forms at the outlet of the nanochannels and escapes into vacuum. Increasing E_+, the pick-off annihilation probability increases and the fraction of escaping Ps decreases. Due to the broadening of the positron implantation profile, at higher E_+ (> 15 keV) a fraction of implanted positrons annihilates into Si beyond the end of nanochannels without forming Ps. One should notice that a small fraction of fast escaping Ps at lower E_+ is not detected or detected with lower efficiency, because it annihilates far from the detector view [18]. In the present samples, for E_+ energies higher than ~ 5 keV, a Ps diffusion model can be applied to fit the data [18] because above this energy the diffusion coefficient D_{Ps} has little variations with the implantation depth and can be considered constant. The continuous lines in figure 1b were obtained with a modified model based on rate equations [29] with an unique fitting parameter L_{Ps}. We found L_{Ps} = 721±13 nm corresponding to a D_{Ps}= 0.16 ± 0.01 cm^2/s. The D_{Ps} in nanochannels results to be four to eight times larger than the one found in disordered porous samples. To cool Ps at meV energies, Ps must be formed deeper in the sample, because of the more linear path of Ps escape, due to the straight channels, and the consequent remarkable L_{Ps}

3.3 Positronium cooling

Ps emitted and confined in a cavity decreases its kinetic energy E_{Ps} from 1-3 eV [17] to few meV by collision with the walls of the cavity. Ps interactions with the atoms of the walls and Ps behaviour depend on the Ps energy and on the size d of the cavity. Two opposite regimes can be distinguished [3, 15, 23].

If the de Broglie wavelength of Ps is less than d, $\lambda_{Ps} = h/\sqrt{4m_0 E_{Ps}} < d$, where m_0 is the e^+ mass, Ps can be treated as a classical particle. This relation requires $E_{Ps} \gg (6\,nm/d)20\,meV$. Up to these energies, Ps collision can be modelled as an elastic scattering between an object of mass $2m_0$ and an atom of mass M of the wall [30, 31].

Differently, when λ_{Ps} becomes comparable to d, quantum effects must be taken into account. i) The Ps scattering depends on the overlap of the Ps wavefunction with several atoms of the wall of the cavity. ii) Ps can be quantum confined. For every given size d, a minimum Ps energy state exists. In an infinite quantum well (modelling a nanochannel) the minimum energy is $E_{Ps} = h^2/(2m_0d^2)$ and the corresponding minimum temperature is $T_m = (2E_{Ps})/(3k_b)$ [15].

The nanochannels of the present sample (d = 5-8 nm) were synthetized to reach a minimum temperature $T_m \cong 100\,K$. Figure 2 shows a Ps kinetic energy spectrum, calculated by multiplying a TOF spectrum measured at 150 K sample temperature by t^3. The TOF spectrum was measured at 7 keV positron implantation energy, corresponding to an e^+ mean implantation depth of about $\bar{z}$=500 nm (density of the sample estimated to be 1.9 gcm^{-3}). For the analysis of the TOF spectra we refer to Ref. 20 and reference therein. The best fit to the $E_{\perp Ps}$ spectrum was obtained with the sum of two distributions, a one-dimensional Maxwellian and a beam Maxwellian:

$$\frac{dN}{dE} = A\frac{{E_{\perp Ps}}^{-1/2}}{\sqrt{\pi k_B T_1}} e^{-\frac{E_{\perp Ps}}{k_B T_1}} + B\frac{1}{k_B T_2} e^{-\frac{E_{\perp Ps}}{k_B T_2}}$$

with T_1 = 142±46 K and T_2= 755±43 K.

The above analysis has pointed out the presence of a fraction of Ps that escapes after thermalization with the sample at 150 K. At this temperature this corresponds to a mean escaping energy of $\langle E_{\perp Ps}\rangle$ = $k_b T$ = 13 meV.

Because of the ordered orientation of the nanochannels, the time spent by the Ps atom in the sample before emission (permanence time) is expected to be short compared to the 142 ns oPs lifetime in vacuum and the Ps time of flight. Assuming a simple stochastic approximation for the random walk of a Ps atom bouncing against the wall of the nanochannel and diffusing towards the open end, the number of collision N is related to the path length L and the channel diameter *d* (single step between collisions) by $N=L^2/d^2$. In the diffusion limit $D_{Ps}=L/2\tau$ where τ, is the mean time for Ps to cover L. For Ps, formed implanting e^+ at 7 keV and escaping into vacuum after cooling from nanochannels with *d*=5 nm, we calculate N=1÷ 2 10^4 and a medium permanence time of τ = 8÷15 ns, considering L in a range between 500 to 700 nm.

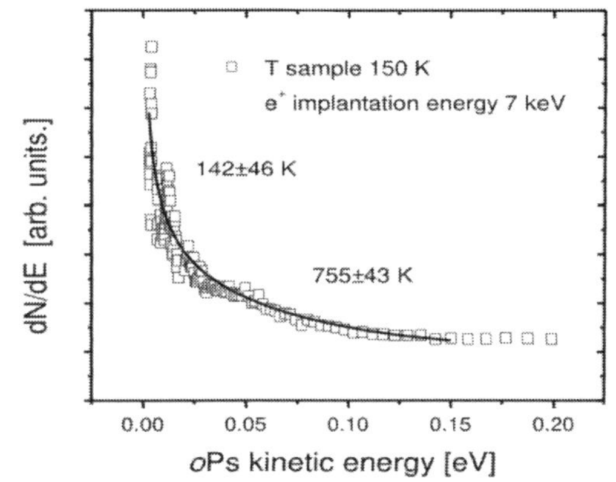

Fig. 2 : Ps kinetic energy spectrum obtained by multiplying by t^3 a TOF spectrum measured at 150 K sample temperature and 7 keV positron implantation energy and 1 cm distance from the target.

4. Conclusions

Oxidized nanochannels in silicon show to be a convenient e^+ → cooled Ps converter for performing advanced experiment. The tunability of the nanochannels size allows one to avoid the quantum confinement and to cool Ps down to cryogenic temperature. In the presented case-study, a Ps thermalized fraction is obtained at 150 K. Preliminary measurements have pointed out a fraction of 50 K cooled Ps in a sample with larger nanochannels.

However, the detailed description of the cooling process is far from being completely understood. Ps starts its cooling history as a classical particle then, progressively, its diffusion and its interaction with the surrounding turn into quantum nature. The possibility of studying the Ps cooling and thermalization in different size nanochannels could help understanding the different cooling regimes. Precise direct measurements of the Ps permanence time before escaping into vacuum would be also of extreme interest for checking the validity of cooling models.

Acknowledgments

The Forschungs-Neutronenquelle Heinz Maier-Leibnitz (FRMII) facility is gratefully acknowledged for allocation of beam time at the NEPOMUC facility.

References

[1] Schultz P J and Lynn K G 1988 *Rev. Mod. Phys.* **60** 701
[2] Mills A P Jr 1995 *Positron and positronium emission spectroscopies* , in *Positron Spectroscopy of Solids*, Dupasquier A and Mills A P eds, (IOS, Amsterdam) p. 209
[3] Brusa R S and Dupasquier A 2010 *Positronium emission and cooling*, in *Physics with many Positrons*, Dupasquier A, Mills A P jr and Brusa R S eds, (North Holland, Amsterdam) p 245
[4] Cassidy D B 2010 *Experiment with many-positron system*, in *Physics with many Positrons*, Dupasquier A, Mills A P jr and Brusa R S eds, (North Holland, Amsterdam) p. 1
[5] Cassidy D B and Mills A P jr 2011 *Phys. Rev. Lett.* **107** 213401
[6] Mills A. P. Jr 2010 *Physics with many Positrons,* in *Physics with many Positrons*, Dupasquier A, Mills A P jr and Brusa R S eds, (North Holland, Amsterdam) p. 77
[7] Cassidy D B, Hisakado T H, Tom H W K and Mills A P jr 2012 *Phys. Rev. Lett.* **108** 133402
[8] Cassidy D B, Hisakado T H, Tom H W K and Mills A P jr 2012 *Phys Rev. Lett.* **108** 043401
[9] Kellerbauer A *et al.*2008 *Nucl. Instrum. Methods Phys. Res.* ***B*** **266** 351
[10] Mariazzi S. et al. 2013 *Nucl. Instrum. Methods Phys. Res. B, submitted*
[11] Mariazzi S, Brusa R S 2011 *J. Phys. : Conf. Series* **262** 012037
[12] Pedersen T S, Danielson J R, Hugenschmidt C, Marx G, Sarasola X, Schauer F, Schweikhard L, Surko C M 2012 *New Journal of Physics* **14** 035010
[13] Mills A P Jr and Leventhal M 2002 *Nucl. Instrum. Methods Phys. Res.* ***B*** **192** 102
[14] Cassidy D B, Hisakado T H, Tom H W K and Mills A P jr 2011 *Phys. Rev. Lett.* **106** 133401
[15] Mariazzi S, Salemi A and Brusa R S 2008 *Phys. Rev. B* **78** 085428
[16] Mazzoldi P, Mattei G, Ravelli L, Egger W, Mariazzi S and Brusa R S 2009 *J. Phys. D: Appl. Phys.* **42** 115418
[17] Nagashima Y, Morinaka Y, Kurihara T, Nagai Y, Hyodo T, Shidara T and Nakara K 1998 *Phys. Rev. B*, **58** 12676
[18] Mariazzi S, Bettotti P, Larcheri S, Toniutti L and Brusa R S 2010 *Phys. Rev. B* **81** 235418
[19] Mariazzi S, Di Noto L, Ravelli L, Egger W, Brusa R S 2013 Journ. of Phys. Conference Series **443** 012061
[20] Mariazzi S, Bettotti P and Brusa R S 2010 *Phys. Rev. Lett.* **104** 243401
[21] Pantel N, Mariazzi S, Toniutti L, Checchetto R, Miotello A, Dirè S and Brusa R S 2007 *J. Phys. D: Appl. Phys.* **40** 5266-5274
[22] He C., Ohdaira T., Oshima N., Muramatsu M., Kinomura A., Suzuki R., Oka T. and Kobayashi, Y. 2007 *Phys. Rev. B* **75** 195404
[23] Cassidy D B, Crivelli P, Hisakado T H, Liszkay L, Meligne V E, Perez P, Tom H W K and Mills A P Jr. 2010 *Phys. Rev.* **A** **81** 012715
[24] Zecca A, Bettonte M, Paridaens J, Karwasz G P and Brusa R S 1998 Meas. Sci. Technol. **9** 409
[25] Egger W 2010 *Pulsed low-energy positron beams in material sciences*, W. Egger in *Physics with many Positrons*, ed. by Dupasquier A, Mills A P jr and Brusa R S eds, (North Holland, Amsterdam) p 419
[26] Hugenschmidt C 2010 *Positron sources and positron beams*, in *Physics with many Positrons* Dupasquier A, Mills A P jr and Brusa R S eds, (North Holland, Amsterdam) p 399
[27] Di Noto L, Mariazzi S, Bettonte M, Nebbia G and Brusa R S 2012 *Eur. Phys. J. D* **66** 118
[28] Brusa R S, Macchi C, Mariazzi S, Karwasz G P, Egger W, Sperr P and G. Kögel G 2005 P*hys. Rev. B* **71** 245320
[29] Cassidy D B, Hisakado T H, Meligne V E, Tom H W K, Mills A P jr. 2010 *Phys. Rev.* **A** **82** 052511
[30] Sauder W C 1968 *J. Res. Natl. Bur. Stand. A* **72** 91
[31] Nagashima Y et al. 1995 *Phys. Rev. A* **52** 258

Journal of Physics: Conference Series **505** (2014) 012039 doi:10.1088/1742-6596/505/1/012039

Possibilities with pulsed polarized high density slow positrons

A P Mills Jr.[1]
Department of Physics and Astronomy, University of California Riverside, Riverside, CA 92521 USA

E-mail: allen.mills@ucr.edu

Abstract. A particularly bright and intense polarized slow positron beam could be formed from isotopically enriched ^{79}Kr produced at a reactor. After moderation with solid Ne, accumulation, compression, and bunching, this type of positron beam would enable a number of experiments including: (1) Long term storage of a neutral polarized electron-positron plasma in a cold box; (2) Pulsed e+ ACAR with a pulsed magnet to measure Fermi surfaces of paramagnetic metals; (3) Single shot measurements of positron annihilation in laser-imploding plasmas; (4) Study of a spin-polarized positronium gas at a density around that of ordinary air to produce a Ps Bose-Einstein condensate at room temperature; (5) High energy polarized positron channelling experiments to study polarized electron spatial wave functions in ferromagnets; and (6) Study of supersonic free expansion spin polarized BEC Ps jets formed from, for example, 10^{11} m=1 triplet Ps atoms created within an open ended 1 μm diameter cylindrical cavity 100 μm in length.

1. Introduction

Positrons are produced with intensities sufficient for multi-user experimentation at nuclear reactors [1, 2, 3, 4] and electron accelerators [5, 6]. Since the positrons are created by pair production from unpolarized photons, the slow positron beams from these sources are unpolarized, i.e. they are produced with no preferential spin orientation. There are many useful and interesting experiments being done with these sources, but one might sometimes wish to use a β^+-decay radioactive source that can produce a beam of polarized positrons [7, 8], for example to examine the state of polarized electrons in a sample [9, 10] and for many other purposes, e.g. [11]. At the present time slow positron beams based on relatively weak (50 mCi) ^{22}Na sources are being used for experiments that require polarized positrons [12]. On the other hand, one may generate an intense polarized slow positron beam starting from positrons pair-produced via backscattering of circularly polarized laser photons [11] from a beam of relativistic electrons [13] or using a radioactive source produced at a reactor [1]. A particularly bright and intense slow positron beam could be formed from isotopically enriched ^{79}Kr produced by neutron activation of ^{78}Kr [14]. In this paper I describe the characteristics of a ^{79}Kr positron source and then briefly discuss some experiments that it would make possible. It is evident that this source could make a valuable addition to the positron facilities at Munich and North Carolina State University [3, 4].

[1] To whom any correspondence should be addressed.

2. Kr-79 polarized slow positron source

^{79}Kr has a half-life of 35 h and yields a 600 keV end-point energy positron in 6.7% of its decays. ^{79}Kr is produced from stable ^{78}Kr which has a thermal neutron capture cross section of ~6 b and a relative abundance of 0.35%. ^{79}Kr has a thermal neutron cross section of ~27 b [15] which implies that a negligible fraction of any ^{79}Kr produced in a reactor will be destroyed. A flux of 10^{13} thermal neutrons per sec per cm^2 at NCSU (or ~10 times more at Munich) and 1 mole of isotopically pure ^{78}Kr would yield 10^{-5} (or 10^{-4}) moles of ^{79}Kr and 10^{10} (or 10^{11}) slow positrons per second assuming a solid Ne moderator efficiency of 0.5% [16]. If the ^{79}Kr were isotopically pure it could be deposited on a 3 mm (or 10 mm) diameter spot to make a bright polarized slow positron source, assuming the moderation process still works at such high intensities. If not, one could use a single crystal W moderator with some loss of yield [17].

The production of ^{79}Kr is very simple, since one needs only to irradiate ^{78}Kr with thermal neutrons by placing it in a sealed container in a suitable location near the core of a nuclear reactor. A hollow container near a reactor core is filled with ^{78}Kr gas via a long pipe. After one or two days, the gas is pumped out and the ^{79}Kr is separated out by the mass-dependence of the sublimation rates [18] and deposited on a 7 K cryogenic surface with a covering of solid Ne to make a slow positron source. The left over Kr gas is pumped back down the pipe to make the next batch of ^{79}Kr. Unfortunately, the deceptive simplicity does not even hint at the real technical and procedural difficulties associated with making a gaseous Kr positron source.

3. Some experiments enabled by pulsed polarized high density slow positrons

This section discusses some experiments that could be enabled by an intense polarized positron source after moderation [19], accumulation [20], compression [21], and bunching [22]. It is remarkable and fortunate that after all of these processes the positrons remain polarized to the same extent that they were when they stopped in the moderator [12]. In this note I am considering a collection of positrons or Ps atoms to be "dense" when

(a) Ps atoms are interacting with other Ps atoms, either via elastic or inelastic scattering, spin exchanging collisions or Ps_2 molecule formation in the presence of a third body or a surface that permits momentum to be conserved.
(b) The positron gas has a measurable Fermi energy or more than one positron is attaching to the same molecule, for example the case of Ps_2O, i.e. positronic water [23].
(c) Single-shot measurements are being performed which require many positrons at the same time, for example experiments requiring the use of a high power pulsed laser, an exploding target or an exploding magnet.

I am using the word "polarized" in the sense that

(a) The positrons have a significant non-zero average spin angular momentum component along some axis.
(b) The positron interaction with the target is dependent on the positron spin direction.

The following are a few things you might think of doing with dense polarized positrons and positronium that might be classified as explorations of interesting composite systems containing antimatter.

3.1. High density unpolarized triplet Ps at low temperature may resist spin exchange

A spin exchange collision between two very slow triplet Ps atoms yields two singlet atoms which move away from each other, each with a kinetic energy equal to the ground state hyperfine interval which is 0.841 meV = $k_B \times 10$ K. Since this energy must be transferred from the Ps internal degrees of freedom to the center of mass the spin exchange rate matrix element will involve an integral over the final state two-body wave function. If there were no Ps-Ps interaction the rate would then be proportional to the temperature of the Ps gas and would vanish at low enough temperature, making it possible to study a vector Ps Bose-Einstein condensate which would have properties analogous to superfluid ^{3}He. However, since the Ps_2 state exists with a binding energy of 0.435 eV, the free particle final state approximation may not be valid and an improved analysis would be recommended before experimenting on this topic. If an experiment were to be done there would be another problem, how to assemble and cool the Ps more rapidly than the spin exchange quenched triplet lifetime at the higher temperatures. One possibility would be to make the cold Ps in a free expansion jet.

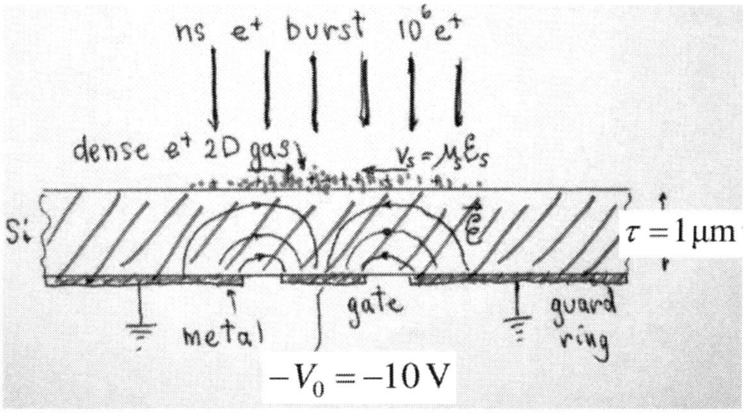

Figure 1. A thin Si sample provided with a back gate and a grounded guard ring to produce an attractive harmonic surface potential for a 2D gas of surface positrons.

3.2. *Polarized* e+ *on a Si FET*

It has recently been found that Ps is formed via what appears to be a surface-excitonic Ps precursor when a Si(100) surface is bombarded with slow positrons [24]. The yield of Ps emitted into vacuum is enhanced when the Si target is irradiated with laser light [25]. It would be interesting to study the mechanism for these processes in detail by observing the likelihood of polarized positrons forming triplet Ps with spin polarized photo electrons produced in selected surface bands. The use of ps laser pulses could reveal something about the formation dynamics [26]. However if dense pulses of positrons were used and the silicon target were supplied with a back-gating structure as shown in Figure 1, one might be able to make a positron super conductor and cause spontaneous positron emission due to the large positronic Fermi energy. Suppose a 1 ns pulse of 2×10^6 positrons is implanted into a Si sample in a ~ 2 μm diameter spot and about half of them diffuse back to the surface to form a 2D surface positrons gas. The positrons are attracted to the surface region over the gate by a nearly harmonic potential energy

$$V(r) = -V_0 \frac{r^2}{\tau^2}$$

in which their oscillation frequency is approximately

$$\omega_0 = 2\pi f_0 = \sqrt{\frac{2eV_0}{m_+\tau^2}} = 250\text{GHz}$$

If the positron surface mobililty is $\mu_+ = 1\ \text{m}^2\text{V}^{-1}\text{s}^{-1}$, the positron damping rate will be

$$\gamma = \frac{e}{m_+\mu_+} = 10^{11}\ \text{s}^{-1}$$

Finally, if the 10^6 positrons are attracted to a 0.1 μm diameter spot, the positron Fermi energy will be on the order of E_F=10 eV, implying the possibility of spontaneous e+ emission and e+ Cooper pairing.

3.3. *Collapse of polarized e+ on a W field emission tip*

The possibility that a dense polarized positron gas moving on the surface of a W field emission tip will collect at the tip to make a 4D overdamped BCS superconductor and a very bright polarized e+ source has been discussed before [27].

3.4. Positronium 2D "BEC"

It is believed that positronium can reside on the surface of a quartz single crystal [28] with a binding energy of about 0.15 eV. A pulse of dense polarized positrons would therefore make a 2D Ps gas which would form a 2D BEC-like state at low temperatures (see Figure 2), with properties very similar to those of the cold 2D polariton gas studied by Kasprzak et al. [29].

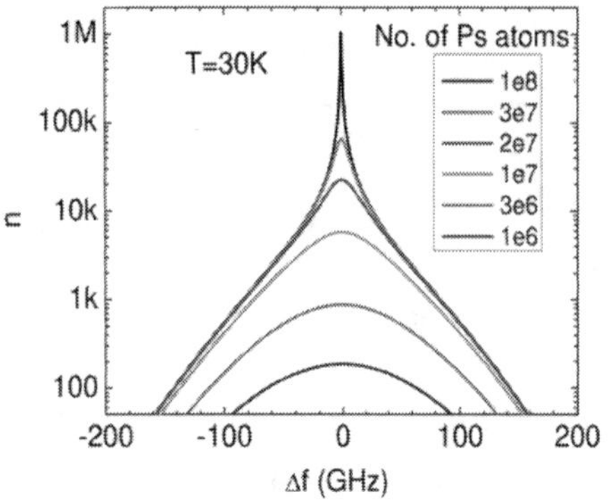

Figure 2. Calculated Doppler distribution of the Ps Lyman-α line at 243 nm for a 2D polarized Ps gas in a 10 μm^2 planar cavity approximating 2D Ps on the surface of quartz. The transition to a 2D BEC (i.e. significant departure from a Maxwell-Boltzmann momentum distribution) is signalled by the appearance of positive curvature in the log of the momentum density.

3.5. Long term storage of a neutral polarized electron-positron plasma

Long term storage of a neutral polarized electron-positron plasma in a cold box [27] could be possible if the walls were coated with solid Ne with perhaps a surface layer of He and the positrons and electrons were both spin polarized so that only triplet collisions would be possible. The Dirac rate is $n\pi r_0^2 c$ so the lifetime of a neutral 100% polarized spin parallel e+ e- plasma is 1 day for density 10^{18} m^{-3} and T=3K. An application of this idea would be for filling a stellarator [30] with a neutral e+e- plasma prior to pulsing the magnetic field.

3.6. Spin-polarized ACAR with a pulsed magnet to measure Fermi surfaces of paramagnetic metals

The theorem of Biasini and Rusz is that one can find the absolute Fermi surface occupancy of a ferromagnetic single crystal with very little distortion by taking the difference between the LCW data for spin up – spin down [31, 32]. An example of the calculated LCW differences between electron spin up and down along two high symmetry directions of a half-metallic ferromagnetic sample is shown in Figure 3. The same technique could be applied to paramagnetic samples which can be significantly polarized in a high magnetic field at low temperature. For example, electrons at the Fermi surface of a paramagnetic metal sample, assuming g=2, have a spin polarization at 4K in B = 10 T:

$$P = \tanh\{\tfrac{1}{2} g\mu \cdot B/kT\} = \sim 0.9$$

One may envision an experiment using a sample in a 10 T field produced by a ~1 mm diameter pulsed magnet firing at 0.1 Hz and probed via 1 ms pulses containing 10^8 spin polarized positrons. The estimated signal rate in this case would be about 10^4 2γ coincidences per pulse detected without significant pile-up by a LYSO PET camera pair 0.5 m in diameter. Using the Biasini-Rusz technique one would be able to measure the undistorted full 3D Fermi surface of any single crystal metal at low temperature, including not only pure metals but also substitutionally disordered alloys [33, 34] which are not amenable to de Haas-van Alphen Fermi surface measurements except at extremely high magnetic fields [35].

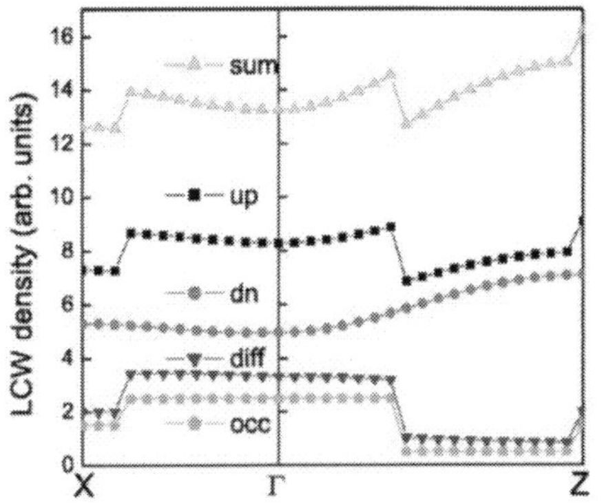

Figure 3. The LCW densities of the up spin, conducting, and down (dn-) spin, insulating, spin-polarized states of CrO_2, together with their sum and difference (diff) along two high-symmetry directions of the Brillouin zone. The occupancy of the up-spin states (occ) is also shown. For the sake of clarity, the occupancy curve is scaled and shifted in the vertical direction. From Ref 32.

3.7. Time-resolved pulsed ACAR for observing the approach to the Ps Bose-Einstein condensate

It has long been hoped that one could produce a spin-polarized positronium gas at a density around that of ordinary air to produce a Ps BEC at room temperature [36, 37]. The Ps BEC would be formed inside a cavity near the surface of a Si wafer as suggested in Figure 4.

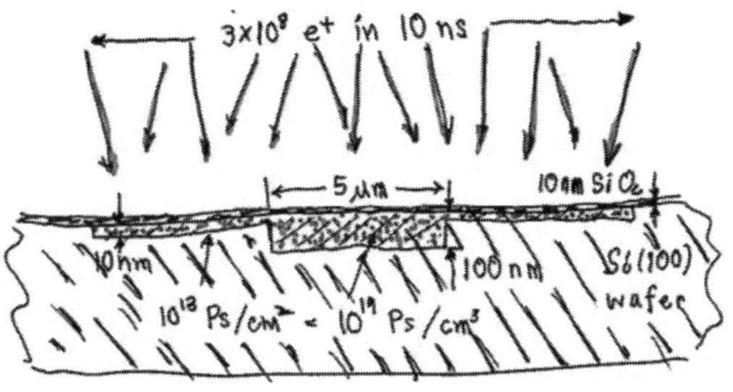

Figure 4. A cavity is formed on the surface of a Si wafer by first forming a depression and then covering it with a thin layer of silica. The design shown here has a top hat structure that might be used to obtain higher densities. The number density hoped for in the drawing is approaching 1 amagat.

The interest would be in obtaining the coldest Ps for spectroscopy, making an atom laser by poking holes in the lid of the cavity, coupling two cavities to obtain Josephson oscillations, observing interesting dynamics in light of the fact that the coherent BEC state is an eigen state of the annihilation operator, and attaining one of the prerequisites for observing stimulated annihilation. The approach to the BEC could be followed using time-resolved angular correlation of annihilation radiation measurements [38, 39].

Acknowledgments

The author is grateful for collaboration with Dr. D. B. Cassidy of University College, London, and with Dr. Adric Jones, Mr. T. H. Hisakado and Prof. H. W. K. Tom of the University of California, Riverside. This work was supported in part by the US NSF grant PHY 1206100.

[1] Lynn K G, Weber M, Roellig L O, Mills A P Jr. and Moodenbaugh A R 1987 A High Intensity Positron Beam at the Brookhaven Reactor in *Atomic Physics with Positrons*, edited by J. W. Humberston and E. A. G. Armour (Plenum, New York, 1987) p. 161.

[2] Sferlazzo P, Berko S, Lynn K G, Mills A P Jr. Roellig L O, Viescas A J and West R N 1988 Evidence for phonon-assisted positronium emission from graphite *Phys. Rev. Lett.* **60** 538.

[3] Hugenschmidt C 2011 The status of the e+ facility at NEPOMUC *J. Phys: Conf. Series* **262** 012002.

[4] Liu M, Moxom J, Hawari A I and Gidley D W 2013 The Intense Slow Positron Beam Facility at the PULSTAR Reactor and Applications in Nano-Materials Study *AIP Conf. Proc.* **1525** 455-459.

[5] Howell R H, et al. 1982 Production of slow positrons with a 100 MeV linac *Appl. Phys. Lett.* **40** 751

[6] Wada K, Hyodo T, Yagishita A, Ikeda M, Ohsawa S, Shidara T, Michishio K, Tachibana T, Nagashima Y, Fukaya Y, Maekawa M and Kawasuso A 2012 Increase in the beam intensity of the linac-based slow positron beam and its application at the Slow Positron Facility, KEK *Eur. Phys. J D* **66** 37.

[7] Jackson J D, Treiman S B and Wyld H W 1957 Possible tests of time reversal invariance in beta decay *Phys. Rev.* **106** 517.

[8] Zitzewitz P W, Van House J C, Rich A and Gidley D W 1979 Spin polarization of low-energy positron beams *Phys. Rev. Lett.* **43** 1281.

[9] Berko S and Zuckerman J 1964 Polarized positron annihilation in ferromagnets *Phys. Rev. Lett* **13** 339.

[10] Kawasuso A, Fukaya Y, Maekawa M, Zhang H, Seki T, Yoshino T, Saitoh E and Takanashi K 2013 Current-induced spin polarization on a Pt surface: A new approach using spin-polarized positron annihilation spectroscopy *Journal of Magnetism and Magnetic Materials* **342** 139.

[11] Hugenschmidt C, Schreckenbach K, Habs D and Thirolf P G 2012 High-intensity and brightness slow e+ source *Appl. Phys. B* **106** 241.

[12] Cassidy D B, Meligne V E and Mills A P Jr 2010 Production of a fully spin-polarized ensemble of positronium atoms *Phys. Rev. Lett.* **104** 173401.

[13] Milburn R H 1963 Electron scattering by an intense polarized photon field *Phys. Rev. Lett.* **10** 75.

[14] Mills A P Jr 1992 ^{79}Kr as a Reactor-based Source of Slow Positrons *Nucl. Sci. Engineer.* **110**, 165.

[15] Kopecky J, Sublet J C, Simpson J A, Forrest R A and Nierop D 1997 *Atlas of Neutron Capture Cross Sections* International Nuclear Data Committee INDC(NDS)-362 (IAEA Nucl Data Section, Wagramerstrasse 5, A-1400 Vienna) p. 84.

[16] Mills A P Jr and Gullikson E M 1986 Solid Neon Moderator for Producing Slow Positrons *Appl. Phys. Lett.* **49** 1121.

[17] Vehanen A, Lynn K G, Schultz P J and Eldrup M 1983 Improved slow positron yield using a single crystal tungsten moderator *Appl. Phys. A* **32** 163.

[18] Bader K and Roth M W 2003 Simulated behavior of krypton/argon mixtures confined between two graphite slabs: new terrain for familiar systems *Surf. Sci* **538** 30-44.

[19] Canter K F, Coleman P G, Griffiths T C and Heyland G R 1972 Measurement of total cross sections for low energy positron-helium collisions *J. Phys. B At. Mol. Phys.* **5** L167.

[20] Surko C M, Leventhal M and Passner A 1989 Positron Plasma in the Laboratory *Phys. Rev. Lett.* **62** 901.

[21] Greaves R G and Surko C M Compression of a positron plasma *Phys. Rev. Lett.* **85**, 1883 (2000).

[22] Cassidy D B, Deng S H M, Greaves R G, and Mills A P Jr 2006 Accumulator for intense positron pulses *Rev. Sci. Instrum.* **77**, 073106.

[23] Schrader D M 1998 Positronic water *Phys. Rev. Lett.* **81** 5113-5116.

[24] Cassidy D B, Hisakado T H, Tom T W K and Mills A P Jr 2011 New Mechanism for Positronium Formation on a Silicon Surface *Phys. Rev. Lett.* **106** 133401.

[25] Cassidy D B, Hisakado T H, Tom T W K and Mills A P Jr. 2011 Photoemission of Positronium from Silicon *Phys. Rev. Lett.* **107** 033401.

[26] Weinelt M, Kutschera M, Schmidt R, Orth C, Fauster T, and Rohlfing M 2005 Electronic structure and electron dynamics at Si(100) *Appl. Phys. A* **80** 995-1003.

Journal of Physics: Conference Series **505** (2014) 012039 doi:10.1088/1742-6596/505/1/012039

[27] Mills A P Jr. 2011 Physics with many positrons *Rivista del Nuovo Cimento* **34**, 151-252, p. 187.
[28] Sferlazzo P, Berko S and Canter K F 1985 Experimental support for physisorbed positronium at the surface of quartz *Phys. Rev. A* **32** 6067.
[29] Kasprzak J, Richard M, Kundermann S, Baas A, Jeambrun P, Keeling J M J, Marchetti F M, Szymańska M H, André R, Staehli J L, Savona V, Littlewood P B, Deveaud B and Dang L S 2006 Bose-Einstein condensation of exciton polaritons *Nature* **443** 409.
[30] Pedersen T T, Danielson J R, Hugenschmidt C, Marx G, Sarasola X, Schauer F, Schweikhard L, Surko C M and E Winkler 2012 Plans for the creation and studies of electron–positron plasmas in a stellarator *New J. Physics* **14** 035010.
[31] Biasini M and Rusz J 2006 Cancellation of probe effects in measurements of spin-polarized momentum density by electron–positron annihilation *J. Phys: Condens. Matter* **18** L289.
[32] Rusz J and Biasini M 2007 Positron wave-function effects in the measurement of the two-dimensional angular correlation of the annihilation radiation of a spin-polarized system *Phys. Rev. B* **75** 235115.
[33] Haghgooie M and Berko S 1979 Studies of the Fermi surface of α-phase CuZn alloys *Proc. 5th Int. Conf. Positron Annihilation* (Japan Institute of Metals, Sendai) p. 291-294.
[34] Berko S 1983 Momentum density and Fermi surface measurements in metals by positron annihilation in *Positron Solid-State Physics, Proc. Int. School of Phys. Enrico Fermi Course LXXXIII* (North-Holland, Amsterdam) p. 64-145.
[35] Mueller F M 1992 De Haas-van Alphen measurements in type II superconductors *Physica B* **177** 47-45.
[36] Platzman P M and Mills A P Jr 1994 Possibilities for BEC of positronium *Phys. Rev. B* **49** 454.
[37] Liang E P and Dermer C D 1988 Laser Cooling of Positronium *Optics Communications* **65** 419.
[38] Nagai Y, Kakimoto M, Hyodo T, Fujiwara K, Ikari H, Eldrup M and Stewart A T 2000 Temperature dependence of the momentum distribution of positronium in MgF_2, SiO_2, and H_2O *Phys. Rev. B* **62** 5531.
[39] Shibuya K, Nakayama T, Saito H and Hyodo T 2013 Spin conversion and pickoff annihilation of orthopositronium in gaseous xenon at elevated temperatures *Phys. Rev. A* **88** 012511.

Gradient correction scheme for bulk and defect positron states in materials: New developments

J Kuriplach[1] **and B Barbiellini**[2]

[1] Department of Low Temperature Physics, Faculty of Mathematics and Physics, Charles University in Prague, V Holešovičkách 2, CZ-180 00 Prague, Czech Republic

[2] Department of Physics, Northeastern University, Boston, Massachusetts 02115, USA

E-mail: jan.kuriplach@mff.cuni.cz

Abstract. As local density approximation positron calculations systematically underestimate positron lifetimes when they are compared with their experimental counterparts, the generalized gradient approximation (GGA) for positrons was introduced in the 1990s, in analogy with the GGA for electrons. New developments in the GGA for positrons are summarized and presented here and it is also discussed how they affect and possibly improve calculated positron lifetimes. In particular, these new GGA approaches are based on the recent perturbed hypernetted-chain and quantum Monte Carlo results.

1. Introduction

The bulk positron lifetime and the affinity are fundamental positron characteristics in materials. They are accessible experimentally and they can be calculated on the basis of the electronic structure of the examined materials. An important aspect of such computations is the type of approach to the electron-positron interactions for positrons experiencing the environment of densely arranged atoms in materials. Though the interaction of a positron with nuclei is of purely electrostatic origin, the interaction of the same positron with electrons requires the consideration of electron-positron correlations in addition to electrostatics. First attempts to describe such complex electron-positron interaction were based on a local density approximation (LDA) using a theory for a positron embedded in a homogeneous electron gas [1]. However, it turned out that, in general, LDA approaches underestimate positron lifetimes, thus some corrections beyond the LDA for positrons are necessary. In analogy with electrons where gradient corrections to the LDA are often successful to cure various LDA deficiencies, a gradient correction scheme for electron-positron correlations was introduced almost 20 years ago [2, 3]. This scheme called generalized gradient approximation (GGA) was quite successful in removing the shortcomings of original positron LDA theories, but still some problems remained (for example, the aluminum calculated lifetime was too long compared to the experiment). The main effect of the GGA was identified to be the reduction of the enhancement factor for core and other more localized electrons.

Here, we present a refinement [4] of the original GGA for positrons. For this purpose, we use the results of a recent quantum Monte Carlo (QMC) study [5] for the homogeneous electron gas with one positron impurity, where accurate parametrizations of the corresponding LDA electron-positron enhancement factor and correlation energy as functions of the electron density were determined. The positron gradient correction is then applied to this LDA based on QMC simulations. We show the results of positron lifetime calculations for selected materials (both metals and semiconductors) and we compare them to available experimental data. We can observe an overall improvement of the agreement between theory and experiment compared to the original GGA scheme. We also compare our new GGA approach to that developed recently by Boroński [6], which is based on the perturbed hypernetted chain (PHNC) approximation [7], and we also notice the similarity between the results obtained with both approaches, except in the case of the alkali metals. For our positron calculations, we make use of precise, self-consistent electronic structures computed without any shape approximation for the electron density and potential. This is important to obtain numerically accurate positron properties. Finally, we demonstrate the capabilities of the new GGA scheme for defect studies and present the positron lifetime and binding energy results for a monovacancy in several materials.

2. Theoretical background and computational methods

The enhancement factor describes the pile-up of electrons around the positron. Enhancement factors are usually expressed in terms of a power series of the density parameter (r_s) where some exponents can be non-integer, like the widely used Boroński and Nieminen (BN) enhancement [8]. In any case, the simplest form to reproduce/fit the results of many-body calculations of the enhancement is the following expression [2]

$$\gamma(r_s) = 1 + 1.23\, r_s + p\, r_s^2 + r_s^3/3\,, \tag{1}$$

where p is the only adjustable parameter, and the linear and cubic terms are fixed (see [2] for details). The density parameter is a radius of a sphere containing one electron, i.e. $4\pi\, r_s^3\, \rho^- / 3 = 1$, ρ^- being the electron density. The fit of QMC data [5] gives $p = -0.22$ whereas the original Arponen and Pajanne (AP) data [1] results in $p = -0.0742$. Interestingly, the Stachowiak and Lach's (SL) enhancement within the PHNC approximation [7] is best reproduced with $p = -0.1375$.

Fig. 1 shows the dependence of the positron lifetime (τ) on the density parameter calculated using the just mentioned LDA enhancements as the inverse of the annihilation rate

$$\lambda = 1/\tau = \pi r_e^2 c\, \rho^-(r_s)\, \gamma(r_s)\,. \tag{2}$$

In this equation, r_e is the classical electron radius and c is the speed of light. One can see that there are quite large differences among approaches investigated, especially for smaller densities (larger r_s), which results in different positron lifetimes for real materials, as demonstrated below. The QMC enhancement [5] (denoted in the following by abbreviation DB) exhibits the longest lifetimes compared to other two approaches: PHNC (or SL) and AP.

The GGA for positrons is introduced using the equation

$$\gamma_{GGA} - 1 = (\gamma_{LDA} - 1) \exp(-\alpha\epsilon)\,, \tag{3}$$

where γ_{GGA} and γ_{LDA} are, respectively, GGA and LDA versions of the enhancement factor. The ϵ quantity characterizes the density gradient and α is an adjustable parameter. In practical calculations, ϵ is chosen to be $(\ln \nabla\rho^- / q_{TF})^2$ where $1/q_{TF}$ is the Thomas-Fermi screening length (see [2] for details). Thus, the original GGA for positrons (denoted further by GC) based on the AP approach was used with $\alpha = 0.22$ [2, 3]. Boroński recently suggested that the SL theory should be used with $\alpha = 0.10$ [6] (marked as SG in the following). And here we suggest [4] that $\alpha = 0.05$ should be used with the QMC (DB) approach (abbreviated as DG hereinafter). In the following, we show how these GGA approaches compare with each other and how they match experimental lifetime results.

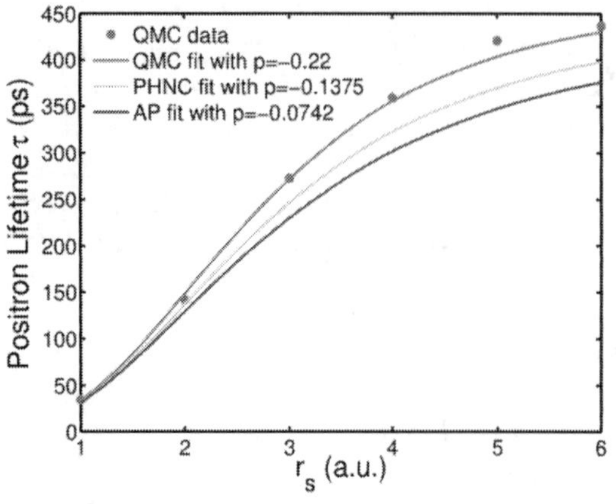

Figure 1. Positron lifetime τ as a function of density parameter r_s for the homogeneous electron gas. Lifetimes calculated using the fitted parameter p in equation (1) for several approaches are shown.

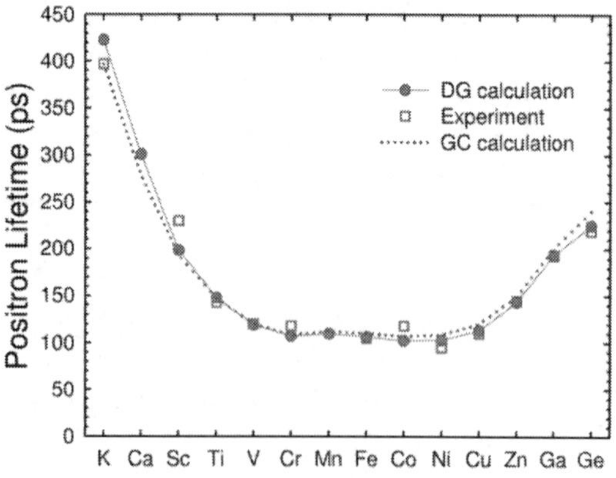

Figure 2. Positron lifetimes for the 4th period of the periodic table of elements. Both calculated (DG and GC approaches) and experimental values are plotted.

The electronic structure calculations for selected materials were carried out using the WIEN2k code [9] which is an implementation of the augmented plane wave plus local orbital (APW+lo) method [10]. This method is considered to be one of the most accurate methods to calculate electronic structure of solids. The WIEN2k program also performs full-potential calculations, which impose no shape restrictions for the electron density and potential. The LDA electron exchange-correlation functional based on QMC simulations by Ceperley and Alder [11] and parametrized by Perdew and Zunger [12] was utilized in the course of the electronic structure calculations.

The positron wave function and energy were obtained using a computer code developed on the basis of a finite difference method [13, 14] to solve the positron Schrödinger equation. The positron potential was constructed using the electron Coulomb potential and a positron correlation potential obtained from the total electron density. Both the Coulomb potential and electron density were taken from self-consistent WIEN2k electronic structure calculations described above. The positron correlation potential in GGA calculations was obtained by scaling [2] the corresponding LDA potential. The positron potential and wave function are calculated on a regular 3D mesh, which covers an appropriately chosen crystallographic unit cell of the studied material. The solver uses a numerical procedure based on the conjugate gradient method and it has been previously used within a non-self-consistent atomic superposition scheme [14]. The positron lifetime is calculated using equation (2) by multiplying its right hand side by the positron density and integrating over the unit cell used in calculations. Various numerical parameters of the WIEN2k code as well as of the positron code were checked and optimized in order to obtain accurate electronic structures and positron wave functions to perform precise positron calculations (the details are described in [4]). As a result, the lifetimes presented in this work are calculated within the precision 0.1 ps. In this way, the comparison with experimental lifetime data is well warranted. It is also possible to test an influence of underlying physical approximations (like electron exchange-correlations) on the positron characteristics with negligible disruption of numerics.

3. Results and discussions

In order to demonstrate the capabilities of the new QMC enhancement and its gradient corrected version, we calculated positron lifetimes for selected elements and compounds. Namely, these are alkali metals, elements of the 4th period of the periodic table and several oxides which attract attention at present.

Alkali metals represent interesting test systems in positron calculations because their conduction electrons behave almost as free, the corresponding electron densities being very low compared e.g. to transition metals. The results of lifetime calculations are presented in table 1. In the third column, we also show the average r_s parameter. The positron lifetime increases with increasing r_s (see equation (2)), i.e. from Li to Rb. The effect of the gradient correction is quite large, especially for the original GGA, and can be attributed to core electrons (see [4]). The best agreement with experiment [15] can be found for GC, LG and DB approaches, the DG values being too long. However, it should be mentioned that these experiments were carried out about 45 years ago and the results might not be very precise considering that no source correction was done, and other effects like the presence of vacancies [16] and impurities and thermalization issues [17] could influence measured lifetimes. Therefore, it would be very desirable to perform new lifetime measurements for alkalies addressing these possible problems.

Being inspired by recent calculations of positron lifetimes for elements of the periodic table [18], we have determined positron lifetimes for the elements of the 4th period. The results are presented in table 2. Fe, Co and Ni were treated as ferromagnetic in calculations whereas an antiferromagnetism of Cr [19] and Mn [20] was not considered. Experimental lifetimes are also given in table 2 and are taken mostly from reviews [16, 21, 22]. To the best of our knowledge, Ca and Mn have not been investigated by means of positron lifetime spectroscopy so far. In table 2, we give most of measured lifetime values in the form "X+", where X is the minimum bulk lifetime detected in experiment. In fact, there is quite big scatter in experimental data for many elements – for instance, lifetimes for Cu typically vary in the (110 – 120) ps range – and specifying the lower limit gives some idea about this uncertainty. The most probable origin of such uncertainties is the source correction which varies among positron laboratories performing measurements (see e.g. [23]).

The lifetime trend across the 4th period can be clearly observed in figure 2. There is a clear minimum in the calculated lifetimes around the center of the period, which corresponds to heavier 3*d* metals having the largest electron densities (smallest volumes per atom [18]). Such a trend was first observed by Brandt *et al.* [24] (see also [18]) and here it is confirmed by more precise calculations. This trend is also seen for experimental values though it is not that apparent due to the scatter in experimental lifetimes. Table 2 also compares the original GGA (GC) with its new version based on QMC results (DG). One can see that the agreement with

Table 1. Lifetimes (in ps) for alkali metals calculated using various approaches to electron-positron correlations. The third column lists the average r_s parameter (in a.u.) based on one conduction *s*-electron per atom. The last column (Exp.) presents experimental lifetimes from [15]. The results for Li and Na were taken from [4].

Metal	Structure	r_s	BN	AP	GC	SL	SG	DB	DG	Exp.
Li	bcc	3.27	299.5	259.4	283.4	277.2	295.0	303.5	315.8	291
Na	bcc	3.99	328.1	290.7	338.6	309.9	341.6	342.8	364.2	338
K	bcc	4.96	373.7	338.5	396.4	360.4	395.0	400.6	422.8	397
Rb	bcc	5.20	376.6	342.0	406.3	364.1	400.5	403.7	426.4	406

Table 2. Lifetime results (in ps) for the 4th period elements (atomic numbers 19–32) calculated using various approaches to electron-positron correlations. The last column indicates experimental values. The results for Fe, Cu and Zn were taken from [4].

Element	Structure	BN	AP	GC	SL	SG	DB	DG	Exp.
K	bcc	373.7	338.5	396.4	360.4	395.0	400.6	422.8	397
Ca	fcc	288.5	250.0	278.3	267.0	283.8	290.7	301.0	–
Sc	hcp	195.1	171.3	192.6	181.7	193.7	191.6	198.3	230
Ti	hcp	144.4	129.7	147.0	136.8	146.2	142.7	147.8	143+
V	bcc	115.1	105.4	120.1	110.5	118.4	115.1	119.3	120+
Cr	bcc	102.6	94.8	108.5	99.2	106.5	103.4	107.2	118+
Mn	α-Mn, A12	104.7	96.7	112.1	101.2	109.3	105.4	109.7	–
Fe	bcc	101.1	93.6	109.9	97.9	106.5	102.1	106.5	105+
Co	hcp	96.6	89.8	106.9	93.8	102.7	97.8	102.5	118+
Ni	fcc	96.9	90.2	108.8	94.1	103.9	98.2	103.2	95+
Cu	fcc	106.4	98.4	120.3	102.9	114.4	107.3	113.2	110+
Zn	hcp	137.1	124.3	149.8	130.7	144.2	136.4	143.6	145+
Ga	α-Ga, A11	187.0	166.2	200.9	175.9	193.2	184.8	194.0	193+
Ge	diamond	214.2	189.8	240.8	201.2	225.2	212.8	225.2	219+

experiment is better for the DG approach. Moreover, the GC approach gives systematically lower lifetimes till Ti whereas in the remaining part of the period the lifetimes are longer. The observed trend through the period clearly shows that positron lifetime spectroscopy is well capable of distinguishing between *sp* and transition elements.

At the beginning of the period, there are two *s* metals (K and Ca) having large atomic volumes (and small electron densities), which results in very long lifetimes (>300 ps). With the increasing atomic number the *d*-block with lighter metals starts, and the atomic volume decreases (electron density increases) in the series, which is reflected in a decrease of the positron lifetime. The last element in this part of the 4th period is Mn having quite complicated cubic structure, which slightly disturbs the decreasing trend. At the beginning of heavier 3*d* metal series, the lifetimes are the shortest with an increase towards the end due to hcp Zn. The next two elements considered, Ga [25] and Ge, confirm the increasing lifetime trend because of their quite open structures (A11 and diamond). Considering the most frequently studied elements Fe and Cu, the lifetimes calculated using the DG (and also SG) approach are very close to experimental ones apart from the scatter mentioned.

Finally, we present the calculated positron lifetimes for several oxides which are frequently studied at present (table 3). So far, we have discussed elements and for these cases metallic or covalent bonding takes place. On the other hand, oxides exhibit ionic bonding (in addition to covalent one) and a large charge transfer occurs between oxygen and other atoms forming the oxides. These effects also influence positron properties since positrons are attracted to oxygen anions and are repelled from cations though it cannot be viewed as a bound state of positrons with oxygen anions; positrons certainly reside in the interstitial space among anions and cations.

In general, experimental positron studies of oxides are rather prone to effects of impurities. For instance, MgO was shown to be affected by unintentional Ga doping [26] which results in the creation of Mg vacancies. When the doping level is decreased under certain limit, the effect disappears and the measured bulk lifetime (see e.g. [27] and references therein) agrees well with calculations, as documented in table 3. Another example is ZnO where H ([28]) and Li ([29])

Table 3. Positron lifetime results (in ps) for selected oxides. Various approaches to electron-positron correlations have been used. The results for MgO and and ZnO were taken from [4].

Oxide	Structure	BN	AP	GC	SL	SG	DB	DG	Exp.
MgO	rock salt	117.8	108.7	145.4	113.8	131.8	119.0	128.4	130
SiO_2	α-quartz	205.8	186.4	297.1	196.7	246.9	207.7	233.2	~261
Cu_2O	cuprite	147.2	133.8	177.2	140.6	161.9	147.2	158.3	~174
ZnO	wurtzite	143.9	131.6	183.3	138.2	162.3	144.6	156.9	151+
ZrO_2	fluorite	128.2	118.0	158.8	123.7	142.5	128.9	138.4	~140
CeO_2	fluorite	138.1	126.7	172.6	133.0	154.0	138.5	149.2	<187

doping plays an important role. Current results (SG and DG) suggest that the lifetime for bulk ZnO is around 160 ps. Furthermore, silica in its α-quartz (or low quartz) form was also studied. Its experimental bulk lifetime is not very well known and was estimated to be about 260 ps in [30]. The calculated lifetimes are apparently shorter (except the GC approach) and this indicates that the mentioned experimental value might require a revision and thereby a new experiment. Cuprite (Cu_2O) is one of the most studied oxides with various practical applications. Like for α-quartz, the bulk lifetime of Cu_2O is not well fixed and recent experiment [31] gives an estimate of 174 ps, which in the light of present results (see table 3) appears to be a bit too long lifetime. Zirconia (ZrO_2), one of the most studied ceramic materials, needs stabilization – e.g. by yttria – to keep the fluorite structure at ambient conditions. The bulk lifetime of ZrO_2 was measured to be approximately 141 ps in [32] and the SG and DG results are well in line with this experimental value. Finally, ceria (CeO_2) exhibiting the same structure as zirconia has many practical applications. It has been only rarely investigated by means of positron annihilation. Therefore, its bulk positron lifetime is not fixed and an estimate based on a study of CeO_2 nanoparticles presented in [33] gives an upper limit of 187 ps, as the shortest defect component detected in samples. Though the authors claim that this component corresponds to a reduced bulk lifetime, it is not very likely when the lifetime results from table 3 are consulted.

Concerning defects studies, we refer readers to our work [4] where we show first results for vacancies in Al, Si and Cu obtained with new GGA enhancements.

4. Conclusions

We have calculated positron lifetimes in selected elements and compounds based on reliable first principles electronic structure calculations taking into account highly precise electronic charge transfers due to an electron self-consistent full potential. The LDA for positrons does not take into consideration charge inhomogeneities due to non metallic charge distribution and the effect of the electron-nuclear interaction which is disrupting the pile-up of electronic charge around the position of the positron. Therefore, we show that gradient corrections to the LDA are still needed in such circumstances despite their intensity is reduced in comparison to the original GGA scheme [2]. On the basis of presented calculations, it is currently difficult to decide – by making comparisons with available experimental positron lifetimes – which is the best GGA approach among the GGA-PHNC (SG) and the GGA-QMC (DG). Thus, further and more precise experiments are highly desirable to answer this important question. Furthermore, we plan to test the new positron GGA's in calculations of the momentum distribution of electron-positron pairs and see whether we can obtain some improvements in addressing the problems outlined e.g. in [34].

Acknowledgments

The authors would like to thank M. Eldrup, D. Keeble and I. Procházka for helpful discussions concerning various aspects of positron lifetime measurements. N.D. Drummond is acknowledged for providing us with numerical results of QMC calculations. This research used resources of the National Supercomputing Center IT4Innovations (Czech Republic), which is supported by the Op VaVpI project number CZ.1.05/1.1.00/02.0070. Partial support by the Czech Science Foundation (project P108/12/G043) is appreciated. B.B. is supported by the US Department of Energy, Office of Science, Basic Energy Sciences Contract No. DE-FG02-07ER46352. He has also benefited from Northeastern University's Advanced Scientific Computation Center (ASCC), theory support at the Advanced Light Source, Berkeley, and the allocation of computer time at NERSC through Grant No. DE-AC02-05CH11231.

References

[1] Arponen J and Pajanne E 1979 *Ann. Phys. (N.Y.)* **121** 343
[2] Barbiellini B, Puska M J, Torsti T and Nieminen R M 1995 *Phys. Rev. B* **51** 7341
[3] Barbiellini B, Puska M J, Korhonen T, Harju A, Torsti T and Nieminen R M 1996 *Phys. Rev. B* **53** 16201
[4] Kuriplach J and Barbiellini B 2013 *submitted to Phys. Rev. B*
[5] Drummond N D, Ríos P L, Needs R J and Pickard C J 2011 *Phys. Rev. Lett.* **107** 207402
[6] Boroński E 2010 *Nukleonika* **55** 9
[7] Stachowiak H and Lach J 1993 *Phys. Rev. B* **48** 9828
[8] Boroński E and Nieminen R M 1986 *Phys. Rev. B* **34** 3820
[9] Blaha P, Schwarz K, Madsen G K H, Kvasnicka D and Luitz J 2001 *WIEN2k: An Augmented Plane Wave and Local Orbitals Program for Calculating Crystal Properties* (Austria: Vienna University of Technology) ISBN 3-9501031-1-2
[10] Sjöstedt E, Nordström L and Singh D J 2000 *Solid State Commun.* **114** 15
[11] Ceperley D M and Alder B J 1980 *Phys. Rev. Lett.* **45** 566
[12] Perdew J P and Wang Y 1992 *Phys. Rev. B* **45** 13244
[13] Puska M J and Nieminen R M 1983 *J. Phys. F: Met. Phys.* **13** 333
[14] Seitsonen A P, Puska M J and Nieminen R M 1995 *Phys. Rev. B* **51** 14057
[15] Weisberg H and Berko S 1967 *Phys. Rev.* **154** 249
[16] Seeger A and Banhart F 1987 *Phys. Stat. Sol. A* **102** 171
[17] Dryzek J, Horodek P and Siemek K 2012 *Nucl. Inst. Methods Phys. Res. B* **291** 22
[18] Campillo Robles J M, Ogando E and Plazaola F 2007 *J. Phys.: Condens. Matter* **19** 176222
[19] Fawcett E 1988 *Rev. Mod. Phys.* **60** 209
[20] Hobbs D, Hafner J and Spišák D 2003 *Phys. Rev. B* **68** 014407
[21] Campillo Robles J M and Plazaola F 2003 *Defect Diff. Forum* **213–215** 141
[22] Saoucha A, Pedersen N J and Eldrup M 1992 *Mater. Sci. Forum* **105–110** 1971
[23] McGuire S and Keeble D J 2006 *J. Appl. Phys.* **100** 103504
[24] Brandt W, Isaacson D and Paulin R 1975 *Phys. Rev. Lett.* **35** 1180
[25] Segers D, Dauwe C, Dorikens M and Dorikens-Vanpraet L 1976 *Appl. Phys.* **10** 121
[26] Tanaka I, Mizoguchi T, Matsui M, Yoshioka S, Adachi H, Yamamoto T, Okajima T, Umesaki M, Ching W Y, Inoue Y, Mizuno M, Araki H and Shirai Y 2003 *Natur. Mater.* **2** 541
[27] Mizuno M, Inoue Y, Sugita K, Araki H, Shirai Y, Mizoguchi T, Tanaka I and Adachi H 2004 *Mater. Sci. Forum* **445-446** 153
[28] Brauer G, Anwand W, Grambole D, Grenzer J, Skorupa W, Čížek J, Kuriplach J, Procházka I, Ling C C, So C K, Schulz D and Klimm D 2009 *Phys. Rev. B* **79** 115212
[29] Johansen K M, Zubiaga A, Makkonen I, Tuomisto F, Neuvonen P T, Knutsen K E, Monakhov E V, Kuznetsov A Y and Svensson B G 2011 *Phys. Rev. B* **83** 245208
[30] Kuriplach J, Anwand W, Brauer G and Skorupa W 2002 *Appl. Surf. Sci.* **194** 84
[31] de la Cruz R M, Pareja R, Díaz L and García-Ramos J V 1989 *Solid State Commun.* **71** 93
[32] Čížek J, Melikhova O, Procházka I, Kuriplach J, Kužel R, Brauer G, Anwand W, Konstantinova T E and Danilenko I A 2010 *Phys. Rev. B* **81** 024116
[33] Chang S, Li M, Hua Q, Zhang L, Ma Y, Ye B and Huang W 2012 *J. Catal.* **293** 195
[34] Laverock J, Haynes T D, Alam M A and Dugdale S B 2010 *Phys. Rev. B* **82** 125127

Journal of Physics: Conference Series **505** (2014) 012041 doi:10.1088/1742-6596/505/1/012041

Positron beam optics for the 2D-ACAR spectrometer at the NEPOMUC beamline

H Ceeh[1], J A Weber[1], C Hugenschmidt[1,2], M Leitner[1] and P Böni[1]
[1] Technische Universität München, Physik Department, Lehrstuhl E21, James-Franck-Straße, D-85748 Garching, Germany
[2] Technische Universität München, Forschungs-Neutronenquelle Heinz Maier-Leibnitz, Lichtenbergstraße 1, D-85748 Garching, Germany

E-mail: hubert.ceeh@frm2.tum.de

Abstract. In the last year a conventional 2D-ACAR spectrometer has been set up and brought to operation at TUM. Once the NEPOMUC beamline is extended to the new experimental hall at the research reactor FRM-II the conventional 2D-ACAR spectrometer will be upgraded with a second sample chamber in order to be integrated to the NEPOMUC beamline facility. This spectrometer will add a complete new quality to 2D-ACAR experiments as it allows to track the evolution of the electronic structure from the surface to the bulk. We present the design features of the positron beam optics and the sample environment.

1. Introduction

Two-dimensional angular correlation of annihilation radiation (2D-ACAR) is a powerful tool to determine the electronic structure of a material. Almost all properties of a material are defined by its electronic structure, in particular its Fermi surface, which is the boundary between occupied and unoccupied states in reciprocal space. 2D-ACAR is used to investigate phenomena which are closely interconnected with the shape and topology of the Fermi surface. Such phenomena are for example the spin density wave anti-ferromagnetism in Chromium or the helical magnetic ordering in GdY. In both of these systems the magnetic ordering is driven by so called nested Fermi surface segments, meaning that different segments of the Fermi surface can be connected by a common spanning vector. These are just two examples among many where nested Fermi surface segments were identified by means of 2D-ACAR investigations[1, 2].

Typically 2D-ACAR studies are performed using β^+-sources such as ^{22}Na [3] with a mean positron energy of 216 keV. This corresponds to material dependent implantation depth of up to 200 μm. Therefore the information of the electronic structure obtained in conventional 2D-ACAR originates almost exclusively from the bulk of the material. However, in many cases it is of high interest to gain insight to the electronic structure at the surface or at the interfaces in layered systems. Once again a prominent example are Chromium thin films and Chromium superlattices which exhibit different properties than bulk Chromium [4]. Therefore, an energy-tunable high intensity positron beam is needed to perform such investigations [5].

In order to operate a 2D-ACAR machine at a positron beamline several design criteria have to be met. To minimize the effect of anisotropic angular resolution the positron beam spot size on the sample has to be as small as possible. Using apertures in the positron beam this could be easily achieved, however at the expense of count rate. Also, the sample needs to be coolable, which allows study of temperature dependent effects and also further improves the angular resolution as the thermal smearing of the positron momentum is decreased.

To meet these criteria two possible options lie at hand: magnetostatic or electrostatic focussing of the positron beam. Though both methods have advantages and disadvantages (see [6] and references therein). We decided to adopt an electrostatic beam focussing because the sample mounting on a cryostat gives rise to several problems, for example, a strong constraint to the magnetic lens system design, or an obstructive magnetic field for the spectrometer. Based on designs of Stadlbauer [7] and Strasser [8] a positron beam optics has been developed which fulfils the requirements to operate a 2D-ACAR spectrometer at a slow positron beam. In the following we present the results of this design study together with an outlook on future measurements.

2. Positron beam optics

In order to focus a positron beam electrostatically the positrons first have to be extracted from the magnetic field. One has to make sure that this process is strongly non-adiabatic, because otherwise the positrons would follow the magnetic flux lines into the field termination. Therefore, two things are required: a long gyration length $l_g = 2\pi\frac{p_l}{qB}$ (with the longitudinal momentum p_l, the magnetic flux density B and the electric charge q) and a steep gradient in the magnetic flux density with $\frac{\nabla B}{B} \gg 1/l_g$. This is achieved by accelerating the positrons and by choosing a small aperture for the field termination. A very elegant way to increase $\frac{\nabla B}{B}$ was implemented by Piochacz [9], he uses magnetic vanes inside the termination aperture to increase the gradient proportional to the number of vanes.

In the present setup a less elaborate scheme is used, the field termination consists of a soft magnetic iron disk with a bore of 25 mm which is placed inside a CF100 flange, which is also soft magnetic (see fig. 1). It is held by three ceramic insulators to create an air gap of 1 mm so an electric potential can be applied to the inner disk. A FEM simulation of the magnetic flux density is shown in figure 3. The NEPOMUC guiding field, which has strength of ≈ 70 G is terminated in the soft magnetic flange ($\mu_r = 5000$) and fed back via mu-metal pipes. The residual flux density after the termination is lower than 0.1 G.

The positron beam is accelerated between two fine meshed grids with a distance of 5 mm about 50 mm before the field termination. In this way it is ensured that the positrons do not acquire additional transverse momentum by $E \times B$ drift, since the electrical and magnetic field lines are parallel within the acceleration gap and the region in the vicinity of the field termination (high $\frac{\nabla B}{B}$) is electrical field free (see fig .2). In this study the accelerating potential was chosen such that the positrons enter the field termination with about 1 keV kinetic energy.

The next component of the beam optics is an Einzel lens which images the beam at the exit of the magnetic field termination. The Einzel lens is designed to operate in deccel-accel mode as this offers higher refractive power compared to accel-deccel mode. The zoom of the Einzel lens is adjusted by ratio $\frac{U_3}{U_2}$ of the potentials at the lens elements 3 and 2 while lens element 1 is kept at the same potential as the magnetic field termination. The lens elements have a tubular geometry with an inner diameter of the lens $D = 50$ mm and a length of $l = 25$ mm. The diameter D is chosen to be about 10 times the expected beam diameter to minimize spherical aerations as the beam travels closer to the optical axis. Also, the focal length f can be tuned in a wider range. For $\frac{eU_2}{E_{e+}} = 1$ (when the potential at lens element 2 matches the beam energy) the lens has its highest refractive power with $f \approx 2D$ (for $U_1 = U_3$). By changing U_3 the focus can be fine tuned.

The drag field which is created by the sample potential and also defines the implantation energy shifts the focal plane towards the sample. The sample itself is elevated 10 mm above a metal plate, which is kept at the same potential. This way the field of view of the detectors remains unobstructed and the electrical field between the lens system and the sample is nearly

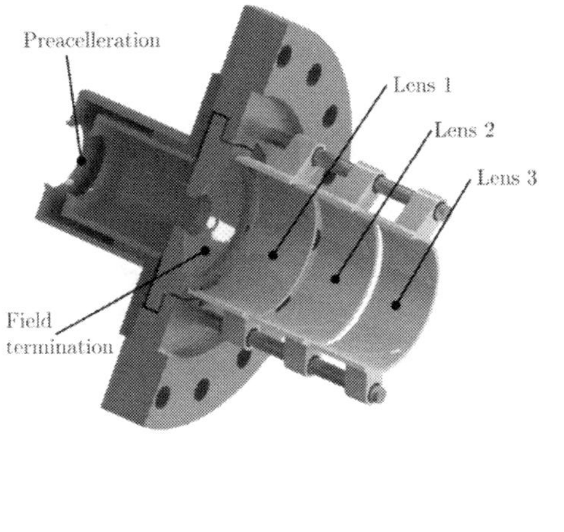

Figure 1. Layout of the beam optics. *Red:* Magnetic field termination with preacceleration. *Green:* Einzel lens for electrostatic foccussing. *Blue:* Insulator elements.

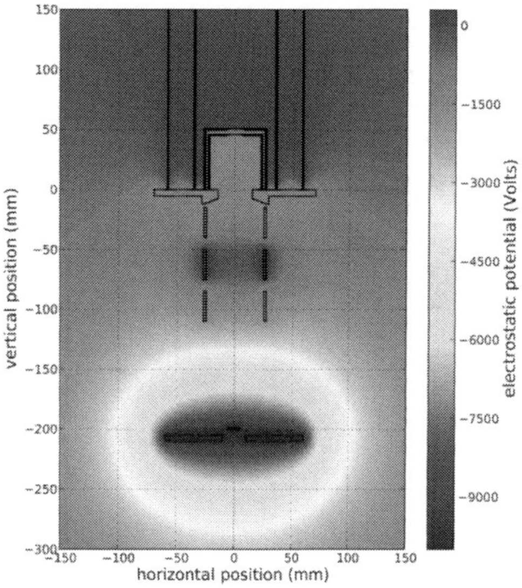

Figure 2. FEM Simulation of the electrostatic potentials exemplary for an implantation energy of 10 keV. The potential at the field termination at the preacceleration is −1 kV.

homogeneous close to the optical axis. With increasing sample potential the focal plane of the Einzel lens has to be moved further away from the sample by decreasing $\frac{U_3}{U_2}$.

3. Performance

Particle tracing calculations using the electric and magnetic fields from the FEM simulation show that a positron beam with realistic parameters ($p_\perp/p_L \leq 0.05$ and $200\,\text{eV} \leq E \leq 1000\,\text{eV}$) can reliably be extracted from the magnetic guiding field (exemplary shown in fig 4 for a 300 eV beam). As a result, a beam spot diameter of (5.0 ± 1) mm for a sample potential of −10 kV is readily achieved, which is sufficient for a 2D-ACAR experiment. The major reason for this limitation is the initial divergence $p_\perp/p_L$ of the beam. In particular the additional transverse momentum that is acquired in the field termination limits the achievable beam focus on the sample. This is due to the fact that in cylindrical symmetric configuration the canonical momentum $p_\Phi = mr^2\dot{\Phi} + qB\frac{r^2}{2}$, which is a conservative of motion, couples the longitudinal and the transverse motion. A positron which is created in a magnetic field with $B = 70\,\text{G}$ at a distance R from the symmetry axis is extracted to $B = 0$ at a distance r will acquire a transverse energy of $E_\perp = \frac{1}{2}m(\omega_g R)^2(\frac{R}{2r})^2$ (with the gyration frequency $\omega_g = \frac{qB}{m}$) which results in this case for $R = r = 3.5\,\text{mm}$ to $E_\perp = 13.2\,\text{eV}$. Therefore, even a monoenergetic beam will experience a smearing of the longitudinal energy which then leads to an imperfect focussing due to chromatic aberrations of the lens system. A way to reduce the amount of additional transverse momentum would be to break the rotational symmetry in the field termination, for example with the magnetic vane design by Piochacz [9].

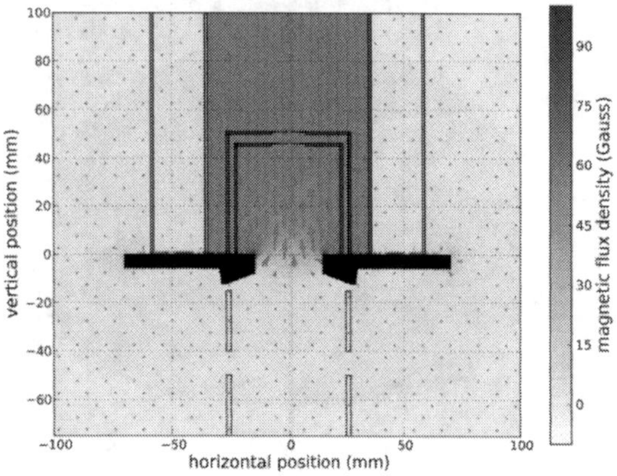

Figure 3. FEM Simulation of the magnetic flux density at the field termination calculated for the nominal operation conditions of the NEPOMUC beamline ($\approx$ 70 G). The absolute value of the magnetic flux density is color coded, the arrows indicate its strength and direction. The strength of the field drops rapidly after the field termination. While the guiding field is homogeneous the field exhibits a significant radial component in the field termination.

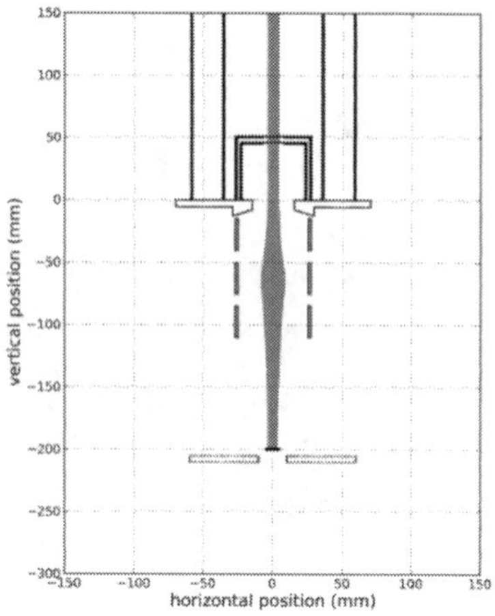

Figure 4. Particle tracing simulation of a 300 eV positron beam with divergence corresponding to $p_\perp/p_L = 0.05$. The diameter of the primary beam is assumed to be 7 mm. The diameter of the beam spot on the sample is in the range of (5.0 ± 1) mm depending on the initial conditions, i.e. the initial beam divergence.

4. Conclusion and Outlook

The extraction and focussing assembly we devised for this study resembles a fusion of different known and applied concepts in positron beam optics. The strength of our design lies in the electrostatic preacceleration prior the extraction of the beam from the magnetic guiding field in combination with a simple yet versatile electrical lens system.

In our simulation we could show that this assembly works under realistic conditions and fulfils the requirements to operate a 2D-ACAR spectrometer at a slow positron beam.

Acknowledgements

This project is funded by the Deutsche Forschungsgesellschaft (DFG) within the Transregional Collaborative Research Center TRR 80 "From electronic correlations to functionality".

References

[1] Hughes R J, Dugdale S B, Major Z, Alam M A, Jarlborg T, Bruno E 2004 *Phys. Rev. B* **69** 174406
[2] Fretwell H M, Dugdale S B, Alam M A, Hedley D C R, Rodriguez-Gonzalez A, Palmer S B 1999 *Phys. Rev. Lett.* **82** 3867
[3] Ceeh H, Weber J A, Hugenschmidt C, Leitner M, Böni P 2013 *Rev. Sci. Instrum.* **84** 043905
[4] Zabel H 1999 *J. Phys-Condens. Mat.* **11** 9303
[5] Falub C V, Mijnarends P E, Eijt S W H, van Huis M A, van Veen A, Schut H 2002 *Phys. Rev. B* **7** 075426
[6] Coleman P 2000 *Positron Beams* (Singapore: World Scientific) pp 22-33
[7] Stadlbauer M, Hugenschmidt C, Schreckenbach K 2008 *Appl. Surf. Sci.* **255** 136
[8] Strasser B 2001 *Dissertation* Technische Universität München
[9] Piochacz C, Kögel G, Egger W, Hugenschmidt C, Mayer J, Schreckenbach K, Sperr P, Stadlbauer M, Dollinger G 2008 *Appl. Surf. Sci.* **255** 98

On the sensitivity of positron annihilation signals to alloy homogeneity in $In_xGa_{1-x}N$

F Tuomisto[1], **V Norrman**[1] **and I Makkonen**[2]

[1] Department of Applied Physics, Aalto University, POB 11100, 00076 Aalto, Finland

[2] COMP Centre of Excellence, Helsinki Institute of Physics and Department of Applied Physics, Aalto University, POB 11100, 00076 Aalto, Finland

E-mail: filip.tuomisto@aalto.fi

Abstract. We present results of theoretical calculations of positron annihilation signals in InGaN alloys with and without vacancies. We demonstrate the sensitivity of the signals to the different configurations of the In/Ga atoms in $In_{1-x}Ga_xN$ supercells.

1. Introduction

The two most used methods in defect studies with positron annihilation are the positron lifetime and Doppler broadening (of the positron-electron annihilation radiation) spectroscopy [1]. These techniques are very efficient in giving important information on vacancy defects in metals and semiconductors: vacancies can be identified (sublattice in compounds, size in the case of vacancy clusters, and decoration by impurities), their charge states can be determined and their concentrations can be evaluated in the technologically important range from 10^{15} to 10^{19} cm^{-3}. Thanks to recent developments in theoretical calculations, computational studies can be directly compared with positron experiments providing possibilities for very detailed interpretations of experimental data [2–6].

The identification of a vacancy defect is at its best when the host lattice has perfect crystalline order. Good examples of such cases include the vacancy-donor complexes in Si [7–10], the Ga and As vacancies in GaAs [11–14], or the Ga vacancy-impurity complexes in GaN [15–18]. In some cases even small substitutional impurities can act as vacancy defects and be identified, such as Li_{Zn} in ZnO [19,20]. As positron trapping to negative defects is temperature dependent, manipulation of the defect charge states through optical illumination can be used to bring additional detail to the identification, as in the case of the EL2 defect in GaAs [21,22] or vacancy clusters in diamond [23–25].

Many of the technologically important semiconductors are in fact alloys instead of simple elemental or compound systems. Detailed identification of vacancy defects has been possible for example in $Si_{1-x}Ge_x$: vacancy-fluorine [26,27] and vacancy-donor [28–30] complexes have been studied quite extensively. However, the random nature of the alloy brings additional complication to the analysis, as the immediate vicinity of the vacancy defect is not well defined. For example, the three group IV atoms surrounding the vacancy in the *V*-Sb complex in $Si_{1-x}Ge_x$ can all be Ge (or all Si). As the core electron shells of Ge and Si are very different, the effect on the annihilation signals is strong.

The main advantage of the III-nitride family of compounds (AlN, GaN and InN) is the possibility to synthesize alloys in order to tune the optical properties (band gap) for the fabrication of opto-electronic devices. Hence it is important to understand the physics of defects in these alloys. In addition, the alloying seems to generate material with elevated point (vacancy) defect concentrations

[31–33]. However, the random alloy nature makes the detailed identification of the detected defects very difficult [34,35], even if it is possible in the respective binary compounds [36–40].

In this paper we present results of theoretical calculations of positron annihilation signals in InGaN systems with and without vacancies. We demonstrate the sensitivity of the signals to the different configurations of the In/Ga atoms. We discuss these results in the light of published experimental data in thin film InGaN alloys and other alloyed materials systems. We also present our view on future developments required for more detailed analysis of defects in semiconductors with less perfect crystalline order.

2. Methods

Our computational scheme [4] is based on the zero-positron density limit of the two-component density functional theory [41]. Valence electron densities are calculated self-consistently using the local density approximation (LDA) and projector augmented-wave method (PAW) [42] implemented in the VASP code [43]. The electronic structure calculations for bulk systems are performed using a 16-atom InGaN wurtzite supercell, with 34 different ways of distributing the In and Ga atoms in the cell. For systems containing a vacancy the supercell size was increased to 128 atoms by multiplying the original 16-atom cell. Ionic positions are relaxed with a convergence criterion of 0.01 eV/Å for forces. Gallium 3*d* and indium 4*d* electrons are included as valence electrons and an energy cutoff of 400 eV is chosen. The Brillouin zone is sampled with a 4×4×4 Monkhorst-Pack **k**-point mesh. After deriving the electron densities in the lattice, the positron densities are solved independently in the calculated Coulomb potential due to electrons and nuclei and the *e-p* correlation potential evaluated with in LDA [41]. This is the so-called "conventional scheme", in which the positron does not affect the average electron density, and further, the *e-p* correlation potential is approximated in the zero-positron-density limit. Momentum densities of annihilating electron-positron pairs are calculated using the state-dependent enhancement model [44] within LDA.

In order to compare the calculated three-dimensional (3D) momentum density to one-dimensional (1D) experimental spectra, we integrate the calculated spectra over the wurtzite *m* plane. To account for the experimental detector resolution, the calculated momentum distributions are additionally convoluted with a Gaussian resolution function of 0.66 a.u. FWHM The line-shape parameters are calculated from the spectra using momentum windows $S < 0.40$ a.u. and $1.53 < W < 3.93$ a.u. The positron lifetime τ is determined as the inverse of the annihilation rate λ obtained from the calculations. For further details on the computational methods, the reader is referred to [3,4,38].

3. Results and discussion

3.1. Bulk InGaN systems

Figures 1–3 show the *S* and *W* parameters and the positron lifetimes calculated in 16-atom InGaN supercells with varying In content and In-Ga distributions. Interestingly, the *S* and *W* parameters (figures 1 and 2) change in different ways with increasing In content. The change in *W* parameter is very close to linear, while the S parameter exhibits strong bowing. The dotted curves are drawn to guide the eye, but also represent a simple model where positrons are considered to be attracted to GaN-resembling and InN-resembling lattices in different ways. For example, in figure 1 the dotted curve is drawn assuming that the relative weights of GaN and InN S parameters are 64% and 36%, respectively. Hence it would seem that from the *S* parameter point of view, positrons are almost twice as likely to annihilate in the GaN-resembling part than in the InN-resembling part of the InGaN lattice. On the other hand, the *W* parameters and the positron lifetimes do not exhibit any significant bowing (or preference), here the relative weights can be modelled as 53% and 47% for GaN and InN, respectively. The bowing seen in figure 4 hence fully originates from the *S* parameter bowing.

Both the S and W parameters for a given In content exhibit scatter as a function of the In-Ga distribution in the supercell, and the largest variations are seen for $In_{0.5}Ga_{0.5}N$. The variations are much smaller in the positron lifetime (figure 3). Interestingly, also the largest variations in *S* and *W* follow the simple model (dotted curve in figure 4), even if they cross the linear behaviour in figures 1 and 2.

The largest variations in the $In_{0.375}Ga_{0.625}N$ – $In_{0.625}Ga_{0.375}N$ supercells (points clearly separated from the others) are due to effective superlattice structure produced by the supercell.

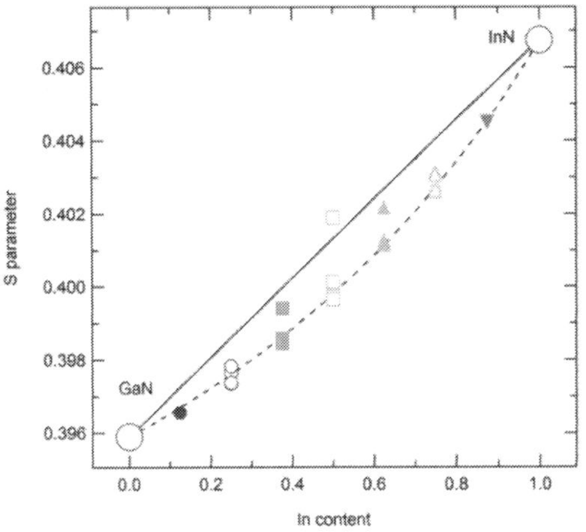

Figure 1. *S* parameters for $In_xGa_{1-x}N$ supercells with varying In-Ga distributions as a function of In content.

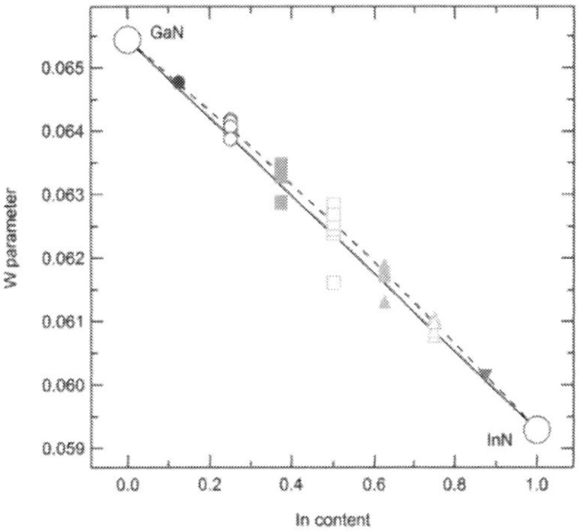

Figure 2. *W* parameters for $In_xGa_{1-x}N$ supercells with varying In-Ga distributions as a function of In content.

It is important to note that the above calculated positron annihilation parameters are not directly comparable on the absolute scale to experimental values. For example, the bulk positron lifetimes for both InN and GaN are 25–30 ps shorter than those determined experimentally. This is in line with earlier comparisons when the LDA approximation is employed [17,34,38]. The relative changes in the positron lifetime can, however, be considered reliable as found in these reports. The situation with the *S* and *W* parameters is more complex. In experiments we have $S_{InN} / S_{GaN} \approx 1.045$ [36], while in the calculations the ratio is 1.028. For the *W* parameter experiments give $W_{InN} / W_{GaN} \approx 0.86$, and the calculated ratio is 0.89. Interestingly, the experiments and theory are closer matched for the *W* parameter.

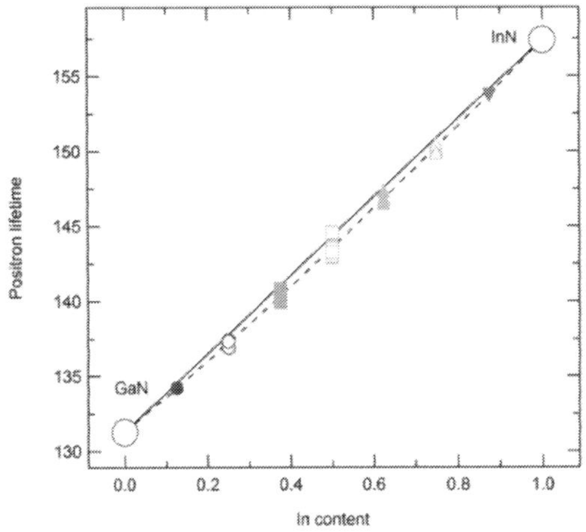

Figure 3. Positron lifetimes for $In_xGa_{1-x}N$ supercells with varying In-Ga distributions as a function of In content.

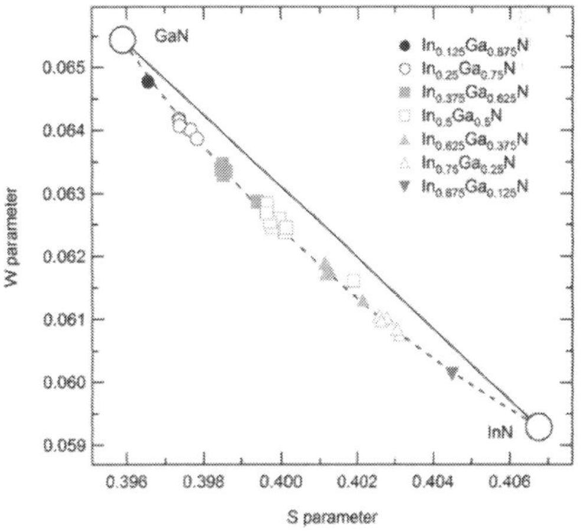

Figure 4. *S* and *W* parameters for $In_xGa_{1-x}N$ supercells with varying In-Ga distributions as a function of In content.

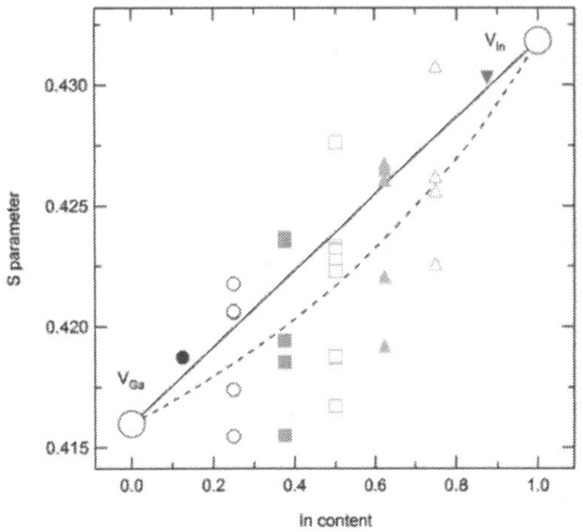

Figure 5. *S* parameters for metal vacancy containting $In_xGa_{1-x}N$ supercells with varying In-Ga distributions as a function of In content.

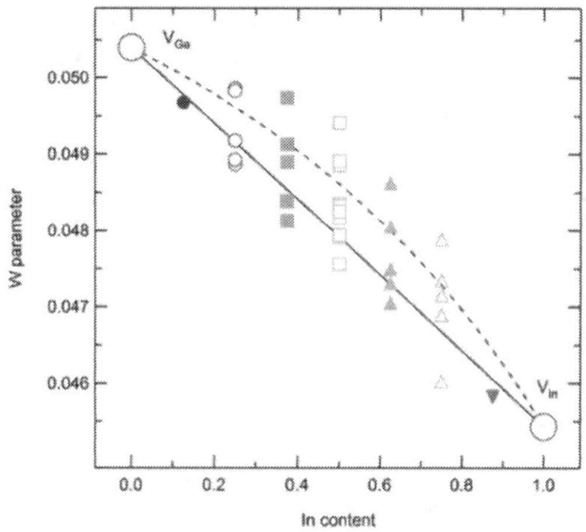

Figure 6. *W* parameters for metal vacancy containting $In_xGa_{1-x}N$ supercells with varying In-Ga distributions as a function of In content.

3.2. Vacancy-containing InGaN systems

Figures 5–7 show the S and W parameters and the positron lifetimes calculated in 128-atom InGaN supercells containing one metal vacancy with varying In content and In-Ga distributions. The removed metal atom was chosen at random in the cell, and only one vacancy for each of the 34 above-descibred systems was considered. Agreement with experiment is similar as in the case of bulk systems.

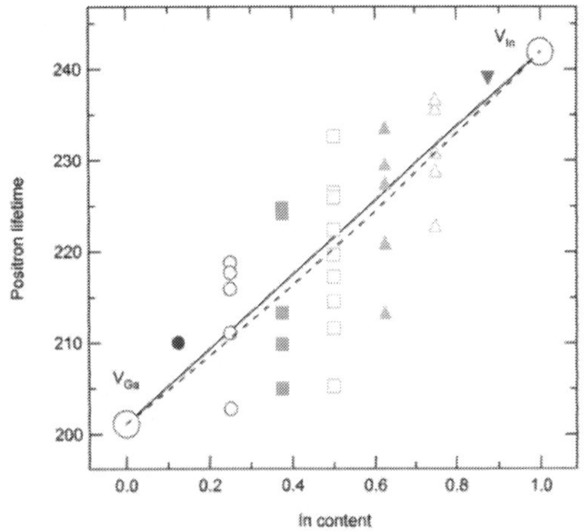

Figure 7. Positron lifetimes for metal vacancy containting $In_xGa_{1-x}N$ supercells with varying In-Ga distributions as a function of In content.

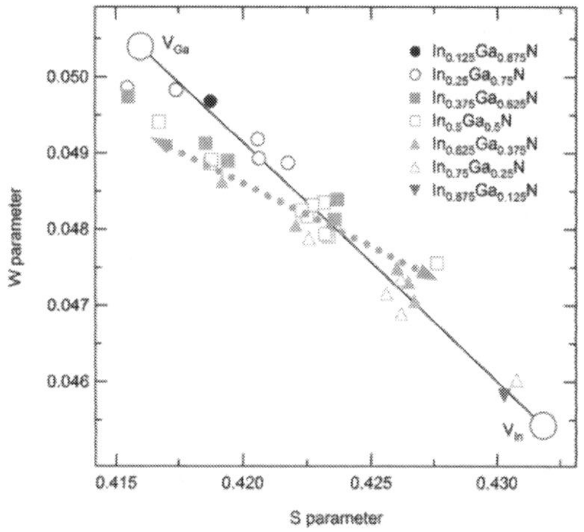

Figure 8. *S* and *W* parameters for metal vacancy containting $In_xGa_{1-x}N$ supercells with varying In-Ga distributions as a function of In content.

The dotted lines in the figures represent a similar model as for the bulk (relative weights for V_{Ga} and V_{In} are 64% and 36%, respectively, in figures 5 and 6), but as the scatter is much more important than in the bulk supercells, the main observation that should be made is that the points are clearly below the linear behaviour in figure 5 and above in figure 6. The scatter is very large also for the calculated positron lifetimes, covering 70% of the total difference between V_{Ga} and V_{In} for the $In_{0.5}Ga_{0.5}N$ supercell (figure 7). Interestingly, as seen in figure 8, the S and W parameters do not deviate strongly from the V_{Ga}-V_{In} line, even if the strong scatter (shown by the green arrow

$In_{0.5}Ga_{0.5}N$) is evident. The number of Ga/In atoms in the 12-atom next-neighbor shell surrounding the metal vacancy has a strong influence on the annihilation parameters too, but detailed correlation of these effects requires further studies.

3.3. Notes and comparison to other alloys

An important observation is that the above considerations suggest that it is very difficult to distinguish the effects of low vacancy concentrations from alloy homogeneity effects in conventional Doppler broadening experiments in $In_{1-x}Ga_xN$ materials. Experimental investigations also in elemental semiconductor alloys, namely $Si_{1-x}Ge_x$, have suggested similar interpretations [28-30]: after introducing vacancy defects by irradiation and then removing them by annealing, the lattice Si/Ge distribution changes in such a way that the S and W parameters are different from the original situation. Further investigations, both theoretical and experimental, are necessary in order to elucidate whether positron annihilation methods could be used to assess, e.g., In clustering in $In_{1-x}Ga_xN$ in a similar manner as in metal alloys [45].

4. Summary

We have performed preliminary theoretical calculations of positron annihilation signals in InGaN supercelles with and without metal vacancies. Our results demonstrate the strong sensitivy of the Doppler broadening S and W parameters as well as of the positron lifetime to the alloy (dis)order in $In_{1-x}Ga_xN$. Related observations can be made in other alloyed semiconductors such as $Si_{1-x}G_xe$.

Acknowledgment
This work was partially funded by the Academy of Finland.

References

[1] Tuomisto F and Makkonen 2013 I *Rev. Mod. Phys.* **85** 1583
[2] Ishibashi S 2004 *Mater. Sci. Forum* **445-446** 401
[3] Makkonen I, Hakala M and Puska M J 2005 *J. Phys. Chem. Solids* **66** 1128
[4] Makkonen I, Hakala M and Puska M J 2006 *Phys. Rev. B* **73** 03513
[5] Ishibashi S, Tamura T, Tanaka S, Kohyama M and Terakura K 2007 *Phys. Rev. B* **76** 153310
[6] Wiktor J, Jomard G, Torrent M and Bertolus M 2013 *Phys. Rev. B* **87** 235207
[7] Lawther D W, Myler U, Simpson P J, Rousseau P M, Griffin P B and Plummer J D 1995 *Appl. Phys. Lett.* **67** 3575
[8] Saarinen K, Nissilä J, Kauppinen H, Hakala M, Puska M J, Hautojärvi P and Corbel C 1999 *Phys. Rev. Lett.* **82** 1883
[9] Ranki V, Pelli A and Saarinen K 2004 *Phys. Rev. B* **69** 115205
[10] Rummukainen M, Makkonen I, Ranki V, Puska M J, Saarinen K and Gossmann H-J L 2005 *Phys. Rev. Lett.* **94** 165501
[11] LeBerre C, Corbel C, Brozel M R, Kuisma S, Saarinen K and Hautojärvi P 1994 *J. Phys.: Condens. Matter* **6** L759
[12] Le Berre C, Corbel C, Saarinen K, Kuisma S, Hautojärvi P and Fornari R 1995 *Phys. Rev. B* **52** 8112
[13] Kuisma S, Saarinen K, Hautojärvi P, Corbel C and LeBerre C 1996 *Phys. Rev. B* **53** 9814
[14] Gebauer J, Lausmann M, Staab T E M, Krause-Rehberg R, Hakala M and Puska M J 1999 *Phys. Rev. B* **60** 1464
[15] Saarinen K, Laine T, Kuisma S, Nissilä J, Hautojärvi P, Dobrzynski L, Baranowski J M, Pakula K P, Stepniewski R, Wojdak M, Wysmolek A, Suski T, Leszczynski M, Grzegory I and Porowski S 1997 *Phys. Rev. Lett.* **79** 3030
[16] Hautakangas S, Oila J, Alatalo M, Saarinen K, Liskay L, Seghier D and Gislason H P 2003 *Phys. Rev. Lett.* **90** 137402

[17] Hautakangas S, Makkonen I, Ranki V, Puska M J, Saarinen K, Xu X and Look D C 2006 *Phys. Rev. B* **73** 193301
[18] Nykänen H, Suihkonen S, Kilanski L, Sopanen M and Tuomisto F 2012 *Appl. Phys. Lett.* **100** 122105
[19] Johansen K M, Zubiaga A, Makkonen I, Tuomisto F, Neuvonen P T, Knutsen K E, Monakhov E V, Kuznetsov A Yu and Svensson B G 2011 *Phys. Rev. B* **83** 245208
[20] Johansen K M, Zubiaga A, Tuomisto F, Monakhov E V, Kuznetsov A Yu and Svensson B G 2011 *Phys. Rev. B* **84** 115203
[21] Krause R, Saarinen K, Hautojärvi P, Polity A, Gärtner G and Corbel C 1990 *Phys. Rev. Lett.* **65** 3329
[22] Saarinen K, Kuisma S, Hautojärvi P, Corbel C and LeBerre C 1994 *Phys. Rev. B* **49** 8005
[23] Dannefaer S, Pu A and Kerr K 2001 *Diamond Rel. Mater.* **10** 2113
[24] Mäki J-M, Tuomisto F, Varpula A, Fisher D, Khan R U A and Martineau P M 2011 *Phys. Rev. Lett.* **107** 217403
[25] Mäki J-M, Kuittinen T, Korhonen E and Tuomisto F 2012 *New J. Phys.* **14** 035023
[26] Pi X D, Burrows C P and Coleman P G 2003 *Phys. Rev. Lett.* **90** 155901
[27] Edwardson C J, Coleman P G, Mubarek H A W E and Gandy A S 2012 *J. Appl. Phys.* **111** 073510
[28] Kuitunen K, Tuomisto F and Slotte J 2007 *Phys. Rev. B* **76** 233202
[29] Kilpeläinen S, Kuitunen K, Tuomisto F, Slotte J, Radamson H H and Kuznetsov A Yu 2010 *Phys. Rev. B* **81** 132103
[30] Kilpeläinen S, Tuomisto F, Slotte J, Lundsgaard Hansen J and Nylandsted Larsen A 2011 *Phys. Rev. B* **83** 094115
[31] Onuma T, Chichibu S F, Uedono A, Sota T, Cantu P, Katona T M, Keading J F, Keller S, Mishra U K, Nakamura S and DenBaars S P 2004 *J. Appl. Phys.* **95** 2495
[32] Slotte J, Tuomisto F, Saarinen K, Moe C G, Keller S DenBaars S P 2007 *Appl. Phys. Lett.* **90** 151908
[33] Uedono A, Tenjinbayashi K, Tsutsui T, Shimahara Y, Miyake H, Hiramatsu K, Oshima N, Suzuki R and Ishibashi S 2012 *J. Appl. Phys.* **111** 013512
[34] Uedono A, Ishibashi S, Watanabe T, Wang X Q, Liu S T, Chen G, Sang L W, Sumiya M and Shen B 2012 *J. Appl. Phys.* **112** 014507
[35] Uedono A, Tsutsui T, Watanabe T, Kimura S, Zhang Y, Lozac'h M, Sang L W, Ishibashi S and Sumiya M 2013 *J. Appl. Phys.* **113** 123502
[36] Tuomisto F, Pelli A, Yu K M, Walukiewicz W and Schaff W J 2007 *Phys. Rev. B* **75** 193201
[37] Tuomisto F, Ranki V, Look D C and Farlow G C 2007 *Phys. Rev. B* **76** 165207
[38] Rauch C, Makkonen I and Tuomisto F 2011 *Phys. Rev. B* **84** 125201
[39] Mäki J-M, Makkonen I, Tuomisto F, Karjalainen A, Suihkonen S, Räisänen J, Chemekova T Yu and Makarov Yu N 2011 *Phys. Rev. B* **84** 081204
[40] Rauch C, Tuomisto F, King P D C, Veal T D, Lu H and Schaff W J 2012 *Appl. Phys. Lett.* **101** 011903
[41] Boroński E and Nieminen R M 1986 *Phys. Rev. B* **34** 3820
[42] Blöchl P E 1994 *Phys. Rev. B* **50** 17953
[43] Kresse G and Furthmüller J 1996 *Phys. Rev. B* **54** 11169
[44] Alatalo M, Barbiellini B, Hakala M, Kauppinen H, Korhonen T, Puska M J, Saarinen K, Hautojärvi P and Nieminen R M 1996 *Phys. Rev. B* **54** 2397
[45] Nagai Y, Hasegawa M, Tang Z, Hempel A, Yubuta K, Shimamura T, Kawazoe Y, Kawai A and Kano F 2000 *Phys. Rev. B* **61** 6574

Journal of Physics: Conference Series **505** (2014) 012043 doi:10.1088/1742-6596/505/1/012043

Positron annihilation in flight: experiment with slow and fast positrons

J Čížek[1], M Vlček[1], F Lukáč[1], O Melikhova[1], I Procházka[1], W Anwand[2], A Wagner[2], M Butterling[2,3], R Krause-Rehberg[3]

[1]Charles University in Prague, Faculty of Mathematics and Physics, V Holešovičkách 2, 180 00 Praha 8, Czech Republic

[2]Institut für Strahlenphysik, Helmholtz-Zentrum Dresden-Rossendorf, Bautzner Landstr. 400, D-01328 Dresden, Germany

[3]Martin-Luther University, Dept. of Physics, 06099 Halle, Germany

E-mail: jakub.cizek@mff.cuni.cz

Abstract. A novel digital coincidence Doppler broadening (D-CDB) spectrometer was employed for energy resolved investigations of two-quantum annihilation-in-flight (TQAF). The TQAF phenomenon was studied using monoenergetic positrons produced in a slow positron beam and also using fast positrons. Because of a low background the measurements on the slow positron beam could be performed in a close geometry and the TQAF contribution in the two-dimensional gamma ray energy spectra fills a 'bowl-like' area delimited by a hyperbolic curve and a kinematical cut-off determined by the kinetic energy of positrons. With decreasing positron energy the area of TQAF contribution becomes smaller and disappears completely for slow positrons with energies below ~ 100 eV. The measurements with fast positrons were restricted to a limited range of angles between the annihilation gamma rays and the TQAF events contribute to a hyperbolic band in gamma ray energy spectrum.

1. Introduction

An energetic positron implanted into a solid loses rapidly its kinetic energy in collisions with electrons and reaches quickly thermal equilibrium with the surrounding medium. As a consequence positrons are annihilated predominantly in thermalized state. However, a small fraction of positrons is annihilated before reaching the thermal equilibrium. These so called annihilation-in-flight (AiF) events differ significantly from annihilations of thermalized positrons because the positron momentum substantially exceeds the momentum of electrons and causes a large Doppler shift of the annihilation radiation. Two-quantum annihilation-in-flight (TQAF) is the dominating AiF channel known for a long time [1]. Coincidence Doppler broadening (CDB) technique introduced by Lynn et al. [2] is based on a precise measurement of the Doppler shift of annihilation radiation and becomes nowadays widely used for characterization of the chemical environment of defects. Weber et al. [3] demonstrated that CDB technique is a suitable tool also for investigation of the TQAF phenomenon. Since TQAF events occur rarely a low background measurement is crucial. A novel (D-CDB) spectrometer [4] developed recently enables to achieve an extremely low background because of detailed shape analysis which eliminates pulses distorted by pile up or by slow charge collection (ballistic deficit). In the present work a D-CDB spectrometer was employed for investigations of the TQAF phenomenon using monoenergetic slow positrons and also fast positrons with continuous energy spectrum.

2. Experimental details

The CDB investigations were performed using a D-CDB spectrometer [4] equipped with two HPGe detectors. To improve the signal-to-noise ratio detector pulses were amplified and sharpened to semi-Gaussian shape in spectroscopy amplifiers. Shaped pulses were sampled by a two channel 12-bit digitizer Acqiris DC 440. Analysis of sampled waveforms was performed off-line by software using the algorithm described in Ref. [4]. The CDB spectra are presented as two-dimensional plots of the sum of energies of annihilation gamma rays $E_1 + E_2 - 2m_0c^2$ (the rest energy of annihilating pair subtracted) versus the difference of these energies $E_1 - E_2$.

TQAF events were investigated using positrons from various sources: (i) monoenergetic positrons with energy adjustable in the range from 0.03 to 36 keV created in a variable energy slow positron beam SPONSOR [5], (ii) fast positrons emitted by $^{68}Ge/^{68}Ga$ positron generator and exhibiting a continuous energy spectrum with the end-point-energy of 1897 keV, and (iii) fast positrons with a continuous energy spectrum up to 16 MeV created from bremsstrahlung radiation by pair production in a target, so called gamma-induced positron spectroscopy (GiPS) [6].

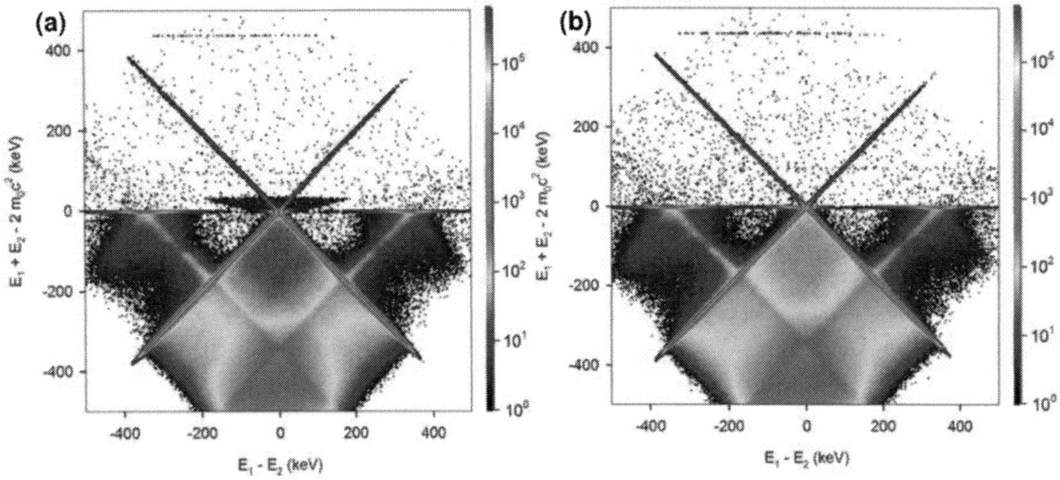

Figure 1. CDB spectra for Fe target obtained using positrons with energy of 35 keV (a) and 27 eV (b).

3. Results and discussion

The CDB spectra measured on the slow positron beam using a 0.5 mm thick Fe target are plotted in Fig. 1. In these measurements HPGe detectors were positioned face-to-face in a distance of 45 mm from the target and perpendicularly to the beam direction. Fig. 1(a) shows the CDB spectrum for positrons with kinetic energy of 35 keV. The TQAF contribution is clearly visible as a 'bowl-like' area located above the main annihilation peak of thermalized positrons. From the conservation of energy and momentum it follows that the sum of energies of gamma rays emitted in a TQAF event is related to the difference of these energies through the relation [7]

$$E_1 + E_2 - 2m_0c^2 = \sqrt{(E_1 - E_2)^2 + \left(\frac{2m_0c^2}{1-\cos\theta}\right)^2} + \frac{2m_0c^2\cos\theta}{1-\cos\theta}, \quad (1)$$

where θ is the angle between the two gamma rays. The sum $E_1 + E_2 - 2m_0c^2$ cannot be higher than the kinetic energy of incident positrons, i.e. 35 keV in Fig. 1(a). This upper limit can be seen as the kinematical cut-off of the TQAF contribution. Since the kinetic energy of a positron implanted into the target is progressively lowered in electronic collisions some positrons lost portion of their kinetic energy prior to the TQAF process and contribute to the 'bowl area' below the kinematical cut-off. The annihilation gamma rays emitted in anti-parallel directions ($\theta = 180°$) exhibit the largest difference of energies. From Eq. (1) it follows that such TQAF events fall on a hyperbolic curve determining borders of the 'bowl-like' area in Fig. 1(a). Projection cuts from the CDB spectra of Fig. 1 are plotted in Fig. 2. The TQAF contribution is clearly visible above the main annihilation peak and exhibits a sharp drop at 35 keV corresponding precisely to the kinetic energy of positrons in the beam. The inset in Fig. 2(a) shows that shape of the TQAF profile is well described by theoretical curve calculated using the TQAF cross section [8] and the positron stopping power [7]. From Eq. (1) one can calculate that for positrons with kinetic energy of 35 keV the TQAF contribution extends to the range $E_1 - E_2 = \pm$ 192 keV, which agrees well with the horizontal projection in Fig. 2(b). The shape of the horizontal projection is again well described by the theoretical curve. The minimum angle between the TQAF gamma rays is related to the maximum kinetic energy of positron $T_{+,max}$ by the relation $(\cos\theta)_{min} = (T_{+,max} - 2m_0c^2)/(T_{+,max} + 2m_0c^2)$ [8], i.e. for $T_{+,max} = 35$ keV, $\theta_{min} = 159^\circ$. Taking into account the geometry of the experiment and dimensions of detectors, if one photon falls into the centre of one detector then the second one can be detected if $\theta > 115^\circ$. Close geometry used here ensures that TQAF events can be detected in the full range of angles $\theta = 180^\circ$-159° allowed for positrons with $T_{+,max} = 35$ keV.

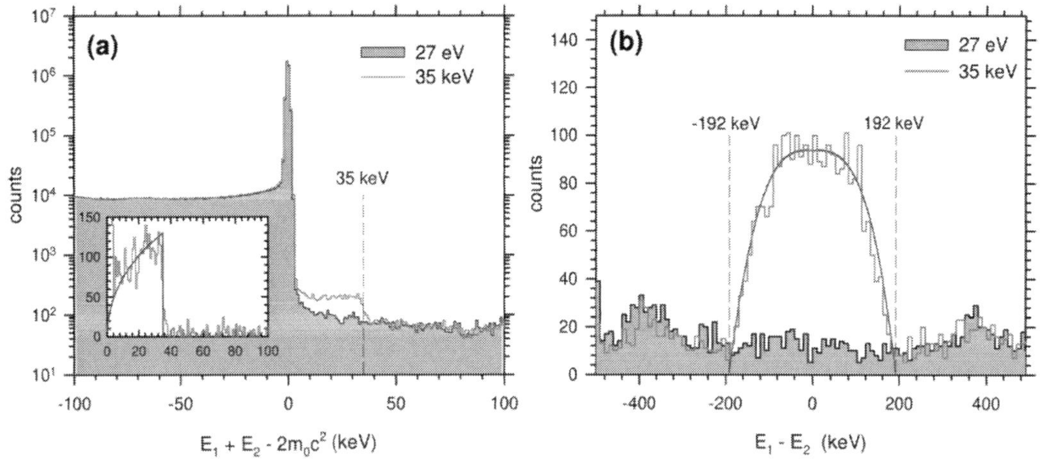

Figure 2. One-dimensional projections from CDB spectra of Fig. 1: (a) vertical cut integrated in the range $-192 < E_1 - E_2 < 192$ keV, inset shows the detail of the TQAF contribution and the theoretical shape calculated within QED, (b) horizontal cut integrated in the range $5 < E_1 + E_2 - 2m_0c^2 < 35$ keV, solid line shows the theoretical shape of the integrated TQAF contribution calculated within QED.

With decreasing energy of incident positrons the kinematical cut-off decreases and the 'bowl' area becomes smaller. In Fig. 1(b) one can see that the TQAF contribution completely disappeared in the CDB spectrum for slow positrons with energy of 27 eV. Hence the present experiment can be used as a benchmark test to verify the absence of fast positrons in the slow positron beam.

Side peaks visible in Fig. 1 on both sides of the main annihilation peak are caused by the events when annihilation gamma ray in one detector is back-scattered into the second detector. If the backscattering angle is 180° the energy difference between the Compton electron (341 keV) registered in one detector and the backscattered gamma ray (170 keV) detected in the second detector together with the second annihilation photon (511 keV) is ± 340 keV which corresponds well to the maxima of the side peaks in Fig. 1. The horizontal line at $E_1 + E_2 - 2m_0c^2 = 438$ keV comes from events when a gamma ray emitted by a ^{40}K radioisotope from background is scattered between the detectors.

The CDB spectra for fast positrons from $^{68}Ge/^{68}Ga$ radioisotope and GiPS are plotted in Fig. 3(a) and 3(b), respectively. The detectors were again positioned in face-to-face geometry but because of a significantly higher count rate the distance of detectors from the target was much higher than in the measurements on the slow positron beam. As a consequence, only the TQAF events with θ in a narrow range 177-180° can be detected and these events fall on a hyperbolic curve which is clearly visible in

Fig. 3 and represents the border of the TQAF 'bowl'. Obviously the TQAF contribution in Fig. 3 extends to significantly higher energies due high energies of incident positrons. Moreover, from comparison of Figs. 1 and 3 it becomes clear that the background above the main annihilation peak in the CDB spectra measured using fast positrons is higher than for slow positrons. In case of ^{68}Ge/^{68}Ga radioisotope the background is caused mainly by Compton scattering of gamma rays with energy of 1078 keV in the detectors. In case of GiPS the background is rather high due to Compton scattering of bremsstrahlung radiation in the target. Asymmetry of background in Fig. 3(b) is caused by the fact that one detector was positioned in the forward direction while the second one in the backward direction with respect to the beam. Since the source of this background is not in pile up events and the corresponding pulses have proper shape they cannot be removed by the shape analysis.

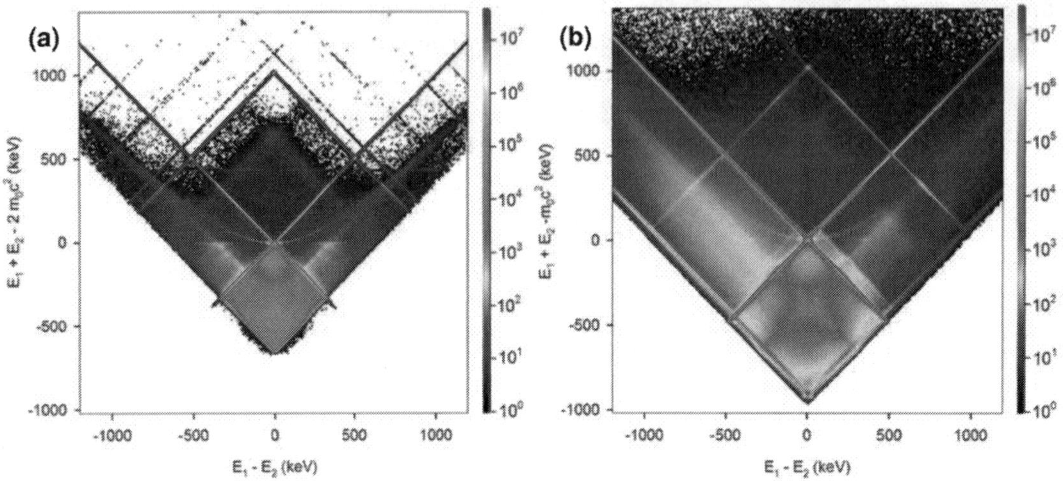

Figure 3. Two-dimensional CDB spectra measured using fast positrons (a) Mg target, positrons emitted by a ^{68}Ge/^{68}Ga positron generator, (b) W target, positrons created by GIPS.

4. Conclusions

A novel D-CDB spectrometer was employed for investigation of the TQAF process. The CDB spectra measured on slow positron beam exhibit low background and a 'bowl' shape TQAF contribution was detected for positrons accelerated to 35 keV. Analysis of the TQAF contribution provides information about positron thermalization in the target. The CDB spectra measured using fast positrons exhibit a remarkably higher background but even here the TQAF contribution was clearly visible.

Acknowledgements

This work was supported by the Czech Science Agency (project P108-13-09436S) and the grant SVV-2013-267303. Availability of beam time at the ELBE, HZDR is gratefully acknowledged.

5. References

[1] Kendall H W and Deutsch M 1956 *Phys. Rev.* **101** 20
[2] Lynn K G, MacDonald J R, Boie R A, Feldman L C, Gabbe J D, Robbins M F, Bonderup E and Golovchenko J 1977 *Phys. Rev. Lett.* **38** 241
[3] Weber M H, Hunt A W, Golovchenko J A and Lynn K G 1999 *Phys. Rev. Lett.* **83** 4658
[4] Čížek J, Vlček M and Procházka I 2010 *Nucl. Instrum. Methods A* **623** 982
[5] Anwand W, Brauer G, Butterling M, Kissenger H-R, Wagner A 2012 *Def. Diff. Forum* **331** 25
[6] Butterling M, Anwand W, Cowan T E, Hartmann A, Jungmann M, Krause-Rehberg R, Krille A, Wagner A 2011 *Nucl. Instrum. Methods B* **269** 2623
[7] Čížek J, Vlček M, Procházka I 2012 *New Journal of Physics* **14** 035005
[8] Dirac P A M 1930 *Proc. Camb. Phil. Soc.* **26** 361

Positron lifetime setup based on DRS4 evaluation board

M Petriska, S Sojak and V Slugeň

Institute of Nuclear and Physical Engineering, Slovak University of Technology, 81219 Bratislava, Slovakia

E-mail: martin.petriska@stuba.sk

Abstract. A digital positron lifetime setup based on DRS4 evaluation board designed at the Paul Scherrer Institute has been constructed and tested in the Positron annihilation laboratory Slovak University of Technology Bratislava. The high bandwidth, low power consumption and short readout time make DRS4 chip attractive for positron annihilation lifetime (PALS) setup, replacing traditional ADCs and TDCs. A software for PALS setup online and offline pulse analysis was developed with Qt,Qwt and ALGLIB libraries.

1. Introduction

Today, digital PALS setups are the usual upgrades from analog PALS systems in positron laboratories. Pulse processing with constant fraction discriminators and time to amplitude convertors are replaced by full software solution applied on digitized detector output signals. To process short nanosecond pulses relatively high bandwidth and sampling rate are needed. For this purpose expensive digital scopes [1], or special digitizer cards [2] were used. Now relatively cheaper cards like PicoScope digitizer with excellent triggering possibilities were used in [3]. Another possibility is to use a digitizer with switched capacitor arrays.

2. Experimental setup

The DRS4 evaluation board (EVB) [4] is based on the DRS4 Switched Capacitor Array chip. In order to simplify the EVB design, only four from eight available inputs are used to digitize the electronic pulses at the sampling speed of 5.12 GS/s. Analog bandwidth of the evaluation board is 700 MHz. This board is equipped with a trigger input and communicates with a computer, controlled by the data acquisition program through an USB interface. Four inputs channels enables to realize not only a standard two detector setup (fig.1), but also a three detector setup (fig.2). An obvious two detectors setup is used in standard PALS measurements. In our setup, the detectors are used in both - "Start" and "Stop" roles at the same time like in [5]. The constant fraction discriminator level for both detectors is set to trigger if pulses are coming from gamma quanta with energy higher than 400 keV. Measured Samples including the ^{22}Na source are placed above the detectors in order to suppress coincidences of the annihilation rays, emitted in opposite directions. Additional lead shield is inserted between these two detectors to minimize unwanted gamma scattering. Final start, stop separation of the pulses is based on the software pulse amplitude discrimination.

The three detector setup is usually used to measure samples containing disturbing Co^{60}. Constant fraction discriminators are set so that the middle detector detects the "Start" pulses from 1275 keV photons, and the side ones the "Stop" pulses from 511 keV photons. The three detector setup has lower counting rate due to the need of tripple coincidence. However better time resolution can be achieved on condition that the resulting lifetime is calculated as the average from the two lifetimes obtained by the three detector system [1]. The BaF_2 cylindrical scintillators with size of 25 mm in diameter and 12 mm in height are coupled to XP2020Q photomultipliers, like in our analog setup and use a high voltage (HV) divider with dynode timing outputs [6]. The fast dynode outputs of all detectors are transmitted to DRS4 evaluation board, that input range is from - 500 mV to 500 mV. The slow dynode outputs are amplified and inverted by Ortec VT120B amplifiers. Signal from these amplifiers feeds the constant fraction discriminators (CFD, Ortec 584). The blocking CFD outputs are summed in signal combiners consisting of a BNC tee connector and 6 dB passive attenuators. The summed signal from all CFD's feeds the leading edge discriminator (LE, Canberra 1428A. According to the LE level and chosen setup, two or three pulse coincidence is used. Standard positive pulse output signal from the LE discriminator is used as a trigger for the DRS4 evaluation module. The trigger rate obtained by this setup in two detector mode with a 1 MBq positron source is around 200 cps. Maximum count rate obtained by evaluation board is about 500 frames per second. One frame means 4 x 1024 measured samples (4 channels, 14 bits waves (2 bytes)). This should be due to limitation of the used USB controller. To obtain higher count rate, the WaveDream board [7] with the same DRS4 chip, but Gigabit Ethernet interface should be considered.

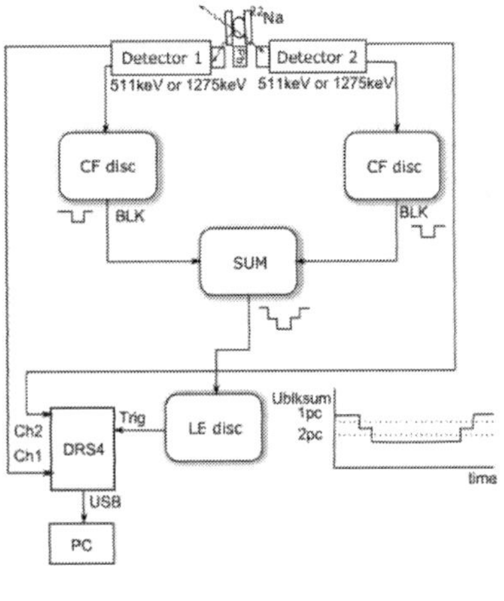

Figure 1. Two detector PALS setup with DRS4 evaluation board.

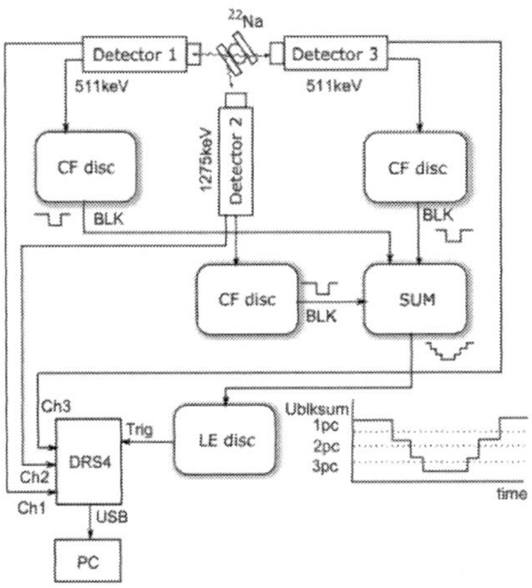

Figure 2. Three detector setup.

3. Waveform collecting

For every trigger event the DRS4 evaluation board provides four waveforms, with length of 200 ns in 1024 samples at 5.2 GS/s sampling rate. These waveforms are transformed to floating values from 14 bit ADCs which are reading data from DRS4 capacitor arrays. Together with measured amplitude arrays, timing information are send to PC through USB2 data interface. The USB connection is implemented with a micro controller (Cypress®CY2C68013A). The high speed modus of the USB 2.0 bus allows for data transfer rates of more than 20 MB/sec.

4. Pulse processing

The data from the DRS4 evaluation board are in nonuniform sampled form. It means that there are small deviation from obvious uniform 192 ps sampling period. This deviation is measured during the calibration procedure provided by the evaluation board libraries and applied to time arrays.

It was reasonable to implement the digital constant fraction, which is one of the most used method to extract timing information from the detector pulse. Principle of the standard digital constant fraction is shown in the figure 3. The method is based on finding the time at which the rising pulse edge crosses 25% of its amplitude, calculated as a difference between the pulse baseline and the minimum of the pulse. Data from the digitizer are in discrete form, therefore to find this time, a data interpolation is required. The Cubic Spline method was found as usable for this purpose [8]. In our software, penalized regression spline provided by the ALGLIB C++ library [9] was applied in the area around 25% of the constant fraction level. The penalized regression spline is a 1-dimensional curve fitting algorithm which is suited for noisy fitting problems. To identify the pulse amplitude we used simple data average value from 10 samples taken before the pulses leading edge, and parabolic least squares fitting around peak to find minimum of the pulse. Finally lifetimes counted as differences between the start/stop times are summed in the PALS histograms.

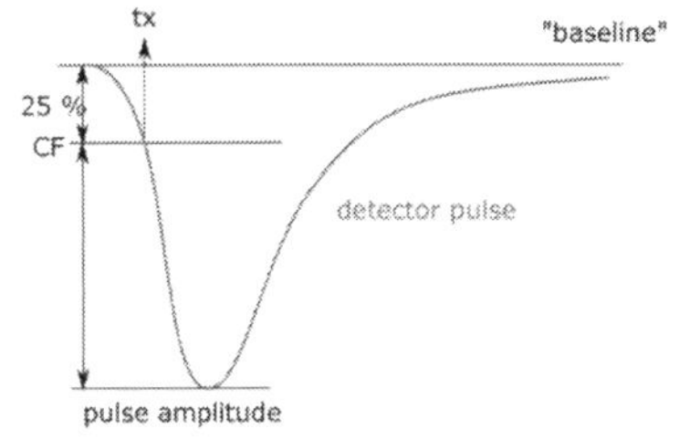

Figure 3. Digital constant fraction principle

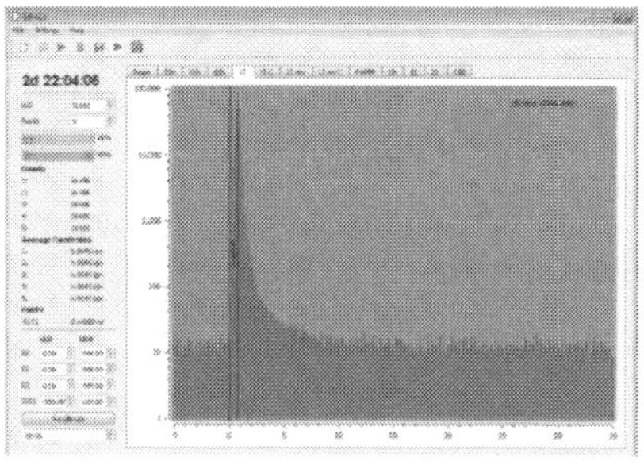

Figure 4. MS windows program QtPALS with user interface using Qt, and Qwt library.

5. QtPALS

To make software user friendly, the standard MS windows program written in C++ using Qt and Qwt libraries was developed. Software is separated into two parts. The first one is used to record samples from detectors, and is capable to do online analysis with collecting two PALS spectra and energy spectra for used detectors. The second part is used for offline analysis, and primarily is used to find optimal parameters for online QtPALS part (fig.4).

6. Results

The developed setup was tested with a reference Si sample and a ^{22}Na source encapsulated in a kapton foil measurement. The best setup resolution (FWHM) in the two detector mode was 190 ps and in the three detector mode with lifetime averaging 148 ps.

7. Conclusion

In our laboratory different digitizers were tested as digital PALS setups in past. In comparison with them (Acqiris DP210, ZT4612), the DRS4 evaluation board is interesting according to its good price and properties which are suitable for processing fast detector pulses. According to the obtained results, there was not significant difference between all of them. To make EVM module even better, delayed coincidence triggering should be delighted. Preliminary QtPALS software is developed under GNU license and is available at http://sourceforge.net/projects/qtpals

8. Acknowledgment

This article was granted by VEGA 1/0204/13 and VEGA 1/0366/12

References

[1] Saito H, Nagashima Y, Kurihara T and Hyodo T 2002 *Nucl. Instrum. Methods A* **487** 612–617
[2] Bečvář F, Čížek J, Lešták L, Novotný I, Procházka I and Šebesta F 2000 *Nucl. Instrum. Methods A* **443** 557–577 ISSN 0168-9002
[3] Korhonen E, Rytsölä K, Kilpelainen S and Tuomisto F 2012 Compact digital positron lifetime spectrometer *Proceedings of Physics Days 2012, University of Eastern Finland*
[4] Ritt S 2008 Design and performance of the 6 GHz waveform digitizing chip DRS4 *IEEE Nuclear Science Symposium Conference Record, 2008. NSS '08* pp 1512–1515
[5] Bečvář F, Čížek J and Procházka I 2008 *Acta Phys. Polonica A* **113** 1279
[6] Čížek J, Bečvář F and Procházka I 2000 *Nucl. Instrum. Methods A* **450** 325–337 ISSN 0168-9002
[7] Friederich H, Davatz G, Gendotti U, Meyer H and Murer D 2012 Real-time data analysis using the WaveDREAM data acquisition system *Real Time Conference (RT), 2012 18th IEEE-NPSS* pp 1–7
[8] Rytsölä K, Nissilä J, Kokkonen J, Laakso A, Aavikko R and Saarinen K 2002 *Applied Surface Science* **194** 260–263 ISSN 0169-4332
[9] Bochkanov S and Bystritsky V ALGLIB URL `http://www.alglib.net`

Recent status of A Positron-Electron Experiment (APEX)

H. Saitoh[1,3], T. S. Pedersen[1], U. Hergenhahn[1], E. V. Stenson[1], N. Paschkowski[1], and C. Hugenschmidt[2]

[1] Max Planck Institute for Plasma Physics, Greifswald and Garching, Germany
[2] FRM-II and Physics Department, Technical University Munich, Garching, Germany
[3] Graduate School of Frontier Sciences, The University of Tokyo, Kashiwa, Japan

E-mail: haruhiko.saitoh@ipp.mpg.de

Abstract. A project is underway to generate an electron-positron plasma by using the NEPOMUC positron source at the FRM-II facility combined with a multicell-type Penning trap (PAX) and a superconducting dipole magnetic field trap (APEX). In the APEX project, proof-of principle experiments are proposed for the development of efficient injection methods of positrons by using a small dipole magnetic field trap with a permanent magnet. Plans for the APEX project and its recent status are reported.

1. Introduction

There has been growing interest in the formation of an electron-positron plasma in a laboratory [1]. Conventional plasmas are characterized by large mass differences between electrons and ions. Many plasma phenomena, such as wave propagation and stability properties, are strongly related to the mass asymmetry and differences in the mobility of electrons and ions. Unlike conventional plasmas, the electron-positron plasma is in the class of pair plasmas, plasmas consisting of equal-mass particles. It is theoretically predicted that pair plasmas exhibit unique physical properties. According to theoretical studies and recent observation in γ-ray astronomy, formation of a large amount of electrons and positrons is predicted in pulsar magnetospheres and active galactic nuclei. Thus the experimental understanding of the electron-positron plasma is also important for astrophysics.

Although theoretical studies have been conducted intensively, there are very few experiments to generate an electron-positron plasma. This is mainly because of (1) the difficulty in the simultaneous confinement of electrons and positrons as plasmas, and (2) the availability of strong positron sources. (1) Concerning the confinement configurations, stable confinement of non-neutral pure electron plasmas has been demonstrated in a toroidal stellarator [2] and in a dipole magnetic field trap [3]. One can confine plasmas at any degree of non-neutrality in the toroidal geometries in principle. Based on these experiments, we will construct A Positron-Electron Experiment (APEX) for the confinement of the electron-positron plasma. (2) The positron source, NEPOMUC [4] at the FRM-II facility is the brightest DC moderated source in the world, with a rate in the order of 10^9 positrons s^{-1}. Moreover, the development of a Positron Accumulation Experiment (PAX) is underway for the accumulation and fast extraction of a large amount of cold positrons. PAX consists of a multicell-type Penning trap and is designed to trap of the order of 10^{11} positrons and extract them in a few milliseconds.

We plan to generate an electron-positron plasma in combination with the NEPOMUC, PAX, and APEX facilities [1]. In this report, we focus on the APEX project and present its plans and recent status. While excellent confinement properties are expected in the closed toroidal geometry, it is not straightforward to transport positrons from the source to the trap region. For this purpose, two methods have been proposed [1]. One is to use external electric fields and the other is to use positronium as intermediate particles, which are generated on single crystal surfaces [5] hit by the DC positron beam. Prior to the pair plasma experiment to be conducted in a superconducting dipole field trap, we plan to conduct proof-of-principle experiments to test the injection methods by using a small dipole field trap with a permanent magnet. The required parameters for the pair plasma formation and plans for the small trap experiment are described in the following sections.

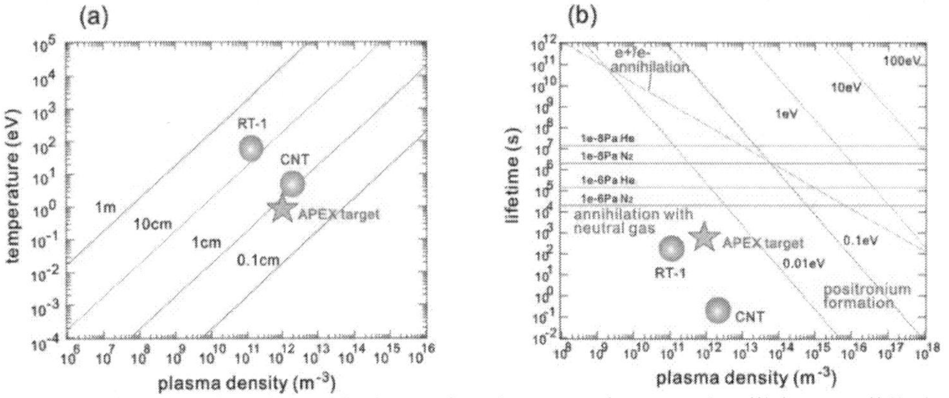

Figure 1. (a) Debye length and (b) life times of positrons set by neutral collision, annihilation with electrons, and positronium formation for various number density and temperature [6].

2. Required parameters for pair plasma production and proof-of-principle experiments

2.1. Target parameters for pair plasma production in APEX

In order to observe collective plasma phenomena of a charged particle cloud, the scale length a of the cloud must be larger (preferably $a > \sim 10\lambda_D$) than the Debye length $\lambda_D = \sqrt{k_B T_e / n_e e^2}$, a typical length of electrostatic shielding [1]. The key techniques are (1) efficient injection methods of a large amount of low temperature positrons into a confinement region with small volume and (2) excellent confinement properties during injection, confinement, and mixing phases. Because the condition $a \sim 10\lambda_D$ is achieved when temperatures $T_e = 1$ eV and a number density $n_e = 10^{12}$ m^{-3} for a realistic scale of $a \sim 10$ cm, we set these parameters as a target (Fig. 1 (a)). By using the DC positron beam, the required confinement time of the trap is ~10 s when the confinement region volume $V \sim 10^{-2}$ m^{-3}. Although the maximum confinement time of a toroidal pure electron plasma exceeds this value, it is difficult to realize such a long confinement by using the relatively weak DC positron beam, clearly showing the importance of the PAX development. The total efficiency (after transport, cooling, and mixing with electrons) above 10% is needed for the injection of 10^{11} positrons from PAX to APEX.

As an antimatter plasma, the effects of (1) annihilation with neutral particles, (2) pair-annihilation with electrons, and (3) positronium formation processes should be considered as well as confinement properties of the trap system. We estimate the lifetimes τ of positrons set by these effects according to Ref. [6] for the above target parameters. (1) As shown in Fig. 1 (b) with solid lines, positron annihilation on neutral gas is negligible in clean UHV environments. For nitrogen gas, $\tau = 2 \times 10^4$ s at pressure 10^{-6} Pa, which is routinely obtained with a standard vacuum system. (2) When mixed with positrons, the dominant two photon annihilation time $\tau[\mathrm{s}] \sim 10^{20}/n_e[\mathrm{m}^{-3}]$, and the annihilation effects are again negligible for low density plasmas, as plotted with a chain line in the figure. (3) Lifetimes set by the three body recombination process are plotted with dot lines for different electron and positron temperatures. It is likely that this effect will not be a problem but may cause significant loss of plasmas at very cold (<0.01eV) and high-density (>10^{14} m^{-3}) cases. In addition to these effects, it is possible that instabilities and enhanced turbulent transport emerge due to the two fluid effects, which should be investigated when positrons are mixed with electrons in future experiments.

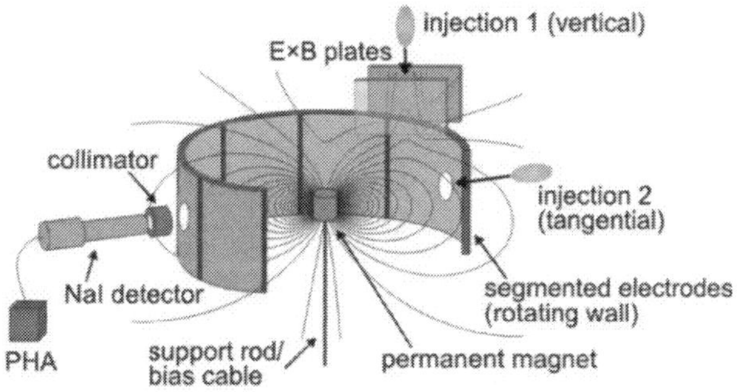

Figure 2. Schematic drawing of a proof-of-principle experiment, including the supported neodymium magnet, E×B plates for vertical injection, rotating wall for tangential injection, and diagnostics.

2.2. *Proposed proof-of-principle experiments in a small dipole field trap*

As a first step experiment, we will develop appropriate injection schemes of positrons (1) by using the effects of external electric fields, and (2) by using the positronium reemission process on solid materials in this proof-of-principle experiment [1,5] . These experiments will be conducted in a small dipole magnetic field illustrated in Fig. 2. The dipole field is generated by a mechanically supported neodymium magnet and the typical field strength in the confinement region is 0.05T.

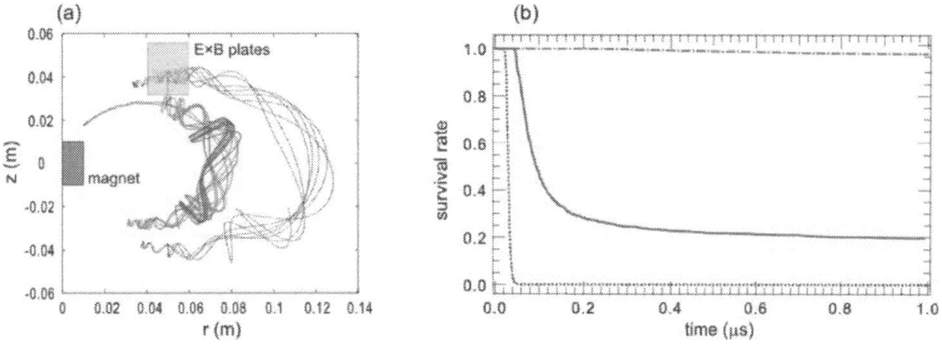

Figure 3. (a) Typical positron orbit projected onto the *r-z* cross section when the electric field is applied (solid line) and not applied (dot line). Electric field of $\mathbf{E} = 1 \times 10^3$ V/m was applied in the marked region from t=0 to 0.1 μs. (b) Ratios of remaining positrons after injection without **E** (dot line), with the application of **E** (solid line), and when the magnet was also biased (chain line).

In order to inject positrons from the guiding field of the beam line into the confinement region, positrons must be transported across closed field lines. As injection methods by using external electric fields, we plan to test two procedures. The first one is a vertical injection scheme by using the **E**×**B** drift motion induced by a local crossed electric field. As shown in Fig. 2, the positron beam is vertically guided to the peripheral region of the dipole field, where a local electric field is applied in the perpendicular direction. Figure 3 (a) shows the typical orbits of a positron with a kinetic energy of 10 eV. Without the application of **E**, the guiding centre motion is trapped on the location of the initial field line (the dot line in the figure). After reflection motions of finite times between the mirror points, positrons are reflected back to the beam line. By applying **E**, positrons are radially transported into the confinement region (the solid line in the figure). Figure 3 (b) shows the calculated remaining ratio of 3086 positrons with vertical energy of 10 eV and perpendicular energy of 0 to 3 eV. When **E** is not applied, as plotted with the dot line, positrons are immediately lost after injection by hitting the **E**×**B** plates. By applying $\mathbf{E} = 1 \times 10^3$ V/m from t=0 to 0.1 μs, as shown by the solid line, the loss on the **E** ×**B** plate is suppressed and approximately 20 % of the injected positrons survived at *t*=1 μs. The main loss channel in this case is the magnet surface, and it is further suppressed by positively biasing the magnet (the chain line), realizing the remaining rate of 98 %. The second scheme with external electric fields is to use a rotating wall (a technique to generate field asymmetry by using segmented electrodes [7]) with tangential positron injection. A charged particle in a poloidal dipole field undergoes a toroidal rotation due to the grad B and curvature drifts. The typical rotation frequency for a 10 eV positron in the present configuration is in the order of MHz. By applying an azimuthal electric field by using segmented electrodes (Fig. 4 (a)) in the dipole field, effective radial transport is induced. Figure 4 (b) shows the typical orbits of a positron. When the rotating wall frequency is synchronized with the rotation frequency of the positron, positrons are effectively transported to the confinement region. After transported inward, positrons are expected to relax into an equilibrium state in the dipole field. For the diagnostics of the injected number of positrons, finally the magnet is negatively biased so that trapped positrons are dumped onto the magnet surface. The γ rays from annihilation are counted by a scintillator detector with a pulse height analysis system.

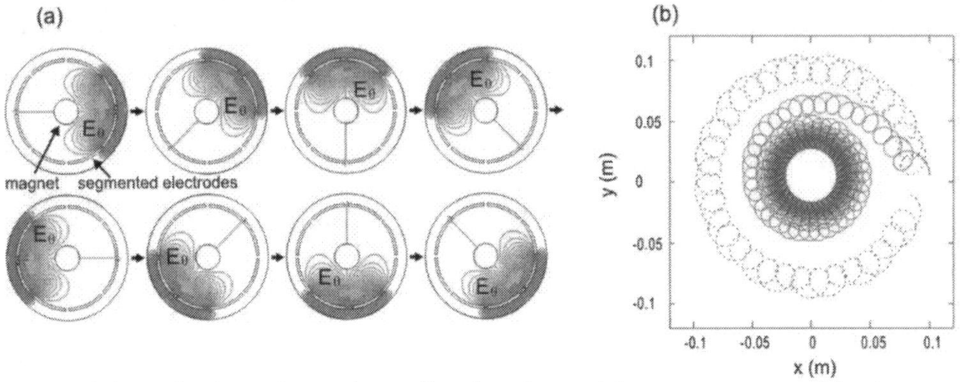

Figure 4. (a) Schematic view of a rotating wall and equipotential contours generated by the segmented electrodes. (b) Typical positron orbits with (dot line) and without (solid line) the rotating wall.

As a second positron injection method [1], positrons are guided to solid state materials and converted into positronium atoms [5]. The neutral positrons are freely transported into the confinement region, where they are photo ionized to generate an equal amount of electrons and positrons. We will study the formation ratio of positrons by injecting the positron beam from NEPOMMUC and access the feasibility for the electron-positron plasma formation. As well as lifetime measurements, a coincident Doppler-broadening spectroscopy will be applied for the measurements [8].

3. Summary and outlook

Aiming for the electron-positron plasma experiment, we have started the APEX project and plan to develop injection methods of positrons by using a small-scale dipole magnetic field trap with a permanent magnet. Based on these proof-of-principle experiments to investigate the injection efficiency of positrons scheduled to be carried out in 2014, we plan to construct a superconducting levitated dipole field trap and simultaneously confine positrons and electrons as a future experiment.

Acknowledgments
This work is partially supported by JSPS KAKENHI 23654201 and 25707043.

References

[1] T S Pedersen *et al.* 2012 *New J. Phys.* **14**, 035010
[2] P W Brenner and T S Pedersen 2012 *Phys. Plasmas* **19**, 050701
[3] Z Yoshida *et al.* 2013 *Plasma Phys. Cntr. Fusion* **55**, 014018
[4] C Hugenschmidt *et al.* 2012 *New J. Phys.* **14**, 055027
[5] D B Cassidy *et al.* 2011 *Phys. Rev. Lett.* **106**, 133401
[6] R G Greaves and C M Surko 2002 in *Non-Neutral Plasma Physics IV*, 10-23
[7] X-P Huang *et al.* 1997 *Phys. Rev. Lett.* **78**, 875-878
[8] C Hugenschmidt *et al.* 2005 *Nucl. Instrum. Meth. Phys. Res. A* **554**, 384-391

Journal of Physics: Conference Series **505** (2014) 012046 doi:10.1088/1742-6596/505/1/012046

Recovering the Fermi surface with 2D-ACAR spectroscopy in samples with defects

SB Dugdale[1] **and J Laverock**[2]

[1]HH Wills Physics Laboratory, University of Bristol, Tyndall Avenue, Bristol BS8 1TL, UK
[2]Department of Physics, Boston University, 590 Commonwealth Avenue, Boston, Massachusetts 02215, USA

E-mail: s.b.dugdale@bristol.ac.uk

Abstract. When two-dimensional angular correlation of positron annihilation radiation (2D-ACAR) experiments are performed in metals containing defects, conventional analysis in which the measured momentum distribution is folded back into the first Brillouin zone is rendered ineffective due to the contribution from positrons annihilating from the defect. However, by working with the radial anisotropy of the spectrum, it is shown that an image of the Fermi surface can be recovered since the defect contribution is essentially isotropic.

Introduction

As a probe of the Fermi surface, positron annihilation occupies an important place in the array of probes which include quantum oscillations [1], angle-resolved photoemission [2] and the closely related technique of Compton scattering [3, 4, 5]. When a positron annihilates with an electron in a metal, the predominant process produces two γ-ray photons which, due to momentum conservation, are emitted almost anti-parallel in the laboratory frame. It can be shown that measurement of the angular distribution of the photons' deviations from anti-collinearity is equivalent to a measurement of the projection of the electron-positron momentum distribution along an axis perpendicular to the mean emission axis of the measurement. The full three-dimensional momentum density can be reconstructed, if necessary, using tomographic techniques [6, 7, 8]. Since the positron has a Maxwell-Boltzmann thermal momentum (with a typical radioisotope source activity, there is only one positron in the sample at any time), the pair momentum is dominated by that of the electron. While the principal strengths of the positron technique, conventionally referred to as two-dimensional angular correlation of positron annihilation radiation (2D-ACAR) spectroscopy, include its bulk sensitivity, the k-resolved nature of the obtained spectra and its insensitivity to the electronic mean-free-path (meaning that substituionally disorded alloys and high-temperature phases are accessible), the fact that the positron is positively charged means that it is energetically favourable for it to become trapped at vacancy-type defects. The fate of a positron on entering a metal containing vacancies depends to a large extent on the concentration and distribution of those vacancies. After rapid thermalization (typically on a timescale of a few ps), in a defect free metal the positron diffuses through the crystal with a diffusion length of the order of ~ 1000Å. Thus whether it encounters a vacancy and consequently becomes trapped depends on the concentration of vacancies. Once trapped, the positron, which was formerly in a delocalized (Bloch) state, is localized at the defect site leading to wavefunctions which are almost isotropic [9]. As pointed out by Prasad *et al.* [10], translational symmetry is destroyed by the presence of a vacancy so that *all* occupied electron states contribute to the momentum distribution. This should be contrasted with the bulk case in which a Bloch-state positron which annihilates with an electron with crystal momentum $\mathbf{k}$ will contribute to the momentum distribution only at momenta $\mathbf{p} = \mathbf{k} + \mathbf{G}$, where $\mathbf{G}$ is a reciprocal lattice vector. In this case the momentum distribution will reflect the anisotropy of the bulk electronic structure and Fermi surface, whereas in the presence of vacancy-type defects this anisotropy is washed out to varying extents depending on what fraction of positrons are trapped. In the case of saturation trapping, when all positrons implanted into the material are trapped, the momentum distribution is very isotropic, leaving no hope of extracting any information about the bulk Fermi surface. In this paper we focus on the intermediate regime where not all positrons are trapped and some anisotropy remains, and show that even at relatively high trapped positron fractions it is still possible to infer the bulk Fermi surface.

For Fermi surface studies, except for the simplest of topologies, a direct interpretation of the measured momentum- (that is real momentum, $\mathbf{p}$-) space spectra is difficult since the signatures of the Fermi surface (the discontinuities in the momentum distribution) are widely distributed across the spectrum since an electron with crystal momentum $\mathbf{k}$ will contribute not only at $\mathbf{p} = \mathbf{k}$, but also, as mentioned above, at $\mathbf{p} = \mathbf{k} + \mathbf{G}$. Moreover, these signatures are themselves difficult to see in the raw spectra, not least because of the projected nature of the measured spectra. In the early days of two-dimensional measurements it was commonplace to subtract an isotropic distribution (either a fitted gaussian or the simple angular average of the spectrum itself) from the measured spectrum, $N(p_x, p_y)$. This is often referred to as the radial anisotropy, $R(p_x, p_y)$, and is defined as

$$R(p_x, p_y) = N(p_x, p_y) - \overline{N(p_x, p_y)}|_{p=\text{const.}} \quad (1)$$

This procedure is still routinely performed during experiments to check that the sample has been properly aligned. However, Lock, Crisp and West had already pointed out that a translationally invariant distribution in crystal momentum- ($\mathbf{k}$-) space could be constructed by 'folding' the real momentum ($\mathbf{p}$)-space distribution back into the the first Brillouin zone by translating those $\mathbf{p}$ points by an appropriate reciprocal lattice vector [11]. For a purely electron momentum distribution within the independent particle model (IPM), this procedure results in the contribution from full bands being constant across the Brillouin zone, and the contributions from partially filled bands reinforcing each other to give step-function-like changes in occupation at the Fermi surface. For positrons, although this 'LCW theorem' is only approximate (even within the IPM), the contributions of the occupied bands nevertheless sum to give a smoothly varying background on top of which the contributions of the partially occupied bands also sum, with the Fermi surface breaks being put back together on top of each other. However, in the situation where some fraction of the positrons are trapped and annihilate from the vacancy, they contribute an (almost) isotropic momentum distribution which is superimposed on the bulk distribution due to untrapped positrons. This extra component makes conventional LCW analysis dangerous, since this peaked defect contribution is likely to dominate the structure of the LCW spectrum. Similar problems can also arise in the reconstruction of the three-dimensional momentum distribution after the application of tomographic procedures, with methods involving an explicit Fourier transform being particularly prone to spurious peaks at low momentum [12, 13]. In fact, this is one of the major reasons for the Bristol group preferring the Cormack method for reconstruction [6, 14, 15, 16].

As mentioned earlier, within the IPM, the LCW 'theorem' applied to the electron momentum distribution is exact since the electronic wavefunctions are orthogonal and normalized. One can immediately grasp what the LCW process is doing in folding back the higher momentum

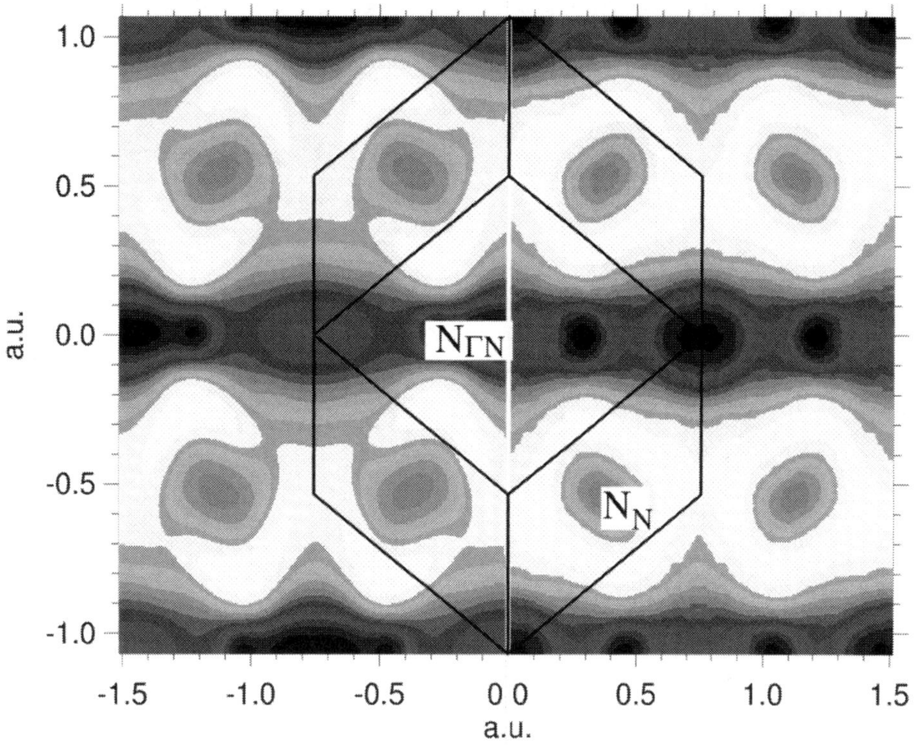

Figure 1. Projected [110] Brillouin zone occupancy of V (horizonal axis is [1$\bar{1}$0] and vertical axis is [001] obtained from an electronic structure calculation (left hand side of figure) and measured LCW spectrum from experiment (right hand side of figure). Some (projected) high symmetry points are labelled and the (projected) Brillouin zone is indicated.

components ('Umklapps') back into the first Brillouin zone. Moreover, in the case of electron-positron momentum distributions, as long as the **k**-dependence of the positron wavefunction is weak, the modulations introduced are likely to be small, and in any case the breaks associated with the presence of the Fermi surface in the electron distribution remain.

Here we show that in fact if one is only interested in seeing the occupied regions of the Brillouin zone, and hence the shape of the Fermi surface, then it is possible to apply the LCW folding procedure not to the momentum distribution itself, but to its radial anisotropy. This has the advantage that if there is a contribution from defects which is (almost) isotropic, then this will not appear in the radial anisotropy distribution and will thus be excluded from the subsequent LCW summation. We illustrate this with reference to a 2D-ACAR experiment on V in which the data are projected down the [110] crystallographic direction (details of the experiment can be found in Ref. [17]). In Fig. 1 the LCW distribution is shown together with an occupancy extracted from a calculation performed using the ELK electronic structure code [18]. As is expected, the LCW distribution broadly reflects the occupancy of the Brillouin zone with the differences principally due to the perturbations introduced by the positron wavefunction [19].

In Fig. 2, a series of LCW distributions are shown to illustrate the effect of adding an (isotropic) gaussian component to represent the presence of a contribution from positrons trapped at defects. At a relatively small defect contribution (with only 1% of positrons being

Journal of Physics: Conference Series **505** (2014) 012046 doi:10.1088/1742-6596/505/1/012046

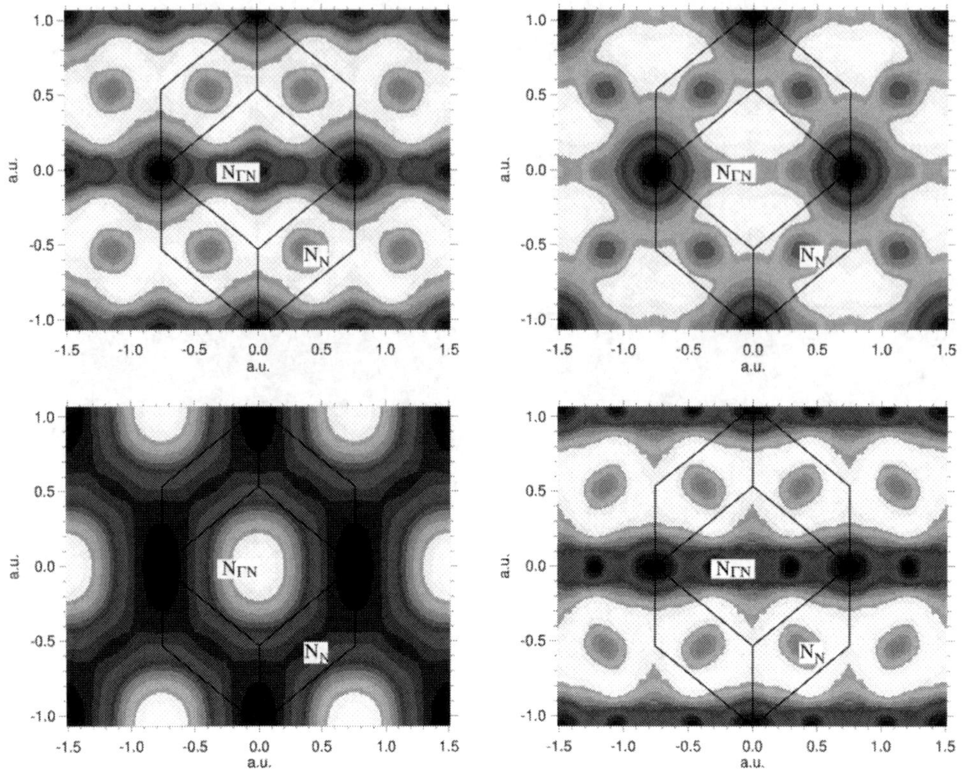

Figure 2. Series of LCW distributions on data with simulated defect component. The fraction of positrons annihilating from the trapped state are 1% (top left), 5% (top right) and 33% (bottom left). The final distribution (bottom right) is the LCW of the radial anisotropy.

trapped), there is little deleterious impact on the LCW distribution (top left). However, with higher positron trapping (5% (top right) and 33% (bottom left)), it is very obvious that the LCW distributions can no longer be relied upon to indicate the Fermi surface, with the 33% distribution unrecognizably different from the occupancy. However, the LCW distribution of the radial anisotropy is, of course, insensitive to the presence of a large isotropic defect contribution (apart from the negative effect on the signal-to-noise ratio). In the bottom right panel of Fig. 2 this radial anisotropy LCW distribution is shown (actually for the 33% trapping) and is clearly very similar to those in Fig. 1. This is because the radial anisotropy procedure emphasizes the anisotropy of the data which is always partly due to the presence of the Fermi surface (in addition to general wavefunction anisotropy) and the LCW folding constructively reinforces these Fermi surface signatures. We believe that it is likely to be applicable whenever the contribution to the momentum distribution from trapped positrons is isotropic.

It is, however, a little surprising just how effective this procedure is. In multiband systems (for it is often in these kinds of materials, for example charge-density-wave systems [20, 21, 22], where the **k**-resolved nature of positron annihilation can make substantial contributions) it is often difficult to disentangle the Fermi surface information from a single projection without the assistance of a theoretical calculation, as pointed out by [23], and when the signals are weak due to either limited positron overlap, or the sheer number of occupied bands it is important to find ways of making the most of your data. The sensitivity of the radial anisotropy to the shape of the Fermi surface (in many cases) was exploited by Major *et al.* [24] who sought to compare the measured radial anisotropies to those from electronic structure calculations [25]. Spurred on by the challenge posed by strong wavefunction effects and small Fermi surface signals in the early days of cuprate superconductivity, several groups were working on image processing methods [26] designed to emphasize the Fermi surface signatures, and it is notable that a kind of 'difference' LCW was used at that time [17], albeit for a different purpose. In the closely related field of Compton scattering, we have also previously used the anisotropic part of the reconstructed momentum distribution in our work on Co-doped $BaFe_2As_2$ [3], and it also appears to be used by other groups [27]. This approach is also likely to be more generally useful for removing Fourier-based-reconstruction artefacts.

Conclusions

In the circumstance of having made a 2D-ACAR measurement on a sample which contains vacancy-type defects in which some fraction of the positrons have been trapped, the LCW folding of the radial anisotropy appears to be a viable route towards accessing the bulk Fermi surface. With new 2D-ACAR spectrometers coming online [28, 29], this technique may prove very useful. Finally we would like to emphasise that we make no claims about these 'radial anisotropy' LCW distributions beyond the statement that this kind of analysis appears to give sensible Fermi surface information in the case of a measurement which contains a contribution from annihilation in defects.

References

[1] Ru N *et al.* 2008 *Phys. Rev. B* **78**(4) 045123
[2] Brouet V *et al.* 2008 *Phys. Rev. B* **77**(23) 235104
[3] Utfeld C *et al.* 2010 *Phys. Rev. B* **81**(6) 064509
[4] Laverock J *et al.* 2007 *Phys. Rev. B* **76**(5) 052509
[5] Dugdale S B *et al.* 2006 *Phys. Rev. Lett.* **96** 046406
[6] Kontrym-Sznajd G 1990 *Phys. Stat. Sol. A* **117** 227
[7] Major Z *et al.* 2004 *Phys. Rev. Lett.* **92** 107003
[8] Haynes T D *et al.* 2012 *New Journal of Physics* **14** 035020
[9] Puska M J and Nieminen R M 1983 *Journal of Physics F: Metal Physics* **13** 333
[10] Prasad R, Benedek R, Robinson J E and Bansil A 1989 *Phys. Rev. B* **40**(13) 8620–8630
[11] Lock D G, Crisp V H C and West R N 1973 *Journal of Physics F: Metal Physics* **3** 561
[12] Fretwell H M *et al.* 1995 *Europhysics Lett.* **32** 771–776
[13] Pecora L M *et al.* 1988 *Phys. Rev. B* **37**(12) 6772–6782
[14] Cormack A M 1963 *Journal of Applied Physics* **34** 2722–2727
[15] Cormack A M 1964 *Journal of Applied Physics* **35** 2908–2913
[16] Crowe S J *et al.* 2004 *Europhysics Lett.* **65** 235–241
[17] Dugdale S B *et al.* 1994 *Journal of Physics: Condensed Matter* **6** L435–L443
[18] URL `http://elk.sourceforge.net/`
[19] Laverock J, Haynes T D, Alam M A and Dugdale S B 2010 *Phys. Rev. B* **82** 125127
[20] Laverock J, Dugdale S B, Major Z, Alam M A, Ru N, Fisher I R, Santi G and Bruno E 2005 *Phys. Rev. B* **71** 085114
[21] Laverock J, Haynes T D, Utfeld C and Dugdale S B 2009 *Phys. Rev. B* **80**(12) 125111
[22] Mansart B *et al.* 2012 *Proceedings of the National Academy of Sciences* **109** 5603–5608
[23] Rabou L and Mijnarends P 1984 *Solid State Communications* **52** 933 – 936
[24] Major Z *et al.* 2004 *J. Phys. Chem. Solids* **65** 2011
[25] Barbiellini B, Dugdale S B and Jarlborg T 2003 *Comp. Mater. Sci.* **28** 287–301
[26] O'Brien K M, Brand M Z, Rayner S and West R N 1995 *Journal of Physics: Condensed Matter* **7** 925
[27] Barbiellini B 2013 *Journal of Physics: Conference Series* **443** 012009
[28] Ceeh H *et al.* 2013 *Review of Scientific Instruments* **84** 043905
[29] Dugdale S B, Laverock J, Utfeld C, Alam M A, Haynes T D, Billington D and Ernsting D 2013 *Journal of Physics: Conference Series* **443** 012083

The effect of regularization on the reconstruction of ACAR data

J A Weber[1], H Ceeh[1], C Hugenschmidt[1,2], M Leitner[1] and P Böni[1]

[1] Technische Universität München, Physik Department, Lehrstuhl E21, James-Franck-Straße, D-85748 Garching, Germany

[2] Technische Universität München, Forschungs-Neutronenquelle Heinz Maier-Leibnitz, Lichtenbergstraße 1, D-85748 Garching, Germany

E-mail: `josef-andreas.weber@frm2.tum.de`

Abstract. The Fermi surface, i.e. the two-dimensional surface separating occupied and unoccupied states in k-space, is the defining property of a metal. Full information about its shape is mandatory for identifying nesting vectors or for validating band structure calculations. With the angular correlation of positron-electron annihilation radiation (ACAR) it is easy to get projections of the Fermi surface. Nevertheless it is claimed to be inexact compared to more common methods like the determination based on quantum oscillations or angle-resolved photoemission spectroscopy. In this article we will present a method for reconstructing the Fermi surface from projections with statistically correct data treatment which is able to increase accuracy by introducing different types of regularization.

1. Introduction

When a positron is implanted into a defect-free sample, it thermalizes within the order of picoseconds. The positron will then annihilate with an electron in such a way that in the centre of mass frame two exactly collinear γ-quanta with an energy of 511 keV each are emitted. However, in the laboratory frame there is a Doppler shift, resulting from the longitudinal momentum of the electron-positron pair. The transverse momentum of the electron-positron pair yields an angular deviation from collinearity. By measuring both Doppler shift and angular deviation from collinearity one would be able to obtain the electron-positron momentum density (ρ_{ep}) directly. This density is continuous except for those points where bands cross the Fermi level; or in other words: It is continuous except at the Fermi surface. In general, the breaks in ρ_{ep} are the only way to identify the Fermi surface.

Due to technical limitations it is only possible to measure either the Doppler shift or the angular deviation from collinearity. With ACAR the latter is the case. This means that an ACAR measurement gives a projection of ρ_{ep}. To determine ρ_{ep} it is necessary to measure a set of projections and to recalculate ρ_{ep} with the help of these. The reconstruction of a 3D density from projections is a well known problem from computed tomography (CT) and is solved by many different approaches (see e.g. [1, 2]). Plenty of these methods have already been used to reconstruct ρ_{ep} from ACAR data [3, 4]. We want to note, however, that there is a fundamental difference between reconstructing from x-ray images of a body and from ACAR data: As the human body has no symmetries, x-ray tomography is a 2D problem. Such is not the case with the

Journal of Physics: Conference Series **505** (2014) 012047 doi:10.1088/1742-6596/505/1/012047

electronic structure of crystalline solids. Here, a projection can provide a lot more information if the full 3D symmetry is taken into account.

Since the measurement time is limited, the experimentalist has to choose whether to collect few projections with high statistics or many projections with few statistics. Different views exist on this question (compare e.g. [5], [6] and [7]). Although e.g. Kontrym-Sznajd et al. gave an rough estimation out of how many projections a 3D density can be reconstructed [8], there has been no detailed study as to which approach is optimal yet.

2. Theory

In this work we will treat the reconstruction problem by minimising χ^2, i.e. the quadratic difference between measurement and reconstruction weighted with the statistical accuracy. Therefore some definitions are needed:

Mathematically a projection of ρ_{ep} in a direction α is expressed by the so called "Radon transformation". In the discrete case it is simply calculated by the multiplication with a projection Matrix $\mathbf{P}^\alpha$. Since the samples investigated with ACAR are crystalline, ρ_{ep} is highly symmetric and it is possible to reduce the complexity of $\mathbf{P}^\alpha$ by restricting the problem into the irreducible wedge of the reciprocal space.

In order to find a density ρ which is in agreement with the measurements, we have to minimize the function $G(\rho)$:

$$G(\rho) = \lambda F(\rho) + \chi^2 = \lambda F(\rho) + \sum_\alpha \sum_i \left(\frac{\mathrm{M}_i^\alpha - (\rho \mathbf{P}^\alpha)_i}{\sigma_i^\alpha} \right)^2$$

The regularization functional $F(\rho)$ is needed, because the problem is highly under-determined. For the reconstruction of ACAR data an entropy-like function $F(\rho) = \sum_i \rho_i \ln \rho_i$ has been used previously [9]. This entropy function treats every point in ρ individually and tries to bring it to its lowest possible value, which minimizes the amount of additional information that is put into the reconstruction. However, we actually do have more information we can put in: ρ should have no sharp steps except for the Fermi surface breaks. A possible way to implement this is given by the sum over the gradient of ρ:

$$F(\rho) = \sum_{|\nabla\rho|_i < a} ((\nabla\rho)_i)^2 + \sum_{|\nabla\rho|_i > a} 2a\,|\nabla\rho|_i - a^2$$

The parameter a has two purposes: It makes the regularization functional differentiable compared to a function which uses only the absolute value. Additionally a defines the strength of the regularization. This regularization is related to the algorithm proposed by Rudin et al. which uses the total variation [10].

Another very general way of putting prior knowledge of ρ_{ep} into the regularization is the histogram method. Therefore, this knowledge has to be coded into a histogram H. For regularization, H is compared to the histogram of the reconstructed density ρ:

$$F(\rho) = \sum_i (H_i - \text{hist}(\rho)_i)^2$$

In practice, instead of calculating the histogram, it is computationally less demanding to sort the point-wise values of ρ. The sorted list can be interpreted as a histogram with infinitesimal resolution. Of course the problem of choosing the right histogram H for regularization still remains. This is non-trivial, but since we will work with simulated data, we can take the histogram of the original underlying density.

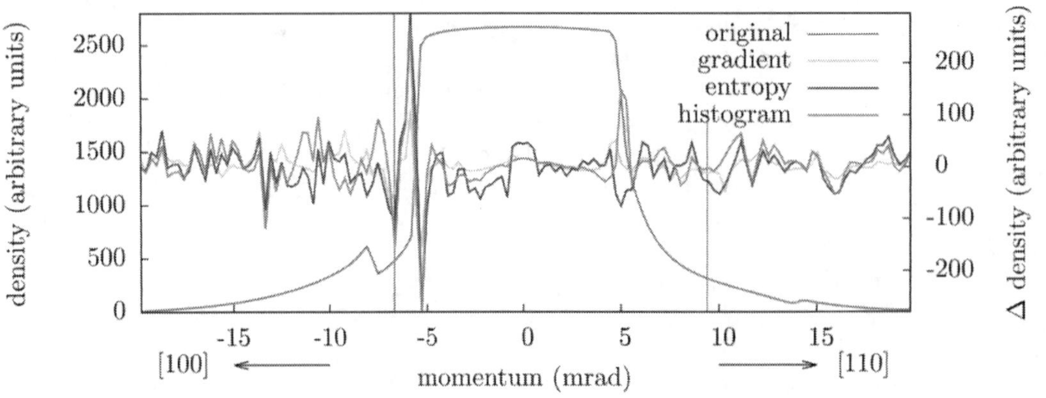

Figure 1: Cut through the original density from [100] to Γ to [110] (red, left axis) and difference between original and reconstruction for different regularization functions (all the other colours, right axis). The vertical lines indicate the zone boundaries.

Table 1: Quadratic difference between original density ρ_{ep} and reconstructed densities ρ_r for various amounts of projections and for different regularization functions. The total number of counts in every set of projections is 275×10^6. For calculating the quadratic differences, the densities were normalized to the original density.

	entropy	gradient	histogram
3 projections	5.3×10^{-7}	1.3×10^{-7}	8.4×10^{-7}
5 projections	4.0×10^{-7}	9.6×10^{-8}	4.6×10^{-7}
17 projections	2.3×10^{-7}	7.3×10^{-8}	3.3×10^{-7}

For χ^2, according to the Poisson statistic, the expected value is $\chi^2 = 1$. This can be achieved with the right choice of λ. When χ^2 reaches $\chi^2 = 1$ the optimization is stopped.

To find the minimum of $G(\rho)$, a physically inspired iterative gradient approach is used: Consider ρ to be the coordinates of a body that moves according to Newton's laws of motion in the potential surface defined by $G(\rho)$, subject to friction. In every iteration step the force $F = -\nabla G(\rho)$ is calculated, which leads to the simple update formulas for v and ρ:

$$v_i = (v_{i-1} + F\Delta t)\, e^{-\mu\Delta t}$$

$$\rho_i = \rho_{i-1} + v_i$$

The friction constant has to be set to $0 < \mu < \infty$ in order to ensure that the minimum will be found.

3. Results

Since the reconstruction algorithm is to be compared quantitatively, projections were created from a simulated ρ_{ep} of Copper (for details see [11]). Three sets of projections with a total amount of counts of 275×10^6 were generated. In every set the projections were calculated by rotating the simulated density around the [110] axis in equidistant steps (22.5°, 15° and 5° respectively). The main symmetry directions at 0°, 54.7° and 90° were avoided, because these provide less information about ρ_{ep} than an arbitrary projection. In a real experiment, however, at least one of this directions will be measured additionally, since it can provides valuable information about the sample orientation and quality. Three reconstructions were created per set, using a different regularization functional for each.

In figure 1 a cut from [100] to Γ to [110] through the original density is shown with the residual (original – reconstruction) for each reconstruction from five projections. Obviously the difference between original and reconstruction is small. Nevertheless it is apparent that the gradient regularization yields the reconstruction with the most efficient noise suppression and with the fewest artefacts.

A more quantitative comparison is given in table 1. There, the quadratic deviations from original density ρ_{ep} and reconstructed density ρ_r are listed ($d^2 = \sum_i (\rho_{ep} - \rho_r)_i^2$) for different methods of regularization and for different number of projections.

With an increasing number of projections d^2 is decreasing for every regularization method. This indicates that the measurement time is used most efficiently if many projections are recorded. But not only is the quality of the reconstruction increasing with the number of projections, the necessary iterations are decreasing. Hence, the problem of reconstruction from many projections is better determined than reconstructing from few, if the total amount of counts is the same.

On the other hand, table 1 shows the influence of regularization: While gradient and entropy regularization can increase the accuracy in every case, surprisingly the histogram regularization is not able to do so, albeit a lot of a priori information (the histogram of the original density) is used for it.

Although the entropy regularization has been the only regularization used in connection with the reconstruction from ACAR data, it is not the one which yields the best results. This is achieved with the gradient regularization in every case. The reason why the entropy regularization has been previously used so often to reconstruct ACAR data may be the simple optimization algorithm proposed by Gull and Daniell [12]. Since it takes much more time to record ACAR data than to calculate a reconstruction, low complexity or fast computation should nowadays not be the reason for choosing an algorithm. However, the optimization algorithm proposed here is nearly as simple and fast as the one of Gull and Daniell. Moreover it is a lot more flexible since the method of regularization can be changed very easily.

4. Conclusion and Outlook

In this article we tested three regularization functions on their effect of reconstructing ACAR data. A novel gradient based regularization was introduced, which was able to increase the quality of a reconstruction compared to common regularizations. Further, we have ascertained here for the first time the fact that measurement time is used most efficiently if many projections as opposed to a few projections with high statistics are recorded.

We want to point out, that the results obtained here are not only restricted to ACAR spectroscopy. In Compton scattering experiments, the measurements are also projections of the electron momentum density density [13]. Hence our results can be also applied to this kind of experiment.

Another important challenge of ACAR spectroscopy is the experimental resolution. To integrate this issue into the reconstruction algorithm is the focus of ongoing work.

Acknowledgements

This project is funded by the Deutsche Forschungsgemeinschaft (DFG) within the Transregional Collaborative Research Center TRR80 "From electronic correlations to functionality".

References

[1] Kak A C and Slaney M 1988 *Principles of Computerized Tomographic Imaging* (New York: IEEE Press)
[2] Cormack A M 1963 *J. Appl. Phys.* **34** 2722
[3] Kontrym-Sznajd G 2009 *Low Temp. Phys.* **35** 599
[4] Weber J A, Böni P, Ceeh H, Leitner M and Hugenschmidt C 2013 *J. Phys. Conf. Ser.* **443** 012092
[5] Haynes T D *et al.* 2012 *New J. Phys.* **14** 035020
[6] Kondo H, Kubota T, Nakashima H, Murakami Y and Tanigawa S 1993 *Phys. Status Solidi (b)* **177** 345

[7] Hamid A S *et al.* 2011 *Phys. Rev. B* **84**(23) 235107
[8] Kontrym-Sznajd G *et al.* 2002 *Phys. Rev. B* **66** 155110
[9] Pylak M, Kontrym-Sznajd G and Dobrzyński L 2011 *Appl. Phys. A* **104**(2) 587
[10] Rudin L I, Osher S and Fatemi E 1992 *Physica D: Nonlinear Phenomena* **60** 259
[11] Leitner M 2013 *Electron momentum reconstruction in ACPAR* Diploma thesis Universität Wien
[12] Gull S and Daniell G 1978 *Nature* **272** 686
[13] Utfeld C *et al.* 2010 *Phys. Rev. B* **81**(6) 064509

AUTHOR INDEX

AUTHOR INDEX